**Conservation Social Science**

# Conservation Social Science

Understanding People, Conserving Biodiversity

*Edited by*

*Daniel C. Miller*
*University of Notre Dame, USA*

*Ivan R. Scales*
*University of Cambridge, UK*

*Michael B. Mascia*
*Conservation International, USA*

*Registered Office*
John Wiley & Sons Ltd, The Atrium, Southern Gate, Chichester, West Sussex, PO19 8SQ, UK

*Editorial Office*
Boschstr. 12, 69469 Weinheim, Germany

For details of our global editorial offices, customer services, and more information about Wiley products visit us at www.wiley.com.

A catalogue record for this book is available from the Library of Congress

Paperback ISBN: 9781444337570; ePub ISBN: 9781119604907; ePdf ISBN: 9781119604921

Cover Image: Oma forest by Agustin Ibarrola. Photograph by Ander Abadia Zallo/CC-BY-SA
Cover design by Wiley

Set in 9.5/12.5pt STIXTwoText by Integra Software Services Pvt. Ltd, Pondicherry, India
Printing and Binding CPI Group (UK) Ltd, Croydon, CR0 4YY

C9781444337570_190123

# Contents

# List of Contributors

**William M. Adams**
Moran Professor of Conservation and
Development
Department of Geography
University of Cambridge
Cambridge
UK

**Arun Agrawal**
Samuel Trask Dana Professor
School for Environment and Sustainability
Professor
Gerald Ford School of Public Policy
University of Michigan
Ann Arbor, MI
USA

**Deborah Blackman**
Professor
School of Business
University of New South Wales
Canberra
Australia

**Steven R. Brechin**
Professor
Department of Sociology
Rutgers, The State University of New Jersey
New Brunswick, NJ
USA

**C. Anne Claus**
Assistant Professor
Department of Anthropology
American University
Washington, DC
USA

**Michael B. Mascia**
Senior Vice President for Strategic Initiatives
Senior Scientist
Conservation International
Arlington, VA
USA

**Daniel C. Miller**
Associate Professor of Environmental
Policy
Keough School of Global Affairs
University of Notre Dame, IN

Adjunct Associate Professor
Department of Natural Resources and
Environment Sciences
University of Illinois at
Urbana-Champaign
Urbana, IL
USA

**Katie Moon**
Senior Lecturer
School of Business
University of New South Wales
Canberra
Australia

**Olin Eugene Myers Jr.**
Professor
Department of Environmental Studies
Western Washington University
Bellingham, WA
USA

**Stephen Polasky**
Regents Professor and Fesler-Lampert
Professor of Ecological/Environmental
Economics
Department of Applied Economics &
Department of Ecology, Evolution and
Behavior
University of Minnesota
St. Paul, MN
USA

**Timmons J. Roberts**
Ittleson Professor of Environmental
Studies and Sociology
Brown University
Providence, RI
USA

**Diane Russell**
President
SocioEcological Strategies, Inc.
Washington, DC
USA

**Ivan R. Scales**
Sir Harvey McGrath Associate Professor
in Human Geography
St Catharine's College
University of Cambridge
UK

**Jennifer Swanson**
PhD Candidate
Department of Sociology
The Maxwell School of Citizenship and
Public Affairs
Syracuse University
Syracuse, NY
USA

# Foreword

Nature—from the majestic blue whale to the humble dung beetle—is the foundation of human existence and society. We rely on functioning ecosystems, and the biological diversity that underpins them, for the food we eat, the water we drink, the air we breathe, and the building blocks of our economies. A healthy planet also contributes to human well-being in less material ways, as a source of inspiration, identity, spirituality, and mental health.

Yet these vital functions are too often overlooked by governments, businesses, and policymakers. As a result, we have allowed the world's biodiversity to become critically imperiled. We are in the midst of Earth's sixth mass extinction, with species being lost at a rate not seen since the last mass extinction 65 million years ago. Along with widespread habitat and biodiversity loss, the global climate is changing rapidly, placing additional stresses on already threatened species, ecosystems, and those who rely on them for their livelihoods and well-being.

Human activities are the driving factor behind both biodiversity loss and climate change. Industry, agriculture, fishing, and transportation place significant pressures on the planet's ecosystems. Wealthier nations and consumers in particular use resources at rates that are simply not sustainable.

While the scale of global environmental challenges can seem daunting, the good news is we have the ability to make change. As individuals, households, communities and nations, together, we can act to stop and reverse biodiversity declines. To do so will require marshaling insights from a range of disciplines. Here, the social sciences hold particular importance. The conservation of biological diversity is, at heart, a social issue, cutting across the political, economic, social, and cultural spheres of human life.

Tackling global biodiversity loss and ecosystem degradation will require applying lessons from the social sciences about human behavior and how we might change and harness it. To identify leverage points, we will need tools from the social sciences to analyze problems and their underlying drivers. We will need to build and test theories of change, as well as ask profound questions about what it means to be human in an era of rapid social, technological, and environmental change. Finally, we will need new ways of engaging with and building knowledge that effectively draw from diverse constituencies, particularly those who have been historically marginalized.

*Conservation Social Science: Understanding People, Conserving Biodiversity* is a vital resource for these tasks. This book provides an easy-to-use overview of the social sciences and what they have to offer both to understanding and tackling global biodiversity loss. Written by leading scholars, it provides a discipline-by-discipline guide to the social sciences and their relevance to conservation.

Too often, social science knowledge and approaches have gone unused in conservation research, policy, and practice. Their absence goes some way to explaining our failure to effectively conserve the planet's natural wealth. But the required integration into the biophysical sciences, engineering, and other disciplines, as well as conservation planning and practice, can be challenging. This book provides a basis for greater integration. It explains and demonstrates key social scientific theories, tools, and ideas through a rich set of case studies drawn from around the world to help students, practitioners, and policy makers understand real-world challenges and develop solutions.

This book is a key reference for the world community as we develop and implement the post-2020 Global Biodiversity Framework for 2030 and beyond. It not only instructs and illustrates. It invites us to draw from the deep history and new developments across the social sciences to develop innovative approaches to tackling one of the most urgent challenges of our time: conserving Earth's rich biological heritage while ensuring thriving human societies. It is now time for us to accept this invitation.

*Inger Andersen*
*Under-Secretary-General of the*
*United Nations (UN) and*
*Executive Director of the UN*
*Environment Programme*

# Acknowledgements

This book has been long in the making. Yet the central idea—to hold up and examine the relationship between a set of core social science disciplines and biodiversity conservation—has remained the same since it was formulated nearly two decades ago. It is, as we believe the pages that follow show, an idea that remains as timely and as important as ever.

Throughout, the contributors to this book have been exceptionally generous and patient. For this and for the scholarly acumen they have brought to these pages, we express our sincere thanks. We also thank Inger Andersen for providing the foreword and Nandita Surendaran for helping to facilitate its creation.

We owe a large debt of gratitude to Joanna Broderick for her outstanding editorial assistance and for creating the index for this volume. It is doubtful that the book would have seen the light of day without her careful and thoughtful contribution. We also thank Katia Nakamura for her help preparing several figures in the book, Sophia Winkler-Schor for writing Box 1 in the introductory chapter, and Mohammed Farrae and Carly Hopkins for help with references and permissions.

Comments by Helen Fox, Louise Glew, Emily McKenzie, Robin Naidoo, Lisa Naughton, Sharon Pailler, Kent Redford, Adena Rissman, Paul Robbins, Chris Sandbrook, Sheri Stephanson, and anonymous reviewers from Wiley allowed us to develop the idea, design, and content of the book. We are enormously grateful for their generosity. We would like to thank the many colleagues, past and present, in the Social Science Working Group of the Society for Conservation Biology, for support over the years.

Several staff at Wiley have helped this book come to life. For their belief in the project and dogged persistence in working with us to see it to fruition, we thank Andrew Harrison, Rosie Hayden, Kelly Labrum, Joss Everett, and Frank Weinreich. Special thanks to Ward Cooper for taking a chance and commissioning this book from an (at the time) early career scholar and Marjorie Spencer for the introduction and vote of confidence that set these wheels in motion.

We are grateful to the John D. and Catherine T. MacArthur Foundation (Grant #05-83705-000-GSS) and the World Wildlife Fund for supporting the early stages of this book with support to one of us (MBM). In particular, three in-person workshops allowed us to develop and refine our overall vision for the book and its corresponding chapters. Support from the Department of Natural Resources and Environmental Sciences at the University of Illinois and the Keough School of Global Affairs at the University of Notre Dame (to DCM) allowed us to bring this book to completion.

**Daniel C. Miller** would like to thank the many colleagues, students and friends at the Universities of Michigan, Illinois, and Notre Dame who have helped inspire, encourage and support the production of this book. Special thanks are due to Arun Agrawal and Mike Mascia for inviting him to be a part of this book, as author and editor. The chapter on Political Science that he and Arun co-wrote has benefitted from insightful comments from Jake Bowers, Elizabeth Gerber, Sourav Guha, Debra Javeline, Mark Lubell, Bob Pahre, Spencer Piston, and a number of other colleagues and students who provided feedback in seminars at Illinois and Michigan.

Finally, most special thanks go to his parents, Steve and Joyce Miller, and his wife, Bea Zengotitabengoa, and his children, Eneko, Maite, and Izei—may theirs be a world in which social sciences, arts and humanities are more insistently fused with the biophysical sciences and other fields to advance the mutual flourishing of people and nature.

**Ivan R. Scales** would like to thank St Catharine's College for the continued support and sabbatical leave that made completing this book possible. He would also like to thank his academic mentors over the years, especially Katherine Homewood and Bill Adams. A very special thank you to his wife Helen Scales for helping him to think and write, and for all the adventures (past, present and future).

**Michael B. Mascia** would like to thank those who initially encouraged him to explore biodiversity conservation through a social science lens and to develop the early thinking that led to this book, including Michael K. Orbach, Margaret McKean, Alan Thornhill, Curt Meine, Gary Meffe, David Hulse, and co-editor Daniel Miller. He would also like to thank friends and colleagues around the world who have broadened and refined his thinking about conservation social science over the years, including Fitry Pakiding, Morena Mills, Art Blundell, Alex Pfaff, and Johan Rockstrom; teammates at the U.S. Environmental Protection Agency, World Wildlife Fund, and Conservation International; and the co-authors of this book. Lastly, and most importantly, he would like to thank his family for their constant encouragement, support, and patience, especially his wife Hannah and children Beatrice and Jonathan.

Of course, any mistakes that remain are our responsibility.

Finally, we thank the reader for putting this book in their hands. We hope it helps inform their thinking—and even action—on the role of the social sciences in a conservation that is equitable, effective, and enduring.

non-material aspects of human well-being, such as identity, inspiration, learning, spirituality, and psychological experience (MEA 2005; Fish et al. 2016). In short, "nature is essential for human existence and good quality of life," as the most comprehensive report on biodiversity and **ecosystem services** yet produced puts it (IPBES 2019). The conservation of biological diversity and ecosystem function therefore stands as one of the biggest challenges facing humanity this century.

Recognition of the importance of biodiversity, as well as mounting threats to it, has spurred a range of different conservation responses around the world. Concerned citizens, conservation scientists, non-government organizations, philanthropic foundations, and government agencies have mobilized to protect vital habitats and take other actions to conserve the Earth's rich natural heritage. From the creation of the first national park in the United States of America in the 1870s to recent attempts to establish payment for ecosystem services schemes, the field of conservation science, policy, and practice has grown remarkably. Conservation now includes a wide array of perspectives (Sandbrook et al. 2019), even as racial, gender, nationality, and other inequalities persist in the conservation movement and among those who study it (Taylor 2014; Wilson et al. 2016; Campos-Arceiz et al. 2018; Jones & Solomon 2019; Bailey et al. 2020).

National parks and other kinds of protected areas perhaps best illustrate the growth of the modern conservation movement. The global protected area estate has increased from a handful of sites at the beginning of the twentieth century to almost 240,000 legally designated protected areas (Figure 1.1). Together, these areas cover over 26 million square kilometers or nearly 15% of the Earth's terrestrial surface and 7.3% of the ocean (UNEP-WCMC 2020). The international community is negotiating much more ambitious targets under a new global biodiversity framework through the Convention on Biological Diversity. Proposals call for setting aside 30% of the Earth for protection with an additional 20% designated as climate stabilization areas outside formally protected areas where carbon-sequestering vegetation is maintained and greenhouse gas emissions prevented (Dinerstein et al. 2019).

In addition to the growth in protected areas, conservation now has its own journals, university departments, international non-government organizations, government agencies, consultancies, and global treaties. Conservation has become a major global enterprise, with tens of billions of dollars spent every year in efforts to protect the planet's biological diversity and ecotourism, and other conservation-related activities, estimated to generate more than one hundred billion dollars annually (Waldron et al. 2013; Deutz et al. 2020).

Conservation science has made major inroads into assessing levels of biodiversity, identifying threats to it, and suggesting where conservation efforts should be concentrated (Myers et al. 2000; Olson et al. 2001; Brooks et al. 2006; IPBES 2019). Almost all countries of the world are parties to the Convention on Biological Diversity (CBD 2019), and conservation actions have been undertaken widely across the globe (IPBES 2019). However, major questions remain concerning the effectiveness of conservation efforts, and how best to focus scarce resources to get the most biodiversity "bang for the buck" (Ferraro & Pattanayak 2006; Waldron et al. 2013; Gerber et al. 2018). Why have some conservation efforts succeeded while others have failed to achieve their aims and sometimes even generated negative social impacts? How might conservation policies and practices be improved to increase the protection of biodiversity, reduce potential negative social impacts, and

# 1

# Introduction: Biodiversity Conservation and the Social Sciences

*Ivan R. Scales, Daniel C. Miller, and Michael B. Mascia*

## 1.1 Global Biodiversity and the Need for Conservation Social Science

Earth's **biodiversity** is under threat. Agricultural expansion, urbanization, industrial pollution, the spread of non-native species, as well as overfishing and overhunting, have led to extinction and continue to place unsustainable pressures on the planet's **ecosystems** (IPBES 2019). From 2000 to 2010, tropical forests were cleared at a rate of over 76,000 square kilometers per year (Achard et al. 2014). Recent studies have revealed dramatic reductions in insect populations around the world, with serious implications for ecosystem function (Sánchez-Bayo & Wyckhuys 2019). Climate change is creating additional stresses for both terrestrial and marine ecosystems and, even with strong mitigation measures, will have profound implications for biodiversity (Seddon et al. 2016; Pecl et al. 2017). Species loss is occurring at a rate not seen since the last planetary mass extinction event more than 65 million years ago (Barnosky et al. 2011). Now, however, this loss is driven not by geological cataclysms or giant meteorites, as in previous epochs, but by human actions (IPBES 2019).

The loss of biodiversity—the variety of living organisms at genetic, species, and higher taxonomic levels—has major implications for our own species. Human well-being is dependent on functioning ecosystems and the biological diversity that underpins them (Diaz et al. 2006; Chivan & Bernstein 2008; IPBES 2019). Natural ecological and evolutionary processes sustain air quality, deliver freshwater, enrich soils, and provide pollination and pest control, among other benefits. For example, more than three-quarters of global food crop types, including fruits, vegetables, and major cash crops like coffee and cocoa depend on animal pollination (Potts et al. 2016). The only planetary sinks for **anthropogenic** carbon emissions are marine and terrestrial ecosystems, which together sequester an estimated 5.6 gigatons of carbon per year—about 60% of total global anthropogenic emissions (IPBES 2019). Forests, grasslands, oceans, and other ecosystems support all dimensions of human health, from reducing disease burden (Herrera et al. 2017) to improving mental health (Cox et al. 2017) and developing new treatments for cancer (Newman & Cragg 2016). A biologically rich and healthy planet also contributes to

*Conservation Social Science: Understanding People, Conserving Biodiversity*, First Edition.
Edited by Daniel C. Miller, Ivan R. Scales, and Michael B. Mascia.
© 2023 John Wiley & Sons Ltd. Published 2023 by John Wiley & Sons Ltd.

| | |
|---|---|
| REDD+ | Reducing Emissions from Deforestation and Forest Degradation |
| RFF | Resources for the Future |
| SC | Structural choice |
| SES | Socio-ecological system |
| SES | Socio-economic status |
| SI | Symbolic interactionism |
| TEK | Traditional Environmental Knowledge |
| TOP | Treadmill of Production |
| USAID | United States Agency for International Development |
| UN | United Nations |
| UNESCO | United Nations Education and Scientific and Cultural Organization |
| VBN | Value-belief-norm |
| VP | Veto player |
| WEIRD | Western, educated, industrialized, rich, and democratic |
| WST | World-systems Theory |

# Abbreviations and Acronyms

| | |
|---|---|
| ABM | Agent-based model |
| AC | Advocacy coalition |
| CBC | Community-based conservation |
| CBNRM | Community-based natural resource management |
| CITES | Convention on International Trade in Endangered Species |
| CPR | Common-pool resource |
| EM | Ecological modernization |
| EPA | Environmental Protection Agency |
| FBS | Folk biological system |
| FAO | Food and Agricultural Organization |
| GDP | Gross domestic product |
| GEF | Global Environmental Facility |
| GIS | Geographical Information System |
| GISc | Geographical Information Science |
| GPS | Global Positioning System |
| HDI | Human Development Index |
| HIV | Human Immunodeficiency Virus |
| IDB | Inter-American Development Bank |
| ICDP | Integrated Conservation and Development Project |
| IEK | Indigenous Environmental Knowledge |
| IRC | Institutional rational choice |
| IUCN | International Union for the Conservation of Nature |
| LEK | Local Environmental Knowledge |
| NGO | Non-governmental organisation |
| NHP | Non-human primate |
| NRM | Natural Resource Management |
| PA | Protected Area |
| PES | Payment for Ecosystem Services |
| PGIS | Participatory Geographical Information System |
| PMT | Protection-motivation theory |
| PPP | Parks and Peoples' Program |
| RCT | Randomized control trial |

(a)

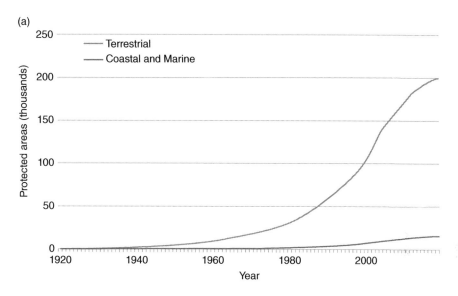

(b)

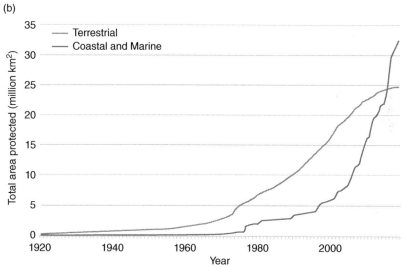

**Figure 1.1** Change in (a) the number and (b) coverage of protected areas globally, 1920–2020. *Data source:* UNEP-WCMC (2020).

contribute to more sustainable economies? How can conservation science, policy, and practice be more inclusive and engage with a broader range of sociocultural values and perspectives?

The central claim of this volume is that the social sciences (Box 1.1) are vital to understanding the drivers of biodiversity loss, the consequences of this loss, and possible solutions to it. To support this argument, the book pulls together insights from six classic social science disciplines—anthropology, economics, human geography, political science, psychology, and sociology—relevant to understanding human–environment interactions

---

**Box 1.1   What are the social sciences?**

The social sciences are academic disciplines that study human societies and the relationships between individuals and groups within those societies. This is a necessarily broad definition. Human thought, behavior, and interaction encompass many different overlapping spheres, including culture, economics, and politics. Furthermore, social processes can be studied at various levels, from the brain functions and psychology of individuals to the actions of households, communities, regions, and nations. Because people can behave and interact in so many different ways, the social sciences draw on a wide range of both **quantitative** and **qualitative** methods to study humans.

For those not familiar with the social sciences, the diversity in methods, approaches, and theories can be overwhelming. Furthermore, as the chapters in this book show, there can often be tensions between different traditions in the social sciences as to how to study social processes. On one end of the spectrum we find approaches that mirror the natural sciences and their emphasis on quantification, large sample sizes, statistical rigor, and hypothesis testing. At the other end of the spectrum, some social science disciplines operate on the assumption that the human condition is something entirely different from biological or physical processes and is best studied through qualitative approaches that emphasize the complexity, richness, individuality, and therefore specificity, of human experience. Given the diversity of the social sciences, it is best not to think of them as a homogenous group but more as a vibrant and sometimes fractious family.

---

generally and conservation more specifically. To introduce the substantive chapters that form the core of this book we first highlight an important set of issues and themes to orient the reader.

The next section describes the intended audience for this book and what readers will gain from the material it contains. We begin by highlighting three major challenges at the heart of the global **biodiversity conservation**. We then summarize the potential contributions of the social sciences to conservation research and policy making. We also consider the barriers to integrating social sciences into conservation research, policy, and practice. Having staked a claim for the social sciences, we make the case for **conservation social science** as a distinct field, defined as the study of the conservation-relevant aspects of human society, including the relationships among humans and between humans and their environment. We finish by providing a brief overview of the chapters that make up the rest of this book.


## 1.2   Whom and What This Book Is For

The goal of this book is to furnish the reader—conservation student, practitioner, scholar, philanthropist, policy maker, or concerned citizen—with a thorough introduction to the diverse approaches that social scientists employ to make sense of conservation problems and conservation itself. We provide tools and knowledge that can inform the myriad forms of conservation policy and practice. To illustrate the theories, tools, and empirical insights discussed in this book, the chapters include case study examples drawn from a range of different countries and ecosystems around the globe. By bringing the rich intellectual traditions of the social sciences to the fore, and by making explicit their collective links to the study and practice of conservation, our ambition is that this primer helps elevate the social sciences to an equal partner to the natural sciences in conservation scholarship and decision-making.

We expect that our audience will be as diverse as the topics and theories that we cover. First and foremost, we are writing for advanced undergraduate and early career graduate students, to provide them with the broad foundation for further scholarship in one or more dimensions of the conservation social sciences. For students with prior training in the social sciences, this book represents an initial foray into conservation-related aspects of six classic social science disciplines and an exploration of the diverse perspectives that disciplines beyond one's own bring to bear on conservation. For students with a background in the natural sciences, this book provides a different way of thinking about and approaching the conservation of biodiversity. For senior scholars, this book will serve as a reference and as a resource to orient one's own work. We believe that an advanced understanding of conservation as a social phenomenon can both translate into a broader scientific understanding and help generate knowledge for science-based conservation policy and practice.

This book is also for practitioners, those involved in the "doing" of conservation: local activists and project managers; grant administrators and philanthropists; concerned citizens, agency staff, and senior officials. For these readers, particularly those primarily trained in the natural sciences, this book will serve as a resource to organize and make sense of personal experiences and observations in novel ways. By providing a new perspective on a topic of long-standing familiarity, this book will help to inform day-to-day conservation decisions and, in the aggregation of these individual choices, inform broader conservation policy and practice.

## 1.3 Challenges for Global Biodiversity Conservation in the Twenty-First Century

The threats driving biodiversity loss are diverse and complex. While humans have a long history of modifying ecosystems and driving species to extinction, the scale and intensity of human environmental impacts is now so great that some argue we are now in the **Anthropocene**, a new geological era where humans are the dominant force (Steffen et al. 2007). Here we identify three major challenges.

### 1.3.1 Understanding Threats to Biodiversity

The major direct threats to biodiversity are well-documented: habitat loss, direct harvesting of organisms, climate change, pollution, and competition from non-native species (Figure 1.2). These processes are themselves underpinned by a wide range of economic, political, cultural, and demographic drivers that shape what and how much humans consume. Responses to these threats have emphasized safeguarding ecosystems and species

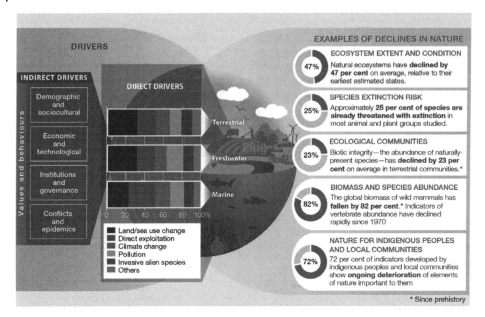

**Figure 1.2** Drivers of global declines in biodiversity. Drivers are listed in order of relative importance according to their contribution to biodiversity loss. Color bands represent the relative global impact of direct drivers on terrestrial, freshwater, and marine ecosystems. Circles show the magnitude of the negative human impacts on a range of different aspects of nature. Figure used with permission from IPBES (2019).

through the creation of protected areas (Watson et al. 2014; Dinerstein et al. 2019). Through community-based conservation schemes and engagement with local stakeholders, conservation efforts have also attempted to change the practices of resource users (for example, hunters, fishers, and farmers) to reduce pressure on ecosystems. Conservation strategies have thus often focused "downstream" on the impacts of human actions on threatened species and habitats. Those advocating and implementing conservation have tended to pay less attention to what happens "upstream"—the broader political, economic, and cultural factors that drive patterns of resource use and the interactions between them. However, recent scientific consensus has begun to present a more holistic picture, with recognition that sustainable conservation will require transformation of the current political–economic systems (IPBES 2019).

Conservationists can be quick to identify human population growth as a major threat to biological diversity (Maurer 1996; Cincotta et al. 2000; McKee et al. 2004). As the subsequent chapters will show, the social sciences are generally skeptical of models and explanations of human natural resource use that focus narrowly on population growth as a driver of environmental degradation. Such **neo-Malthusian** models tend to underplay the critical role of high levels of consumption in wealthier countries in driving biodiversity loss. Simplistic models along these lines also often ignore the capacity of human societies to change how they manage resources as population densities increase, including through the use of technology (Tiffen et al. 1994; Boserup 2014 [1965]).

When discussing threats to biodiversity, conservation thinking and practice have been dominated by certain stories and cautionary tales. Many of these will be familiar to readers: environmental "collapse" on Easter Island, the extinction of the dodo, the destruction of the Amazon rainforest. Such stories are powerful and help conservationists increase awareness of important issues and raise funds to tackle them. But they can also create problems by oversimplifying complex issues, leading to misguided policies. The "tragedy of the commons," popularized by ecologist Garrett Hardin (1968), represents an especially prominent example of how an oversimplified story can capture the imagination of scientists and policy makers and persist despite deep flaws (see Box 6.3 in Chapter 6, Political Science and Conservation). While we do not wish to downplay the magnitude of conservation challenges and threats to biological diversity, conservationists need to pay closer attention to the stories they tell each other and the world.

Studies of the environmental and socioeconomic interactions between distant regions of the world, known as telecoupling (Liu et al. 2013), provides a promising example of research seeking to provide a fuller picture of threats and opportunities to biodiversity conservation. In an increasingly interconnected world, telecoupled forces like international trade of agricultural products and wildlife are unprecedented in their speed, extent, and intensity (Carrasco et al. 2017). These forces underpin many of the threats to biodiversity like habitat loss, direct exploitation of valuable biota, and invasive species. Production of beef, palm oil, soybean, and other commodities, driven by consumer demand in increasingly affluent societies, has led to tropical deforestation (Newton et al. 2013) with negative impacts on biodiversity (Lenzen et al. 2012; Moran & Kanemoto 2017).

At the same time, however, telecoupling has also brought potential opportunities for conservation "in distant supermarkets, corporation boardrooms, stock markets, and the Internet" (Carrasco et al. 2017, p. 7). For instance, shifting consumer demands have created pressures on multinational corporations and governments to support more sustainable commodity production. Certification of commodities like coffee, paper, seafood, and wood as sustainably sourced is increasingly widespread and can help advance conservation objectives (Tayleur et al. 2017; Lambin et al. 2018), although critics suggest that consumers have limited power to influence large-scale commodity chains in comparison to large businesses and governments (Scales 2014). Conservation research and practice are increasingly recognizing that a global perspective is needed that trains attention on consumers, corporations, and governments in wealthy countries as much as on small-scale producers in poorer yet often biologically rich ones.

## 1.3.2 The Effectiveness of Conservation Policy and Practice

Given the growing threats to biodiversity and the expanding roll call of species on the brink of extinction, there is an urgent need to better understand what determines the effectiveness of conservation interventions. Historically, conservation policy has tended to focus on establishing protected areas. In reality, many protected areas are little more than "paper parks" that exist only on maps and in policy documents, with resource extraction and environmental degradation often continuing to occur. Global studies of protected area performance have found that less than half of protected areas are effectively managed (Leverington et al. 2010; Watson et al. 2014) due in significant part to insufficient capacity (Gill et al. 2017) and funding (Coad et al. 2019). Legal changes that reduce the protections and extent of protected areas further challenge the effectiveness of conservation efforts (Golden Kroner et al. 2019). Conservation policy has also often proved faddish, with organizations chasing the latest policy fashions and funding trends (Redford et al. 2013). It seems that the urgency of the problem leads to an ever faster policy treadmill. There have been too few efforts to take stock of what has worked (or not worked) and why (Ferraro & Pattanayak 2006; Miteva et al. 2012; Burivalova et al. 2019; Wardropper et al. 2022).

In response to shortcomings and unintended consequences of many conservation projects, more rigorous and systematic assessments of the effectiveness of different conservation tools and practices are being published (Sutherland et al. 2004). A recent boom in gathering and analyzing conservation evidence has been driven by a desire to make conservation policy and practice more rigorous and objective. While we welcome efforts to improve the success of conservation interventions, as Adams and Sandbrook (2013) note, two important questions must still be answered: What counts as evidence? and How does evidence count?

With regard to the first question, it is crucial that conservation decision makers resist the temptation to draw only on forms of knowledge with which they are familiar (e.g. **quantitative** data from the natural sciences) and engage with different kinds of social scientific data (both quantitative and **qualitative**) as well as indigenous and other relevant forms of knowledge (Charnley et al. 2017). In terms of how evidence is used to inform policy, it is not simply a case of getting conservation experts to gather "better" data to hand over to decision makers so they can make the "right" decision. This sort of conservation decision-making privileges certain individuals, groups, and forms of knowledge and excludes others. The production, distribution, and use of knowledge are processes intimately tied to the exercise of power. What counts as knowledge, how it is generated, and who gets to make decisions on whose behalf should be at the heart of any discussion of conservation evidence and policy making.

Efforts to conserve biodiversity have brought mixed results for people living in and around protected areas. Evidence of positive impacts on livelihoods and other aspects of human well-being (McKinnon et al. 2016) coexists with studies finding a range of negative impacts, including evictions, loss of access to natural resources, and exclusion from decision-making (Brockington & Igoe 2006; West et al. 2008; Dressler et al. 2010; Oldekop, Holmes et al. 2016). Conservationists have often been poor at understanding the different worldviews and priorities of other stakeholders, leading to antagonism and conflict (Scales 2012; Parathian 2019). The fact that some of the poorest people on the planet pay the highest costs for the conservation of global biodiversity is morally unacceptable (Martin 2017). The principal problem is that the conservation of biological diversity largely remains something that is done by conservation experts rather than a process that engages with diverse interest groups. This reduces the chances that conservation policies will succeed.

### 1.3.3 The Search for Sustainable Conservation Solutions in an Uncertain Future

Given the urgency of the global conservation challenges, there has been a tendency in policy circles to look for panaceas—magic bullets that will solve all problems (Ostrom et al. 2007). However, as will become apparent in many of the subsequent chapters, win-win solutions can be elusive in conservation and context is key. Conservation challenges are often the result of the complex interactions among various social and environmental factors, which preclude simple one-size-fits-all solutions.

In addition to the issues surrounding evidence-based conservation highlighted in the previous section, we lack knowledge of which kind of intervention is most effective in what context. Do incentive programs like paying landowners for habitat conservation work better than government-run protected areas? Are information-based approaches like those to inform consumers about sustainably harvested seafood or zero-deforestation beef, palm oil, or soybean more effective than encouraging conservation through ecotourism? Under what conditions do approaches implemented successfully in one country or ecoregional context work in another? Recent reviews are beginning to synthesize available evidence to answer such questions (Miteva et al. 2012; Agrawal et al. 2018; Burivalova et al. 2019), but relevant research remains scant. A national-scale study from Mexico (Sims & Alix-Garcia 2017) comparing the effectiveness of setting aside land for conservation versus paying landowners to protect it provides a notable exception. The authors find that protected areas and payments for ecosystem services approaches had about the same positive effects in conserving forests, but that the latter was more successful in also alleviating poverty. They conclude that interventions combining sustainable financing, flexible zoning, and recognition of local aspirations are more likely to deliver conservation gains without compromising local livelihoods.

Beyond considering the relative efficacy of different conservation approaches and devoting closer attention to local social context, conservation policy and practice also need to do more to recognize and address tradeoffs between different outcomes (McShane et al. 2011). Biodiversity conservation involves difficult decisions, especially in the context of limited resources: which species to focus on; which geographical areas and ecosystems to prioritize; how to balance the needs of humans and non-human species; and how to balance the demands and priorities of different groups and stakeholders. Once again, questions of power emerge. Who gets to decide on the tradeoffs that are made, as well as how the costs and benefits of different projects and actions are shared? The social sciences have an important role in addressing such questions.

To date, conservation actions have largely been reactive, responding to threats and attempting to slow the loss of habitats and species. There have been notable attempts to "horizon scan" and identify future trends and priorities (Oldekop, Fontana et al. 2016; Sutherland et al. 2019). However, conservation policy and practice need to go further in thinking about the future. Climate change will complicate efforts to manage biodiversity. The rate and scale of projected climate changes in the twenty-first century are likely to have profound impacts on the functioning of Earth's ecosystems. It is still unclear how this will unfold and which ecosystems and species will be most affected (Seddon et al. 2016). As well as the important biological questions of how different species and ecosystems will react, there are also major questions about how humans will adapt and what this will mean for biodiversity and human well-being (Maxwell et al. 2015; Pecl et al. 2017; Marselle et al. 2019). A greater emphasis on prediction and learning from other fields like finance, military studies, and public health promise to help conservation advocates to anticipate shocks and pre-empt their impacts in an increasingly uncertain world (Travers et al. 2019).

Nevertheless, it is also important to remember that even when conservation strategies work for a time, there is no guarantee that they will endure. Studies of protected area downgrading, downsizing, and degazettement have shown, for example, how conservation policies can quickly be undone (Mascia & Pailler 2011; Golden Kroner et al. 2019). Research on the long-term impacts of conservation interventions remains rare, however (Miller et al.

2017), and this situation will need to change to enable more informed efforts to foster just and sustainable conservation.

## 1.4 Opportunities and Challenges for Conservation Social Science

Traditionally, biodiversity conservation has been viewed primarily through the lenses of the discipline of biology, especially genetics, population biology, and biogeography. Given that the term *biodiversity* refers to the variety of living organisms, it is tempting to see the natural sciences (especially the various biological disciplines) as the most relevant to supporting conservation policy and practice. This is indeed the way much of modern conservation has developed, including the discipline of conservation biology. Conservation research has mainly focused on measuring extinction rates, compiling data on biological diversity at various levels, assessing threats to species and ecosystems, and more recently on calculating the economic values of ecosystem services in the hope of convincing businesses and policy makers that biodiversity matters. So while Soulé (1985, p. 727) labeled conservation biology as "multidisciplinary," "synthetic," and "eclectic" (and explicitly noted the need for insights from the social sciences), the reality has been a concentration on biological processes rather than relevant, but often harder-to-measure indicators from many of the social sciences (Hicks et al. 2016).

However, a growing number of conservation researchers and practitioners are realizing that conservation is in fact not primarily about biology but about people and the choices they make (Balmford & Cowling 2006; Amel et al. 2017). It is clear that **biodiversity conservation** is a social phenomenon. Threats to biological diversity are influenced by a wide range of social factors. The conservation of biodiversity is conceived and carried out by people. Biodiversity conservation is a manifestation of human beliefs and values. In every corner of the planet, formal and informal social **norms** establish expectations and standards for protecting genes, species, ecosystems, and the relationships among them. Written laws and unwritten taboos govern hunting, fishing, logging, recreation, agriculture, and human settlement. Choices about which species and habitats to conserve, how to prioritize efforts, and how to conserve them are inherently political. Environmental education programs attempt to provide individuals with sufficient information to make informed decisions (Box 1.2) about how they interact with the environment while marketing, advocacy, and lobbying campaigns promote specific conservation agendas. Government agencies, non-profit organizations, for-profit corporations, and individuals invest billions of dollars and spend countless hours designing and implementing these and other conservation actions.

It is clear that the conservation of biological diversity is a social process, with consequences that affect humans and other species. The choice to conserve is a human one. The various ways of doing it are social initiatives. The impacts of how it is done are felt by people (as well as other species), and yet conservation policy and practice have been dominated by various branches of the biological sciences. This book argues that the social sciences have much to offer. In developing our argument, the chapters that follow extend and deepen previous efforts that have sought to show how social science contributions are vital to understand conservation and to the field's overall success (Mascia et al. 2003; Kareiva & Marvier 2012; Bennett, Roth, Klain, Chan, Christie et al. 2017).

---

**Box 1.2   Crossing boundaries: changing consumer behavior to reduce wildlife trade in Asia. Author: Sophia Winkler-Schor, University of Wisconsin**

Influencing the choices people make is crucial to conservation and thus, conservationists must understand human behavior to achieve global conservation goals (Balmford et al. 2021). Conservation marketing is a burgeoning discipline and is defined as "the ethical application of marketing strategies, concepts and techniques to influence attitudes, perceptions and behaviors of individuals, and ultimately societies, with the objective of advancing conservation goals" (Wright et al. 2015, p. 46). Advertising and marketing techniques in commercial sectors have seen great success in identifying segments of a population who are most persuadable and then developing techniques to influence their preferences and behavior. During the last few decades, conservation marketing experts have increasingly adopted these techniques to identify subpopulations and frame campaign messages in a way that speaks to the values, norms, and attitudes of the people (Veríssimo 2019). From the protection of endemic endangered St. Vincent parrot (*Amazona guildingii*; Jenks et al. 2010) to reducing lawn watering by North American homeowners (McKenzie-Mohr 2000), conservation marketing has helped change human behavior and contributed to conservation success.

In recent years, conservation marketers have turned their attention to tackle the problem of elephant ivory consumption in China and other countries in Asia, which threatens the existence of global elephant populations. Various campaigns have been designed and implemented to dissuade consumers from buying ivory, with conservation marketing as a core strategy of these campaigns (Greenfield & Veríssimo 2019). The spike in ivory consumption over the past 15 years has been spurred by increasing affluence in China and other Asian countries, and so conservation marketing experts have sought to understand what would influence ivory consumers' behavior through focus groups, interviews, and surveys (see, e.g. Lee et al. 2016). Results indicate that:

1) People were unaware of the basic facts of where ivory comes from (many did not know that elephants are killed for their tusks) and the future implications of the illegal ivory trade (overall elephant extinction).
2) There was confusion surrounding legal and illegal ivory. Ivory consumption was not banned in China until December 31, 2017.
3) People wanted to reduce government corruption, and ivory consumption is linked to government corruption and bribery. Ivory is largely used as "gifts" for government officials in China.
4) People wanted to combat organized crime, and ivory consumption is illegal as of 2018.
5) Ivory is highly ingrained into Chinese tradition and seen as a wise financial investment. Thus, these cultural norms must be uncoupled from ivory.

Conservation marketers compiled the findings from focus groups and interviews to develop persuasive campaign messages featuring local celebrities. Preliminary results suggest that since the ivory ban took effect in 2018 and the public campaign, only 12% of respondents claim to have purchased ivory in the past six months compared to 26% of respondents who reported doing so in a similar 2017 pre-ban survey (GlobeScan 2021), a 54% decline. The entirety of the campaign success has not yet been evaluated as it is still in its implementation phase. However, while such behavior change

*(Continued)*

---

**Box 1.2 (Continued)**

campaigns are becoming increasingly popular, very few define clear, time-bound objectives or a control group to enable rigorous assessment of success (Veríssimo & Wan 2019). Such measures are necessary for this type of conservation action to effectively address the unsustainable trade in wildlife.

---

Conservationists have long recognized the important role social sciences have to play in advancing conservation objectives (Soulé 1985; Leopold 1987 [1949]). The conservation literature is full of strong calls for their greater use and integration (Mascia et al. 2003; Fox et al. 2006; Cowling 2014; Bennett, Roth, Klain, Chan, Clark et al. 2017) and for interdisciplinarity (Schultz 2011; Guerrero et al. 2018; Stern 2018). The social sciences are increasingly better integrated into conservation science (Teel et al. 2018; Hintzen et al. 2019), but their incorporation into the mainstream of conservation policy and practice lags behind (Mascia et al. 2003; Adams 2007; Bennett, Roth, Klain, Chan, Christie et al. 2017; Nature Editorial Board 2022).

The challenges to integrating the social sciences into conservation research and practice are manifold. Most readers will be aware that the natural and social sciences have different vocabularies and different **methodologies**. But as will become apparent through this book, differences between the natural and social sciences go even deeper. They can be based on very different philosophies of what research is for and even what counts or does not count as valid knowledge (Chapter 2, Social Science Foundations). To some, the immensity and diversity of social science theory, research foci, methods, and philosophical foundations represent substantial barriers unto themselves: "To the uninitiated, the social sciences can seem like the 'Tower of Babel'" (Phillipson et al. 2009).

Beyond these linguistic and philosophical barriers, more mundane and bureaucratic barriers often inhibit more integrated conservation knowledge. Professional incentives tend to discourage interdisciplinary collaboration and applied problem-solving and push researchers down ever narrower subfields of specialization (Fox et al. 2006). In addition, social scientists (like natural scientists) often struggle with the tension among the roles of scholar (to document, explain, and critique), practitioner (to identify problems and implement solutions), and advocate (to encourage specific goals and actions). Indeed, some scholars fear that engaging too deeply in the policy process hinders one's ability to observe and critique (Lackey 2007) and that one's knowledge or expertise might be misused (Chapin 2004). Others counter that specialization is essential to rigorous scholarship or that social scientists lack sufficient conservation knowledge to contribute effectively to conservation science and policy (Fox et al. 2006).

Despite these considerable barriers, there is a growing trend toward interdisciplinarity in conservation research and policy making (Bennett, Roth, Klain, Chan, Clark et al. 2017; Charnley et al. 2017). We very much welcome this development but with two important caveats. The first is that conservationists need to draw on a wider range of social science methods and approaches. To date, conservationists have tended to engage with a relatively narrow subset of the social sciences, favoring quantitative approaches from economics, political science, and behavioral sciences (Moon et al. 2019). The reasons for this are explored in Chapter 2 (Social Science Foundations) but mainly relate to the fact that these approaches fit well with the quantitative scientific traditions within the conservation sciences. In contrast, conservation scientists drawing on the social sciences have tended to be less engaged with questions of values and power (Hicks et al. 2016).

The second caveat follows from the observation that conservation practitioners and policy makers have tended to have a rather instrumentalist view of the social sciences. In other words, they have seen the social sciences as a means to help conservationists achieve their desired goals. However, the social sciences are not simply at the service of conservation science or conservation policy and practice. For example, many social scientists work *on* conservation rather than *for* conservation (Sandbrook et al. 2013). That is, their main interests are to study and critique conservation science, policy, and practice. Such social scientists view conservation as an important object of study itself, capable of yielding more general insights about human behavior and meaning.

While conservationists often can find it uncomfortable to be under the gaze of social scientists—to be the object of academic study—this form of conservation social science can add significantly to biodiversity conservation. It can help researchers and practitioners reflect on values and beliefs, as well as the power relations, that are often taken for granted. For example, research has shown that even within the world of conservation research, policy, and practice, there is a wide range of contrasting and even conflicting views of what conservation is for and how it should be carried out (Sandbrook et al. 2011, 2019). Through studying conservation as a social process, the social sciences can help conservation policy and practice with dialogue, discussion, and debate. Only when we can acknowledge and recognize different viewpoints can we begin constructive dialogue.

The social sciences have a long history of studying human interactions with nature. However, this book serves as a platform for taking this engagement further and for moving from social science approaches to studying human–environment interactions to conservation social science. This book is organized according to the classic disciplines within the social sciences. It is distinctive in its in-depth treatment of these different social science disciplines as opposed to exploration of more applied, cross-cutting social sciences and humanities as reviewed elsewhere (Bennett, Roth, Klain, Chan, Christie et al. 2017). Nevertheless, it will become apparent that the boundaries between the core social science disciplines are often fuzzy. Our hope is that conservation social science will develop into a mature field that transcends these boundaries as it also helps break down others between expert knowledge, citizen science, and indigenous knowledge; between Western and non-Western values; and between research, policy, and practice.

## 1.5 Plan of the Book

Given the diversity of our audience, we have chosen a straightforward and consistent organization. Before we get to the discipline-based chapters that form the bulk of this book, and are ordered alphabetically, there is an important chapter that we have titled "Social Science Foundations." While it is perfectly possible to read each chapter individually and in no particular order, we encourage readers to start with this overview chapter. As we have already alluded, some of the biggest barriers to bridging the natural and social sciences involve the different ways in which many social scientists approach knowledge: what it is, what counts as valid knowledge, and how it should be collected. These differences in turn lead to unfamiliar and often daunting terminology. Chapter 2 is designed to guide the reader through the different ways in which social scientists think and form knowledge about the world, especially nature and conservation.

The core of this book is then comprised of six chapters, each focusing on one of six major social science disciplines—anthropology, economics, human geography, political science, psychology, and sociology—and the conservation-relevant insights that each has to offer. Each chapter is written by experts and provides a synthesis of how the discipline studies and understands conservation; how it can contribute to conservation; and how conservation may, in turn, contribute to the development of the social sciences. Each chapter opens with a brief history of the discipline and its major foci before turning to the key conceptual lenses and analytic frameworks that each offers to the study of biodiversity conservation. The chapters then conclude by sketching out future directions for their respective disciplines in relation to conservation research and practice. We use text boxes to illustrate our message. "Applications" boxes highlight the practical application of social theory in the real world. "Debates" boxes highlight areas of scientific uncertainty or debate. "Methods" boxes explore the tools and approaches that scholars use to study conservation questions, while "Crossing boundaries" boxes examine topics at the interface of multiple disciplines. Each chapter features at least one of each kind of box, which together present a range of illustrative examples from around the world. We conclude the book with a final chapter that articulates a vision for the future, highlighting the components of what could become a new paradigm for conservation.

We recognize that many ideas, debates, and areas of inquiry are common between social science disciplines. At the same time, however, each discipline has a unique history, the legacy of which is a particular body of scholarship (theories, empirical insights, methodological approaches). Rather than develop our own synthesis or fusion of these diverse approaches, we prefer to lay out this diversity for the reader. In so doing, we provide an entry point to each of the diverse histories of scholarship on conservation and encourage the reader to weave these diverse threads together in a meaningful tapestry to inform the study and practice of conservation.

## For Further Reading

**1** Thinking like a human (Adams 2007, *Oryx* 41: 275–276).

**2** Conservation social science: understanding and integrating human dimensions to improve conservation (Bennett, Roth, Klain, Chan, Christie et al. 2017, *Biological Conservation* 205: 93–108).

**3** Money for nothing? A call for empirical evaluation of biodiversity conservation investments (Ferraro & Pattanayak 2006, *PLoS Biology* 4: e105).

**4** Perceived barriers to integrating social science and conservation (Fox et al. 2006, *Conservation Biology* 20: 1817–1820).

**5** Achieving the promise of integration in social–ecological research: a review and prospectus (Guerrero et al. 2018, *Ecology and Society* 23 (3): 38).

**6** Engage key social concepts for sustainability (Hicks et al. 2016, *Science* 352 (6281): 38–40).

**7** *Summary for Policymakers of the Global Assessment Report on Biodiversity and Ecosystem Services of the Intergovernmental Science-Policy Platform on Biodiversity and Ecosystem Services* (2019, Intergovernmental Science-Policy Platform on Biodiversity and Ecosystem Services, Bonn, Germany).

**8** Conservation and the social sciences (Mascia et al. 2003, *Conservation Biology* 17: 649–650).

**9** *Environmental Social Science: Human-Environment Interactions and Sustainability* (Moran 2010, Wiley-Blackwell, Malden, MA).

**10** Social research and biodiversity conservation (Sandbrook et al. 2013, *Conservation Biology* 27: 1487–1490).

# References

Achard, F., Beuchle, R., Mayaux, P. et al. (2014) Determination of tropical deforestation rates and related carbon losses from 1990 to 2010. *Global Change Biology* 20: 2540–2554.

Adams, W.M. (2007) Thinking like a human: social science and the two cultures problem. *Oryx* 41: 275–276.

Adams, W.M. & Sandbrook, C. (2013) Conservation, evidence and policy. *Oryx* 47: 329–335.

Agrawal, A., Hajjar, R., Liao, C. et al. (2018) Editorial overview: forest governance interventions for sustainability through information, incentives, and institutions. *Current Opinion in Environmental Sustainability* 32: A1–A7.

Amel, E., Manning, C., Scott, B. et al. (2017) Beyond the roots of human inaction: fostering collective effort toward ecosystem conservation. *Science* 356: 275–279.

Bailey, K., Morales, N. & Newberry, M. (2020) Inclusive conservation requires amplifying experiences of diverse scientists. *Nature Ecology & Evolution* 4: 1294–1295.

Balmford, A., Bradbury, R.B., Bauer, J.M. et al. (2021) Making more effective use of human behavioural science in conservation interventions. *Biological Conservation* 261: 109256.

Balmford, A. & Cowling, R. (2006) Fusion or failure? The future of conservation biology. *Conservation Biology* 20: 692–695.

Barnosky, A.D., Matzke, N., Tomiya, S. et al. (2011) Has the Earth's sixth mass extinction already arrived? *Nature* 471: 51–57.

Bennett, N.J., Roth, R., Klain, S.C., Christie, P. et al. (2017) Conservation social science: understanding and integrating human dimensions to improve conservation. *Biological Conservation* 205: 93–108.

Bennett, N.J., Roth, R., Klain, S.C., Clarke, D.A. et al. (2017) Mainstreaming the social sciences in conservation. *Conservation Biology* 31: 56–66.

Boserup, E. (2014 [1965]) *The Conditions of Agricultural Growth: The Economics of Agrarian Change under Population Pressure*. London: Routledge.

Brockington, D. & Igoe, J. (2006) Eviction for conservation: a global overview. *Conservation and Society* 4: 424–470.

Brooks, T.M., Mittermeier, R.A., Da Fonseca, G.A. et al. (2006) Global biodiversity conservation priorities. *Science* 313: 58–61.

Burivalova, Z., Allnutt, T.F., Rademacher, D. et al. (2019) What works in tropical forest conservation, and what does not: effectiveness of four strategies in terms of environmental, social, and economic outcomes. *Conservation Science and Practice* 1 (6): e28.

Campos-Arceiz, A., Primack, R.B., Miller-Rushing, A.J. et al. (2018) Striking underrepresentation of biodiversity-rich regions among editors of conservation journals. *Biological Conservation* 220: 330–333.

Carrasco, L.R., Chan, J., Mcgrath, F.L. et al. (2017) Biodiversity conservation in a telecoupled world. *Ecology and Society* 22 (3): 24.

CBD (Convention on Biological Diversity) (2019) *List of Parties.* https://www.cbd.int/information/parties.shtml (accessed September 16, 2019).

Chapin, M. (2004) A challenge to conservationists. *World Watch* 17 (6): 17–31.

Charnley, S., Carothers, C., Satterfield, T. et al. (2017) Evaluating the best available social science for natural resource management decision-making. *Environmental Science & Policy* 73: 80–88.

Chivan, E. & Bernstein, A. (2008) *Sustaining Life: How Human Health Depends on Biodiversity.* Oxford, UK: Oxford University Press.

Cincotta, R.P., Wisnewski, J. & Engelman, R. (2000) Human population in the biodiversity hotspots. *Nature* 404: 990–992.

Coad, L., Watson, J.E., Geldmann, J. et al. (2019) Widespread shortfalls in protected area resourcing undermine efforts to conserve biodiversity. *Frontiers in Ecology and the Environment* 17: 259–264.

Cowling, R.M. (2014) Let's get serious about human behavior and conservation. *Conservation Letters* 7: 147–148.

Cox, D.T.C., Shanahan, D.F., Hudson, H.L. et al. (2017) Doses of neighborhood nature: the benefits for mental health of living with nature. *BioScience* 67: 147–155.

Deutz, A., Heal, G.M., Niu, R. et al. (2020) Financing nature: closing the global biodiversity financing gap. The Paulson Institute, The Nature Conservancy, and the Cornell Atkinson Center for Sustainability.

Diaz, S., Fargione, J., Chapin, F.S., III et al. (2006) Biodiversity loss threatens human well-being. *PLoS Biology* 4 (8): e277.

Dinerstein, E., Vynne, C., Sala, E. et al. (2019) A global deal for nature: guiding principles, milestones, and targets. *Science Advances* 5 (4): eaaw2869.

Dressler, W., Büscher, B., Schoon, M. et al. (2010) From hope to crisis and back again? A critical history of the global CBNRM narrative. *Environmental Conservation* 37: 5–15.

Ferraro, P.J. & Pattanayak, S.K. (2006) Money for nothing? A call for empirical evaluation of biodiversity conservation investments. *PLoS Biology* 4: e105.

Fish, R., Church, A. & Winter, M. (2016) Conceptualising cultural ecosystem services: a novel framework for research and critical engagement. *Ecosystem Services* 21: 208–217.

Fox, H.E., Christian, C., Nordby, J.C. et al. (2006) Perceived barriers to integrating social science and conservation. *Conservation Biology* 20: 1817–1820.

Gerber, L.R., Runge, M.C., Maloney, R.F. et al. (2018) Endangered species recovery: a resource allocation problem. *Science* 362: 284–286.

Gill, D.A., Mascia, M.B., Ahmadia, G.N. et al. (2017) Capacity shortfalls hinder the performance of marine protected areas globally. *Nature* 543: 665–669.

GlobeScan (2021) Demand under the ban: China ivory consumption research 2020. Washington, DC: World Wildlife Fund.

Golden Kroner, R.E., Qin, S., Cook, C.N. et al. (2019) The uncertain future of protected lands and waters. *Science* 364: 881–886.

Greenfield, S. & Veríssimo, D. (2019) To what extent is social marketing used in demand reduction campaigns for illegal wildlife products? Insights from elephant ivory and rhino horn. *Social Marketing Quarterly* 25 (1): 40–54.

Guerrero, A.M., Bennet, N., Wilson, K.A. et al. (2018) Achieving the promise of integration in social–ecological research: a review and prospectus. *Ecology and Society* 23 (3): 38.

Hardin, G. (1968) The tragedy of the commons. *Science* 162: 1243–1248.

Herrera, D., Ellis, A., Fisher, B. et al. (2017) Upstream watershed condition predicts rural children's health across 35 developing countries. *Nature Communications* 8: 811.

Hicks, C.C., Levine, A., Agrawal, A. et al. (2016) Engage key social concepts for sustainability. *Science* 352: 38–40.

Hintzen, R.E., Papadopoulou, M., Mounce, R. et al. (2019) Relationship between conservation biology and ecology shown through machine reading of 32,000 articles. *Conservation Biology* 00 (00): 1–11. https://conbio.onlinelibrary.wiley.com/doi/pdf/10.1111/cobi.13435.

IPBES (Intergovernmental Science-Policy Platform on Biodiversity and Ecosystem Services) (2019) Summary for Policymakers of the Global Assessment Report on Biodiversity and Ecosystem Services of the Intergovernmental Science-Policy Platform on Biodiversity and Ecosystem Services. Bonn, Germany: IPBES.

Jenks, B., Vaughan, P.W. & Butler, P.J. (2010) The evolution of Rare Pride: using evaluation to drive adaptive management in a biodiversity conservation organization. *Evaluation and Program Planning* 33 (2): 186–190.

Jones, M.S. & Solomon, J. (2019) Challenges and supports for women conservation leaders. *Conservation Science and Practice* 1: e36.

Kareiva, P. & Marvier, M. (2012) What is conservation science? *BioScience* 62: 962–969.

Lackey, R. (2007) Science, scientists, and policy advocacy. *Conservation Biology* 21: 12–17.

Lambin, E.F., Gibbs, H.K., Heilmayr, R. et al. (2018) The role of supply-chain initiatives in reducing deforestation. *Nature Climate Change* 8: 109–116.

Lee, R., Workman, C. & Whan, E. (2016) New National Geographic Society study on ivory demand in five key consumption countries. *Oryx* 50 (2): 203–204.

Lenzen, M., Moran, D., Kanemoto, K. et al. (2012) International trade drives biodiversity threats in developing nations. *Nature* 486: 109–112.

Leopold, A. (1987 [1949]) *A Sand County Almanac, and Sketches Here and There.* New York: Oxford University Press.

Leverington, F., Costa, K.L., Pavese, H. et al. (2010) A global analysis of protected area management effectiveness. *Environmental Management* 46: 685–698.

Liu, J., Hull, V., Batistella, M. et al. (2013) Framing sustainability in a telecoupled world. *Ecology and Society* 18 (2): 26.

Marselle, M.R., Stadler, J., Korn, H. et al. eds. (2019) *Biodiversity and Health in the Face of Climate Change.* Cham, Switzerland: Springer.

Martin, A. (2017) *Just Conservation: Biodiversity, Wellbeing and Sustainability.* New York: Routledge.

Mascia, M.B., Brosius, J.P., Dobson, T.A. et al. (2003) Conservation and the social sciences. *Conservation Biology* 17: 649–650.

Mascia, M.B. & Pailler, S. (2011) Protected area downgrading, downsizing, and degazettement (PADDD) and its conservation implications. *Conservation Letters* 4: 9–20.

Maurer, B.A. (1996) Relating human population growth to the loss of biodiversity. *Biodiversity Letters* 3: 1–5.

Maxwell, S.L., Venter, O., Jones, K.R. et al. (2015) Integrating human responses to climate change into conservation vulnerability assessments and adaptation planning. *Annals of the New York Academy of Sciences* 1355: 98–116.

McKee, J., Sciulli, P., Fooce, C.D. et al. (2004) Forecasting global biodiversity threats associated with human population growth. *Biological Conservation* 115: 161–164.

McKenzie-Mohr, D. (2000) Fostering sustainable behavior through community-based social marketing. *American Psychologist* 55 (5): 531–537.

McKinnon, M.C., Cheng, S.H., Dupre, S. et al. (2016) What are the effects of nature conservation on human well-being? A systematic map of empirical evidence from developing countries. *Environmental Evidence* 5: 1–25.

McShane, T.O., Hirsch, P.D., Trung, T.C. et al. (2011) Hard choices: making trade-offs between biodiversity conservation and human well-being. *Biological Conservation* 144: 966–972.

MEA (Millenium Ecosystem Assessement) (2005) Summary for decision-makers. In: *Ecosystems and Human Wellbeing: Biodiversity Synthesis*, 1–16. Washington, DC: World Resources Institute.

Miller, D.C., Rana, P. & Wahlén, C.B. (2017) A crystal ball for forests? Analyzing the social–ecological impacts of forest conservation and management over the long term. *Environment and Society* 8: 40–62.

Miteva, D.A., Pattanayak, S.K. & Ferraro, P.J. (2012) Evaluation of biodiversity policy instruments: what works and what doesn't? *Oxford Review of Economic Policy* 28: 69–92.

Moon, K., Blackman, D.A., Adams, V.M. et al. (2019) Expanding the role of social science in conservation through an engagement with philosophy, methodology, and methods. *Methods in Ecology and Evolution* 10: 294–302.

Moran, D. & Kanemoto, K. (2017) Identifying species threat hotspots from global supply chains. *Nature Ecology & Evolution* 1: 0023.

Myers, N., Mittermeier, R.A., Mittermeier, C.G. et al. (2000) Biodiversity hotspots for conservation priorities. *Nature* 403: 853–858.

Nature Editorial Board (2022) Biodiversity faces its make-or-break year, and research will be key. *Nature* 601: 298.

Newman, D.J. & Cragg, G.M. (2016) Natural products as sources of new drugs from 1981 to 2014. *Journal of Natural Products* 79: 629–661.

Newton, P., Agrawal, A. & Wollenberg, L. (2013) Enhancing the sustainability of commodity supply chains in tropical forest and agricultural landscapes. *Global Environmental Change* 23: 1761–1772.

Oldekop, J.A., Fontana, L.B., Grugel, J. et al. (2016) 100 key research questions for the post-2015 development agenda. *Development Policy Review* 34: 55–82.

Oldekop, J.A., Holmes, G., Harris, W.E. et al. (2016) A global assessment of the social and conservation outcomes of protected areas. *Conservation Biology* 30: 133–141.

Olson, D.M., Dinerstein, E., Wikramanayake, E.D. et al. (2001) Terrestrial ecoregions of the world: a new map of life on earth: a new global map of terrestrial ecoregions provides an innovative tool for conserving biodiversity. *BioScience* 51: 933–938.

Ostrom, E., Janssen, M.A. & Anderies, J.M. (2007) Going beyond panaceas. *Proceedings of the National Academy of Sciences of the United States of America* 104: 15176–15178.

Parathian, H. (2019) Understanding cosmopolitan communities in protected areas: a case study from the Colombian Amazon. *Conservation and Society* 17: 26–37.

Pecl, G.T., Araújo, M.B., Bell, J.D. et al. (2017) Biodiversity redistribution under climate change: impacts on ecosystems and human well-being. *Science* 355: eaai9214.

Phillipson, J., Lowe, P. & Bullock, J.M. (2009) Navigating the social sciences: interdisciplinarity and ecology. *Journal of Applied Ecology* 46: 261–264.

Potts, S.G., Imperatriz-Fonseca, V., Ngo, H.T. et al. (2016) Safeguarding pollinators and their values to human well-being. *Nature* 540: 220–229.

Redford, K.H., Padoch, C. & Sunderland, T. (2013) Fads, funding, and forgetting in three decades of conservation. *Conservation Biology* 27: 437–438.

Sánchez-Bayo, F. & Wyckhuys, K.A.G. (2019) Worldwide decline of the entomofauna: a review of its drivers. *Biological Conservation* 232: 8–27.

Sandbrook, C., Adams, W.M., Buscher, B. et al. (2013) Social research and biodiversity conservation. *Conservation Biology* 27: 1487–1490.

Sandbrook, C., Fisher, J.A., Holmes, G. et al. (2019) The global conservation movement is diverse but not divided. *Nature Sustainability* 2: 316–323.

Sandbrook, C., Scales, I.R., Vira, B. et al. (2011) Value plurality among conservation professionals. *Conservation Biology* 25: 285–294.

Scales, I.R. (2012) Lost in translation: conflicting views of deforestation, land use and identity in western Madagascar. *The Geographical Journal* 178: 67–79.

Scales, I.R. (2014) Green consumption, ecolabelling and capitalism's environmental limits. *Geography Compass* 8: 477–489.

Schultz, P.W. (2011) Conservation means behavior. *Conservation Biology* 25: 1080–1083.

Seddon, A.W., Macias-Fauria, M., Long, P.R., Benz, D. & Willis, K.J. (2016) Sensitivity of global terrestrial ecosystems to climate variability. *Nature* 531: 229–232.

Sims, K.R.E. & Alix-Garcia, J.M. (2017) Parks versus PES: evaluating direct and incentive-based land conservation in Mexico. *Journal of Environmental Economics and Management* 86: 8–28.

Soulé, M.E. (1985) What is conservation biology? *BioScience* 35: 727–734.

Steffen, W., Crutzen, J. & Mcneill, J.R. (2007) The Anthropocene: are humans now overwhelming the great forces of Nature? *Ambio* 36: 614–621.

Stern, M.J. (2018) *Social Science Theory for Environmental Sustainability: A Practical Guide*. Oxford, UK: Oxford University Press.

Sutherland, W.J., Fleishman, E., Clout, M. et al. (2019) Ten years on: a review of the first global conservation horizon scan. *Trends in Ecology & Evolution* 34: 139–153.

Sutherland, W.J., Pullin, A.S., Dolman, P.M. et al. (2004) The need for evidence-based conservation. *Trends in Ecology & Evolution* 19: 305–308.

Tayleur, C., Balmford, A., Buchanan, G.M. et al. (2017) Global coverage of agricultural sustainability standards, and their role in conserving biodiversity. *Conservation Letters* 10: 610–618.

Taylor, D. (2014) *The State of Diversity in Environmental Oragnizations*. Washington, DC: Green Diversity Initiative (Green 2.0).

Teel, T.L., Anderson, C.B., Burgman, M.A. et al. (2018) Publishing social science research in *Conservation Biology* to move beyond biology. *Conservation Biology* 32: 6–8.

Tiffen, M., Mortimore, M. & Gichuki, F. (1994) *More People, Less Erosion: Environmental Recovery in Kenya*. Chichester, UK: John Wiley & Sons.

Travers, H., Selinske, M., Nuno, A. et al. (2019) A manifesto for predictive conservation. *Biological Conservation* 237: 12–18.

UNEP-WCMC (United Nations Environment Programme World Conservation Monitoring Centre) (2020) Protected areas map of the world. www.protectedplanet.net. (accessed January 5, 2021).

Veríssimo, D. (2019) The past, present, and future of using social marketing to conserve biodiversity. *Social Marketing Quarterly* 25: 3–8.

Veríssimo, D. & Wan, A.K. (2019) Characterizing efforts to reduce consumer demand for wildlife products. *Conservation Biology* 33 (3): 623–633.

Waldron, A., Mooers, A.O., Miller, D.C. et al. (2013) Targeting global conservation funding to limit immediate biodiversity declines. *Proceedings of the National Academies of Science of the United States of America* 110: 12144–12148.

Wardropper, C.B., Esman, L.A., Harden, S.C. et al. (2022) Applying a "fail-fast" approach to conservation in US agriculture. *Conservation Science and Practice* 4: e0619.

Watson, J.E., Dudley, N., Segan, D.B. et al. (2014) The performance and potential of protected areas. *Nature* 515: 67–73.

West, P., Igoe, J. & Brockington, D. (2008) Parks and peoples: the social impact of protected areas. *Annual Review of Anthropology* 35: 251–277.

Wilson, K.A., Auerbach, N.A., Sam, K. et al. (2016) Conservation research is not happening where it is most needed. *PLoS Biology* 14: e1002413.

Wright, A.J., Veríssimo, D., Pilfold, K. et al. (2015) Competitive outreach in the 21st century: why we need conservation marketing. *Ocean & Coastal Management* 115: 41–48.

# 2

# Social Science Foundations

*Katie Moon and Deborah Blackman*

## 2.1  Introduction

One of the main premises of this book is that the conservation of biological diversity is a social phenomenon. As such, understanding the threats to **biodiversity** and offering possible solutions to conservation problems requires input from the social sciences, as well as dialogue between disciplines. Although there are many benefits to such interchange, important differences exist in how researchers from different disciplines conduct their research (e.g. Moon, Adams, et al. 2019).

This chapter focuses on how knowledge is discovered and formed in the social sciences, and the assumptions that influence research and practice. As subsequent chapters will reveal, the social sciences can differ significantly in terms of what is studied (the object of study), how it is studied (research design), and how knowledge generated is valued and used (application). Any attempt to establish a distinct field of **conservation social science**, or even to develop effective interdisciplinary research projects that include a social science discipline, requires at least a basic understanding of the fundamentals that underpin the broad discipline of the social sciences.

One of the most obvious differences between academic disciplines is in *what* is being studied. In the natural sciences, the object of study tends to be physical; in the social sciences, it can be physical and non-physical. Using money as a simple example, a coin or a bank note can be described in terms of its physical properties, such as its color, weight, material, and size. But to understand precisely what money is and how it functions, it must also be described in terms of its non-physical properties, for example, what money can buy (or indeed not buy) in a society (e.g. food, the ownership of land, an individual's labor) and who creates money and guarantees its value (e.g. the monarchy, the state, banks).

A forest provides a conservation-relevant example of how natural and social science disciplines might approach a specific object of study differently. Biologists might measure various biological parameters along a transect (e.g. canopy cover, forest structure, the diversity of various flora and fauna within the forest **ecosystem**). They can then convert these data into various indices (e.g. of biodiversity or of human impacts), which they could use to describe a forest (e.g. according to type: tropical, dry-deciduous), and record the

*Conservation Social Science: Understanding People, Conserving Biodiversity*, First Edition.
Edited by Daniel C. Miller, Ivan R. Scales, and Michael B. Mascia.
© 2023 John Wiley & Sons Ltd. Published 2023 by John Wiley & Sons Ltd.

levels of threat and conservation importance (e.g. various prioritizing tools such as ecoregions and biodiversity hotspots).

Objects of study in the social sciences can also be physical. For example, researchers could focus on the land-use practices of people living in the forest, including the choice of crops, adoption of technologies, and the influence of income on conservation practices. Other social science approaches might focus on non-physical aspects, such as beliefs that influence attitudes and behaviors toward nature. Such approaches might look at how beliefs, taboos, and values shape forest use. Other researchers might be interested in the power struggles over forest resources, for instance, between landholders, conservation organizations, and governments. The objects of study in these examples are immaterial and non-physical, in that they cannot be seen or touched, but instead exist within human consciousness. Of course, they nevertheless can lead to human action and ecological change and have real material effects.

Thus, a forest can be understood as a physical entity made up of organic and non-organic materials (e.g. plants, animals, fungi, soils, water), a landscape that is perceived subjectively in different ways by different individuals and groups (e.g. as wild, beautiful, frightening, sacred, a source of profit), and as a site for political struggles (e.g. between logging companies, conservation organizations, and indigenous groups), among others. The ways in which we seek to generate knowledge about different aspects of a system are underpinned by different philosophical assumptions.

This chapter serves as a guide to different ways of thinking about conservation social science research (Figure 2.1). We begin by examining two very different perspectives of reality: realism and relativism. These **ontologies** (philosophical positions on the nature of reality) include assumptions about the fundamental nature of reality, what the real world is made of, and whether we can ever, in fact, objectively experience and know a "real world." Next, we turn our attention to how we create knowledge about reality, which is the focus of **epistemology**. Epistemology is the bridge between thinking and reality—whether we can create knowledge that corresponds with something that is "real." We explore three of the many possible epistemological positions: objectivism, constructionism, and subjectivism. We then examine different theoretical perspectives. These perspectives, also known as paradigms or worldviews, represent a person's system of values, which guides their (research) action. Finally, we provide an overview of **methodology**—the rationale and strategy for creating knowledge—used in the different social sciences. We focus on how different research designs are underpinned by different views of how to produce, justify, and apply knowledge and the implications for conservation social science and practice. In so doing, this overview points to key references where interested readers can learn more about the wide range of research design elements used in the social sciences (including methods that produce both **qualitative** and **quantitative** data) (e.g. Patton 2002; Denzin & Lincoln 2011; Bernard 2013; Bryman 2016).

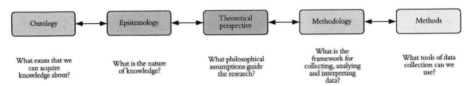

**Figure 2.1** Overview of the relationships among ontology, epistemology, theoretical perspective, methodology, and methods. Adapted from Crotty (1998) Figure 1, p. 4.

It is important to note that this chapter engages primarily with "Western" ontologies and epistemologies, that is, those derived from the historical context of Western Europe but that, through colonialism and other means, have an influence that extends beyond the boundaries of that original geography. Such ontologies, epistemologies, and related methods underpin much of contemporary social science and are therefore our focus here. We recognize and support, however, the need for engagement with indigenous and non-Western ontological and epistemological approaches to conservation social science. Works by Agrawal (1995) and Chilisa (2020), among others (e.g. Wilson 2008; Tuhiwai 2021), provide a starting point for such engagement for the interested reader. Importantly, such engagement should seek to participate in processes of **decolonization** (Parsons & Fisher, 2020); acknowledge the **social construction** and plurality of knowledge/s (Castleden et al., 2017); practice respect, reciprocity, and responsibility (Kirkness & Barnhardt 1991); and work alongside indigenous people and scholars to build partnerships for knowledge sharing (e.g. Woodward et al., 2020).

## 2.2 Ontology

**Ontology** is the study of what is considered to be "real," or what exists in the world that humans can acquire knowledge about. Those who study ontology ask, "How sure can we be about the nature of reality? What does it mean for something to be 'true' or 'real'?" These broad questions are important to consider because our experience of reality is limited by our senses, as well as our interpretation of that sensory information, which can differ between individuals. Therefore, we can ask, "What 'truth claims' can a researcher make about reality? How might researchers deal with different and conflicting ideas of reality?"

Making ontological assumptions explicit helps others understand how and why research is conducted in certain ways. In practical terms, revealing these assumptions can help with understanding why individuals and groups might disagree about the need for, and nature of, conservation in different spatial and temporal contexts. For example, restoration ecologists might make recommendations for a deforested landscape to be replanted with certain plant species, based on particular criteria and objectives (e.g. the recovery of endangered species or carbon sequestration). Here, the landscape is perceived and managed according to the categories of ecological science, which are assumed to correspond to the way the world "really is." Yet these recommendations might conflict with the perceptions of nature and landscapes held by local communities, for example, the notion of "country" as a cultural–spiritual landscape in which plants play many important roles that are ecological, spiritual (e.g. for storytelling), and cultural (e.g. providing weather knowledge), as is the case within Aboriginal Australian **culture** (Kinnane 2005; Stocker et al. 2016). Thus, thinking about ontology helps us understand how people define the nature of their reality and the consequences that arise from those definitions.

Many ontological positions exist (Feyerabend 1981; Morton 1996; Stokes 1998; Johnson & Gray 2010; Tashakkori & Teddlie 2010). One of the main axes of difference across these positions is the dichotomy between realism and relativism (Figure 2.2). This dichotomy can be useful to consider in understanding the importance of ontology and its influence on the application and outcomes of social science.

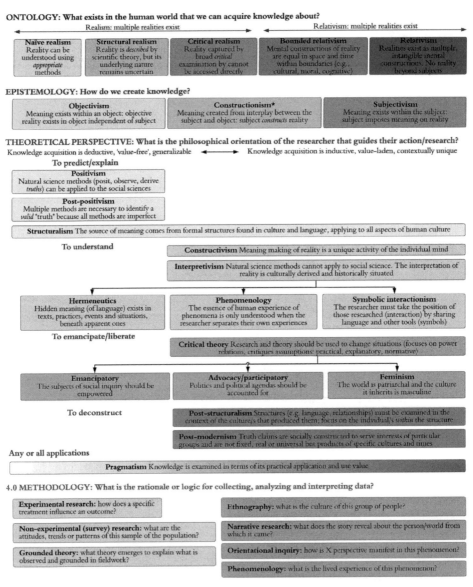

**Figure 2.2** Social science research guide consisting of ontology, epistemology, and theoretical perspective. When read from left to right, elements take on a more multidimensional nature (e.g. epistemology: objectivism to subjectivism). The elements within each branch are positioned according to their congruence with elements from other branches so when read from top to bottom (or bottom to top), elements from one branch align with those from another (e.g. critical realist ontology, constructionist epistemology, and interpretivist theoretical perspectives). Note that subcategories of elements (e.g. emancipation, advocacy/participation, and feminism) are to be interpreted as positioned under the parent category (i.e. critical theory). *Source:* Adapted from Moon & Blackman 2014 Figure 1, p. 1169.

## 2.2.1 Realism versus Relativism

Realist ontology holds that one single reality exists that can be studied, understood, and experienced as a "truth." In other words, it assumes that a real world exists independent of, or external to, human experience (Moses & Knutsen 2012); an external reality exists independent of our minds. This ontological position dominates in the natural sciences, where the purpose of research is to describe and explain an objective "reality" (Evely et al. 2008).

In contrast to realism, relativist ontology holds that reality is constructed within the human mind, so that no one "true" universal reality exists. Instead, reality is "relative" according to each individual who experiences it. Bounded relativists argue that a shared reality exists within a bounded group, but that different realities exist across groups (Moon & Blackman 2014). For instance, in some cultures, beliefs in the healing properties of certain (parts of) animals have led to their continued harvest, even when these animals are endangered (Graham-Rowe 2011). The belief in the "reality" of the benefit is shared among those within the cultural group. In another culture, people might not hold beliefs about the healing properties of animals but might hold beliefs about animal sentience. As a result, they might consider it unacceptable to harvest animals, especially when harvest poses a risk to the species' well-being as individuals or the survival of the species (e.g. Biggs et al. 2013).

Unlike the natural sciences, the social sciences can align with either, or both, relativism and realism. Many social scientists, like natural scientists, carry out research based on the principle that the things they study exist independently of human minds. However, other social scientists work, to varying degrees, on the assumption that "reality" is relative. Using a hypothetical example of tropical deforestation, an ecologist might consider the description of an ecosystem as "degraded" to be a simple factual statement that can be supported with measurements of ecological function (e.g. vegetation cover or species diversity) while a relativist might consider "degradation" to be less a statement of fact and more a product of seeing landscapes from a culturally influenced (and thus subjective) viewpoint. In Madagascar, for instance, ecologists and conservation practitioners have tended to assume that the island was once entirely forested and that its extensive grasslands are therefore degraded landscapes and the product of human action. Research has shown, however, that the island was never entirely forested and that not all of Madagascar's grasslands are **anthropogenic** (see Box 5.1 in Chapter 5, Human Geography, for more details on the "island forest" debate). Perceptions of "reality" for these ecologists and practitioners were influenced by assumptions of what they expected to see on the island. A relativist ontology means recognizing and engaging with different views of reality and relativist research is typically "people-centered" (Brown 2003). This includes research that elicits mental models (see Box 2.1.) to reveal individuals' knowledge, values, and beliefs that frame how they view the world (Kolkman et al. 2007; Moon, Guerrero, et al. 2019).

---

**Box 2.1  Applications: Mental models**

A mental model exists in the mind as a small-scale model of how (a part of) the world works (see Moon & Browne 2021). Moon et al. (2015) elicited the mental models of scientists and policy advisors to understand the extent to which the two groups considered that scientific evidence should be used in policy-making processes for invasive species

---

*(Continued)*

---

**Box 2.1 (Continued)**

management in Tasmania, Australia. The two groups described different policy-making realities. Policy advisors' mental models showed policy making as a multifaceted process that needs to accommodate multiple roles, for both scientific evidence and community members, and incorporate the knowledge and insights of communities. In contrast, scientists' mental models revealed a preference for evidence-based policy making, where the role for community members was limited to assisting the government in implementing the eradication program (Moon et al. 2015). The scientists explained that they considered community stakeholders' knowledge to be of limited value in informing policy because it was based on misinformation, misconceptions, and mistrust in science. This example shows that different cultures (i.e. a scientific culture and a government culture) can form their own sense and definitions of what is true or false. While policy advisors appeared to accept multiple views of reality, scientists appeared to consider that one "true" reality existed, which was based on scientific evidence.

---

## 2.3 Epistemology

Epistemology is concerned with "providing the philosophical grounding for deciding what kinds of knowledge are possible and how we can ensure that they are both adequate and legitimate" (Maynard 1994, p. 10). In other words, it asks questions about the nature of knowledge—how do we know what we know? Epistemology considers all aspects of the validity, scope, and methods of acquiring knowledge of reality, such as (i) what constitutes a knowledge claim; (ii) how can knowledge be produced or acquired; and (iii) how can the extent of its applicability be determined?

A wide range of epistemologies are described, discussed, and used in the social science literature (e.g. Audi 1998; Crotty 1998; Patton 2002; Evely et al. 2008; Denzin & Lincoln 2011). To assist in understanding the role of epistemology in conservation social science, we examine three different positions commonly found in the social sciences: objectivism, constructionism, and subjectivism (Figure 2.2) (Crotty 1998). We use these three positions simply to provide some clear differentiations between epistemologies and how they influence knowledge creation. Reflecting the focus of this book, the attention of this section is on the "researcher"—how do they seek to create knowledge?

### 2.3.1 Objectivism

Objectivist epistemology is based on the assumption that it is possible to discover an objective "truth" that is empirically verifiable, valid, generalizable, and independent of social thought and social conditions (Crotty 1998). This epistemological position underpins the natural sciences. It assumes that the researcher can remain detached from the subjects they are researching and that their interests, values, or interpretations do not bias the generation of knowledge (Pratt 1998). Objectivism assumes that "people can rationally come to know the world as it really is; the facts of the world are essentially there for study" (Pratt 1998, p. 23). The value of objectivist research to conservation is in its external validity

**Figure 2.3** Abandoned dwelling, Western Province, Solomon Islands. Objectivist research can be used to identify the dominant factors that explain people's behaviors and motivations, such as fishing practices, which could be transferred to similar contexts. Photo: K. Moon.

(results can be applied to other contexts) and its reliability (results are consistently produced) (Evely et al. 2008).

An example of objectivist social science research is one that focuses on a defined system in which conservation takes place, for instance, fisheries (Figure 2.3). To help reduce fishing in depleted fisheries, researchers assume that they can objectively discover the factors that affect fishers' decisions and that lead to overfishing. They might start by conducting a survey of a random sample of a defined population of fishers. The resultant data could be analyzed to identify dominant factors that explain how and why fishers make decisions. Isolating these factors could then be used to try to change fisher behavior, for example, by investing in creating alternative employment opportunities for those fishers who are predicted to be most likely to continue fishing (e.g. Cinner et al. 2009). Successful interventions could be transferable to other communities that are socioeconomically similar. The aim of objectivist research is to describe or "test" reality by collecting and analyzing evidence to explore assertions, corroborate claims, and provide correspondence with the "real world" (Patton 2002).

### 2.3.2 Constructionism

Constructionist epistemology assumes that "truth," or meaning, comes into existence in and out of our engagement with the realities in our world; no "real world" pre-exists that is independent of human activity or symbolic language (Bruner 1986). In other words, "knowledge is not passive—a simple imprinting of sense data on the mind—but active; mind does something with these impressions, at the very least, forms abstractions or concepts" (Schwandt 1994, p. 125). This epistemological position assumes that different individuals can construct meanings of the same object or phenomenon in different ways. How an individual engages with and understands their world is based on their cultural, historical, and social perspectives and meaning arises as a result of being human and inter-acting with other humans (Crotty 1998; Creswell & Creswell 2017).

Constructionist research is often context-specific, since it does not necessarily assume that findings will be universal. For example, Weeks and Packard (1997) looked at resource dependency and the willingness of two different resource-dependent communities—west Texas ranchers and Gulf Coast oyster fishers—to accept different scientific management prescriptions. They found that scientists viewed resource management according to the scientific enterprise and valued factors associated with scientific integrity, such as method-ological rigor. In contrast, the resource-dependent communities viewed resource management according to their own context: their historical relationship with the management agency, the match between scientific explanations and local experience (knowledge), and the conceptual fit between managers' and communities' views on resource management. For instance, ranchers considered their own local knowledge to be a better predictor of threats to ecosystem function than scientific models.

Constructivist research can therefore generate *contextual* understandings of a defined conservation topic or problem (Box 2.2). Findings of this type of research can be used to enable governments and stakeholders, for instance, to design contextually relevant responses to conservation problems, which have been demonstrated by some researchers to have a higher likelihood of success (e.g. Waylen et al. 2010).

---

**Box 2.2  Application: Constructed knowledge of a fishery**

Brewer (2013) sought to understand perceptions of the ecological state of a coral reef fishery in the Solomon Islands (Figure 2.4) by exploring the dominant **discourses** among fishers and fish traders on the topic of reef fish decline, and then comparing these discourses to scientific knowledge. His research adopted a constructionist epis-temology to inform local conservation initiatives in ways that were meaningful to the fishers and fish traders in the region.

Brewer found that overall, both fishers and fish traders tended to identify fishing as the main cause of fish decline. These perceptions agreed with scientific findings and explanations. However, there was no single dominant discourse. In other words, not all fishers and traders agreed on the drivers of the fish decline. There were some-times clear dichotomies in perceptions. The most pronounced dichotomy was between those respondents who blamed fish decline on economic self-interest and laziness (with fishers using modern fishing gear and dynamite), and those who attributed fish decline to poverty and lack of alternatives (forcing people to overfish). There were also dichotomies in terms of the management strategies proposed to increase fish stocks. Respondents tended to perceive that either command-and-control (i.e. laws and pen-alties) or community-based management would increase fish stocks. Traders were more likely than fishers to be supportive of market regulations that restricted fish size and fishing effort, while fishers tended to be more supportive of spatial restrictions on fishing through community cooperation. Differences in views and perceptions could be explained, in part, by local knowledge networks; discourses tended to be place-based. Brewer (2013, p. 252) suggests that fishers and fish traders are likely to "have developed a site-specific market culture relating to the fishery, including a shared understanding of causality of fish stock variability." These findings have important implications for

**Box 2.2  (Continued)**

**Figure 2.4**  Fishing boats amid runoff from deforestation activities, Western Province, Solomon Islands. Coral reef fisheries in the Solomon Islands, where dominant discourses within the fishery are constructed on the basis of a range of social–ecological factors (e.g. knowledge and level of wealth). Photo: K. Moon.

managing natural resources. The similarity between local perceptions of fishery decline and scientific understandings suggests that local resource users are aware of, and thus might support, fishery management strategies based on scientific evidence. However, management strategies must pay close attention to differences in perceptions between resource users, as perceptions are likely to differ between locations and because threats to resources and preferred management strategies are likely to be context specific.

### 2.3.3  Subjectivism

Subjectivist epistemology assumes that people impose meanings and values on the world and interpret it in ways that make sense to them (Crotty 1998; Pratt 1998). A lot of risk-perception research adopts subjectivist epistemology, because people tend to perceive risk on the basis of very personal experiences (Burgman 2005; Moon, Adams et al. 2019). For example, people who lived through the 2019/2020 Australian "megafires"—which burned more than 24 million hectares and saw the loss of an estimated 3 billion animals (Celermajer et al. 2021)—will have had pre-existing mental models (see Box 2.1) that would have shaped the way they perceived and experienced the fires (Moon 2020). To illustrate, Australians in fire-affected communities who assumed the government (and firefighting service) would provide a certain level of protection had the potential to feel abandoned when they had to face the fires without the assistance of the emergency services, as happened in many cases. In contrast, those who did not expect the government to provide protection were arguably less likely to experience these social-psychological effects, and potentially be more prepared. Assumptions about the predictability and controllability of the fires also affected the ways in which people prepared for and responded to the fires.

Whereas the motto of objectivism might be "seeing is believing," the motto of subjectivism might be "believing determines what is seen" (Pratt 1998, p. 25). To illustrate, a shadow in the water could be interpreted differently by a person scuba diving according to whether she or he was waiting for a boat, was alerted to a shark in the area, or was expecting a change in the weather—the same shadow could be seen as a boat, a shark, or a cloud. For subjectivists then, "we see the world as we are; that which we have inside, we see outside" (Pratt 1998, p. 24).

Subjectivists assume that each individual observes the world from a specific place of purpose and interest. For example, a researcher taking a subjectivist approach to climate change might explore how emotion, values, worldviews, trust, and imagery influence perceptions of the risks of climate change (Slovic 2000). They would not necessarily be looking for what is happening external to the person (e.g. whether or not the person is likely to experience an extreme weather event) but rather what is happening internally to the person (e.g. how a person assesses risk). The thought processes in which an individual engages could be intimately tied to her or his external environment, or completely separate from it. The aim of the research might be to uncover important insights into the factors that contribute to individual behavior (Fishbein & Ajzen 1975), revealing how individuals' experiences shape their perceptions of, and behaviors within, the world.

We can return to the example of tropical deforestation to contrast the role of different epistemologies in informing approaches to conservation research and policy. An objectivist social scientist working on tropical deforestation might focus on studying the behavior of individual farmers at the forest frontier and reducing their motivations for clearing forest to a discrete set of (testable) criteria (e.g. poverty or changes in crop prices). The researcher might discover a number of dominant motivations for forest clearance and trial interventions that respond to those motivations (e.g. alternative sources of income).

In contrast, a constructionist working on tropical deforestation might focus on how culture, history, and experience create meaning around various forest-clearance activities, such as slash-and-burn agriculture, within a defined social context. The researcher might find that forest clearance is deeply linked to farmers' cultural identities, presenting unique challenges for changing practices (e.g. Scales 2012).

The subjectivist, adopting a third epistemology, might focus on interpretation, seeking to understand what the forest means to different people and determine how widely held those meanings are and how they correspond between people with different levels of experience. These subjective meanings could be useful in understanding the likelihood that various interventions would achieve their intended outcomes.

## 2.4 Theoretical Perspectives

Theoretical perspectives represent a system of values to which people adhere (Evely et al. 2008), and which guide action (Guba 1990). They are also called paradigms (Guba & Lincoln 1994; Morgan 2007), perspectives (Patton 2002), and worldviews (Creswell & Creswell 2017). Whereas epistemology is about beliefs around knowledge, theoretical perspectives can be considered as a set of assumptions that structure an approach to research, something personal that influences what, how, and why research is conducted (i.e. its purpose). For example, some authors think that "conservation biology is science in advocacy for certain normative agendas ... characterizing habitat loss and reduction of biodiversity as crises, asserting the intrinsic value of biodiversity, and acknowledging our responsibilities to effect positive change or prevent harm" (Roebuck & Phifer 1999, p. 444). This **normative** (i.e. based on value judgments) or mission-oriented (see Soulé 1985) position can influence how conservation science is conducted, particularly in terms of what questions are asked, how data is interpreted, and what recommendations are made. Thus, theoretical perspectives are important to conservation science because, when made explicit, they reveal some of the assumptions that researchers bring to their research, influencing choices of methodology and methods (Crotty 1998) (see Table 2.1).

In the social sciences, research designs are not necessarily aligned with a single theoretical perspective and all its associated characteristics (Bietsa 2010). It is common for more than one theoretical perspective to resonate with researchers, and even for researchers to change their perspectives (and potentially the related epistemological and ontological positions) toward their research over time (see Moses & Knutsen 2012; Moon, Adams et al.

**Table 2.1** Examples of research questions, and their associated assumptions, on forest use and deforestation to illustrate different research approaches that could be taken to studying the same topic according to different philosophical perspectives of the system being studied.

| Theoretical perspective | Example research question | Researcher's assumption(s) |
| --- | --- | --- |
| Structuralism | What is the purpose of the (social) structural relationships in this community (e.g., social classes, governments) and how do they influence forest clearance practices here and elsewhere? | Once I can understand the systematic structure of social classes and relationships, I can generalize the knowledge and apply it to all aspects of human culture (in space and time). |
| Constructivism | What currently motivates individuals in this community to clear forest? | I know that each individual defines and frames problems in his or her own way, and these differences must be understood to evaluate the system. |
| Hermeneutics | Why don't individuals stop clearing forest or cutting down trees for timber, when they said they would? | I can interpret the (hidden) meanings of a text/event from the perspective of the author/participant, within its social and historical context. |
| Phenomenology | Why do people clear forest? | I believe that researchers can put their "system of meaning" (of reality) to one side, so as to interpret the immediate personal experience of a phenomenon, to give rise to a new, refreshed, or richer meaning of the phenomenon. |
| Symbolic interactionism | How do different individuals' descriptions, definitions, and metaphors of the trees/forest affect land use outcomes in this community? (e.g., are the trees considered to be part of a "forest" or are they considered to be a "resource?") | I believe that the *meaning* of objects arises out of social interaction (*language*) between people, and that people interact with and interpret objects on the basis of the meanings those objects have. People are conscious of their role in interaction (*thought*) and are able to change their behavior. |

*(Continued)*

**Table 2.1** (Continued)

| Theoretical perspective | Example research question | Researcher's assumption(s) |
|---|---|---|
| Emancipation | How can we ensure that the community shares in the benefits of logging, or alternatives to logging? | I want to create a mutual interdependence between the research participants and me to transform structures that exploit people. |
| Advocacy/ participation | How can we garner support and develop effective forest governance structures to enable sustainable livelihoods in this community? | I want to collaborate *with* the people in the system, rather than conduct research *on* them, to create an agenda for active change/political reform. |
| Feminism | Does examining deforestation from a feminist perspective offer alternative understandings of the dynamics and power relations among and between the stakeholders? | I believe that forest clearance is a masculine activity and reflects a patriarchal world and culture. Exploring deforestation solely from a traditional scientific (i.e., non-feminist) perspective limits our opportunities to understand behavior and create change. |
| Post-structuralism | What are the narrative structures within this system that describe how a deforestation debate has arisen in this historical context? | I need to understand not only what the system appears to be, but also how it emerges from the history and culture of the people who comprise the system. In understanding the history and culture, I can come to understand whether or not what I have learned about this system can be applied to other systems. |
| Postmodernism | Why is it assumed that deforestation is a problem? | I am skeptical of approaches to generating knowledge and want to scrutinize, contest, deconstruct, and make visible the (invisible) origins, assumptions, and effects of meaning. |

*Source:* Adapted from Moon & Blackman 2014.

2019). For example, Moon, Adams et al. (2019) share aspects of their research journeys, explaining how their experiences within the social sciences changed the ways in which they approached their research practice. They identified the value that comes from "understanding that social science is not just answers, but stories; not just data, but meaning; and that place is a critical part of understanding socio-ecological phenomena and processes" (Moon, Adams et al. 2019, p. 427). They came to learn that knowledge is partial, context is critical, and that **positionality** (i.e. personal characteristics, such as gender, race, personal experiences, beliefs, biases, and political and **ideological** stances) influence perspective (see also Berger 2015 and Box 2.4).

Many generalized philosophical perspectives have been defined, some of which can also be viewed as, and interchanged with, epistemological or ontological positions (Tashakkori & Teddlie 2010; Denzin & Lincoln 2011). We include in Figure 2.2 some perspectives we believe are relevant to conservation social science, plus others that are not commonly used in conservation social science but could play a more important role in expanding and extending approaches to scientific inquiry. In Sections 2.4.1–2.4.5, we elaborate on five common perspectives and present additional perspectives in Table 2.1 (see also Figure 2.2).

### 2.4.1  Positivism and Post-positivism

**Positivism** is based on a conviction that only knowledge gained through the scientific method (i.e. deriving predictions from hypotheses) through unprejudiced use of the senses is accurate and "true" (Crotty 1998). Post-positivism is similar but is based on the premise that humans can never know reality perfectly. Rather than "proving," they seek to "falsify" their theories or laws, whereby a genuine counter-instance would act to falsify the theory (Popper 1963).

   We can use our tropical deforestation example once again to provide a simple distinction between these two perspectives. A positivist approach would involve testing a hypothesis, for example, that providing farmers at the forest frontier with a financial payment equal to the value of the crops they would have grown on deforested land will prevent deforestation. A post-positivist approach would test both a null hypothesis—providing farmers with a financial payment to stop deforestation will have no effect on deforestation activities—and an alternative hypothesis—providing farmers with a financial payment to stop deforestation will result in a net reduction in deforestation activities. Both positivist and post-positivist social sciences assume that people (including their culture and behavior) and the social systems of which they are a part can be studied in the same way as natural processes and systems (e.g. Evely et al. 2008).

### 2.4.2  Interpretivism

Interpretivism emerged in "contradistinction to positivist attempts to understand and explain human and social reality" (Crotty 1998, pp. 66–67; see also Schwandt 1994). Interpretivism assumes that the object of study in the social sciences is fundamentally different from the natural sciences, and thus requires a "different logic of research procedure, one that reflects the distinctiveness of humans as against the natural order" (Bryman 2016, p. 15). Instead of seeking to identify regularities or establish "laws" that explain human behavior, interpretivist approaches broadly seek to understand events or phenomena, often qualitatively (Crotty 1998). Returning to the tropical deforestation example, an interpretivist approach might explore how culture explains why farmers clear forest. For instance, in Madagascar, slash-and-burn agriculture is part of a cultural system of taboos, where forest can be cleared as long as offerings are made to the ancestral spirits that inhabit the forest (Scales 2012). Interpretivist research outcomes emerge from the researcher's interaction with the participants, and all the (different) interpretations are considered contextually dependent on the history and culture that influences how each individual interprets and makes meaning of their world.

### 2.4.3  Critical Theory

Critical theory aims to reveal conflict and oppression and/or bring about change (Crotty 1998; Evely et al. 2008). According to Horkheimer (cited in Bohman 2016), critical theory is adequate only if it is (i) explanatory (explains what is wrong with current social reality),

(ii) practical (identifies those actors who can change that reality), and (iii) normative (provides clear **norms** for criticism and achievable and practical goals that support social transformation), all at the same time.

Researchers who adopt a critical theory approach begin with an explicit ideological perspective that dictates how a chosen theoretical framework will direct data collection and interpretation (Patton 2002). For example, a researcher might start from the position that power imbalances exist within a community (Figure 2.5.) and then explore those imbalances. A critical theorist might ask what power imbalances manifest when Western scientists seek and use traditional ecological knowledge to inform conservation interventions (Box 2.3; see also Chapter 3, Anthropology and Conservation, for more on local environmental knowledge). Critical theory has an important role to play in conservation social science, particularly in bringing about positive change for minority or oppressed groups. This perspective aligns with the idea that "the philosophers have only *interpreted* the world in various ways; the point is to *change* it" (Marx 1845[1992], p. 423).

---

**Box 2.3   Application: Critical theory and traditional (or indigenous) ecological knowledge**

Conservation researchers and policy makers have increasingly sought to draw on traditional or indigenous ecological knowledge (TEK or IEK) to improve conservation outcomes (Berkes et al. 2000; Bohensky & Maru 2011; Althaus 2020). For example Jackson et al. (2014) found IEK was different yet complementary to scientific knowledge on freshwater fishes and provided nuanced data on fish predation, fish habitat, and seasonal and inter-annual variation in abundance that affected conceptual models of flow ecology and influenced the structure of risk assessment tools.

At first glance, inclusion of TEK or IEK might appear to make conservation more inclusive. Yet, highly unequal power relations are often involved (e.g. Parsons & Fisher 2020). In reviewing the literature, Shackeroff and Campbell (2007) identified three main power relations regarding Western science and TEK. First, the legacy of colonialism has positioned Westerners as superior and non-Westerners as inferior. Here, TEK can be considered to "fill in some knowledge gaps, but cannot challenge Western knowledge" (Shackeroff & Campbell 2007, p. 346). Second, Western scientists usually hold the power over research processes, which then typically benefit those conducting it. Meanwhile, research subjects' power depends on how scientists decide to incorporate them, including how their TEK is used. The authors cite the work of Nadasdy (1999), who suggests that TEK is usually expected to conform to Western knowledge systems, irrespective of its cultural context. Consequently, "data are incorporated into existing bureaucracies and acted upon by scientists and managers, [serving] to concentrate power in administrative centres rather than indigenous people" (Shackeroff & Campbell 2007, p. 347). Third, knowledge is power itself: "once a researcher collects TEK data, he/she often controls that knowledge through the interpretation of results and deciding how, when, where, and to whom conclusions are presented" (Shackeroff & Campbell 2007, p. 347).

**Figure 2.5** Village in the Sierra Nevada de Santa Marta, Northern Colombia. Positionality and power relations can affect the ways, and extent to which, traditional ecological knowledge is accessed and used by researchers and decision makers. Photo: K. Moon.

### 2.4.4 Pragmatism

Pragmatism involves the judgment of knowledge with respect to how well it serves human purpose(s). In other words, knowledge is good not because it is entirely accurate but because it is useful. For example, the Mercator map projection (a cylindrical projection that allows the Earth's surface to be represented on a two-dimensional page) has no truth as a representation of the planet because it distorts the shapes of objects, yet it is the best map available for navigation. For pragmatists, truth claims, cultural values, and ideas are explored in terms of consequences and application or use-value (Crotty 1998; Scott & Marshall 2009). Pragmatists consider that research should be contextually situated and not necessarily committed to any one philosophical position, instead using a diversity of methods to understand a given problem (Creswell & Creswell 2017). Returning to the tropical deforestation example, a pragmatist might ask, "How can I understand what is really happening 'at this point in time,' so that the different needs of the community, non-government organizations, and other stakeholders can be balanced to reduce the negative effects of deforestation?" Pragmatists are typically comfortable with knowledge continually evolving and changing as an ongoing practice.

### 2.4.5 Postmodernism

Postmodernists argue that there is no single "truth," but rather societies and cultures construct knowledge within different disciplines, potentially to serve particular interests. **Postmodernism** questions the structures, values, and norms that we assume to be "real" and universal in a given society or culture. It typically assumes that the basic structures upon which our societies and cultures are built are largely "social constructs," which can be defined as the meanings and values that have been given to things by human societies

**Figure 2.6** A "Wilderness Park" in The Kimberley, Western Australia. Through a postmodern lens, the term *wilderness* is not an objective description of place, but rather as a socially constructed one that reflects specific elements of the social context, and which can lead to power imbalances and oppression. Photo: K. Moon.

(e.g. money and the economy, goods and services, markets, nations, sovereignty, races, gender, governments, marriage, religion, education, culture, norms). Postmodernists do not see these **social constructs** as fixed, real, or universal. They see them instead as products of specific cultures and specific times.

An example of a postmodern view of conservation is the large body of work exploring the idea of wilderness (e.g. Cronon 1996). While many conservation biologists would use the term *wilderness* as a supposedly objective term that describes places without human influence (Figure 2.6.), postmodernists see wilderness as an idea that emerged out of specific contexts (for example the American frontier in the nineteenth century) and reflects romantic values of nature rather than reality (see Chapter 5 for more on the social construction of wilderness).

## 2.5   Methodology

While ontology and epistemology relate to what reality is and how we create knowledge about it, methodology is concerned with the strategies we use to discover or create that knowledge. As such, the methodology we use in our research reflects our philosophical approach to research and practice. Social sciences draw on a wide range of methodologies (Table 2.2). Methodology is important to conservation science because it provides a rationale, or logic, for how to collect, analyze, and interpret data (Denzin & Lincoln 2011; Creswell & Creswell 2017). In selecting a methodology, the researcher is also establishing the framework for identifying the type(s) of data (quantitative and/or qualitative) necessary to answer their research question(s) (McCaslin & Scott 2003).

Methodology provides an overarching framework for undertaking a program of research; it explains why and how the research is being undertaken and guides the choice of methods (Creswell & Creswell 2017). Methodologies shape the design phase wherein the researchers

**Table 2.2**   Illustrative examples of methodological variety in social science research.

| Methodology | Central questions |
| --- | --- |
| Experimental | How does a specific treatment influence an outcome, by providing a specific treatment to one group and withholding it from another and examining differences between the two groups? |
| Nonexperimental | What are the attitudes, opinions, trends, or patterns of this sample of the population? |
| Ethnography | What is the culture of this group of people? |
| Grounded theory | What theory emerges from systematic comparative analysis and is grounded in fieldwork so as to explain what has been and is observed? |
| Narrative analysis | What does this narrative or story reveal about the person and world from which it came? How can this narrative be interpreted to understand and illuminate the life and culture that created it? |
| Orientational (e.g. critically theoretical, feministic) inquiry | How is X perspective manifested in this phenomenon? |
| Phenomenology | What are the meaning, structure, and essence of the lived experience of this phenomenon for this person or group of people? |

*Sources:* Adapted from Patton (2002, pp. 132–133) and Creswell & Creswell (2017).

decide what it is that they want to do, while methods represent how they want to do it, or the "doing" phase. It is the combination of, and logical connections between, methodology and methods that establishes research quality (i.e. an assessment of the trustworthiness of the research and its outputs) (see Table 2.3). Assessment of research quality is challenging and contested, but relates to issues such as subjectivity and bias, recruitment and sampling, data analysis and interpretation, and reflexivity (Box 2.4) (Belcher et al. 2015).

---

**Box 2.4   Methods: Positionality and reflexivity**

Positionality is a practice that involves researchers' explaining how they are positioned relative to their research, in terms of their experience, knowledge, values, and theoretical perspective, and how that position influences their research design, including the interpretation of results. The practice of reflexivity allows researchers to disclose the position they took in their research and to reflect on how that position might have caused them to interpret, or even miss, certain important aspects of the research context (e.g. Koch & Harrington 1998; Horsburgh 2003; D'Cruz et al. 2007; Hammersley & Atkinson 2007). Reflexive statements reveal how another person, coming from a different position and with a different experience, could have developed a completely different research program that generated different research outcomes. Understanding how positionality affects the outcomes of research is an important consideration of research design, particularly in terms of assessing the "trustworthiness" of the research outcomes (Guba 1981). Reflexivity can encourage a perspective that acknowledges multiple authorities with different insights that should be heard, with the researcher offering just one.

*Source*: adapted from Moon, Blackman, Adams et al. 2019, p. 299

**Table 2.3** Measures and definitions of research quality.

| Quantitative data measures | Questions to assess quality | Qualitative data measures | Questions to assess quality |
|---|---|---|---|
| Reliability | Does the method, applied to the same units, consistently yield similar measurements over and over? (Reliability is a precursor for validity; an unreliable measure cannot be valid.) | Dependability | How can one determine whether the findings of an inquiry would be consistently repeated if the inquiry were replicated with the same (or similar) subjects (respondents) in the same (or similar) context? |
| Internal validity | Can the variations in the outcome (dependent) variable be attributed to controlled variation in an independent variable? Can we infer truth or falsity of cause and effect between two variables? | Credibility | How can one establish confidence in the "truth" of the findings of a particular inquiry for the subjects (respondents) with which, and the context in which, the inquiry was carried out? |
| Objectivity | Would multiple observers agree on the phenomenon of cause and effect? | Confirmability | How can one establish the degree to which the findings of an inquiry are a function solely of the subjects (respondents) and conditions of the inquiry and not of the biases, motivations, interests, perspectives, and so on of the inquirer? |
| External validity/ generalizability | To what extent can we infer that the causal relationship can be generalized across other persons, settings, and times? | Transferability | How can one determine the degree to which the findings of a particular inquiry may have applicability in other contexts or with other subjects (respondents)? |

*Sources:* Adapted from Guba 1981, Lincoln & Guba 1985 and Moon et al. 2021.

Researchers should be encouraged to describe their methodology, because a description of the method alone is insufficient to allow a researcher to make strong claims about what they have done and why it was appropriate (Wolcott 2002). In other words, before we begin on a research journey, we must first ask what methodological routes might lead us there (Waltz 2010).

To assist the reader in understanding methodology, we discuss eight methodologies that are either common or have value to add in conservation social science: experimental research, nonexperimental research, **ethnography**, grounded theory, narrative research, orientational research, phenomenology, and participatory research. We are not suggesting that these are the only methodologies that should be considered, nor are we attempting to simplify the important diversity of views about methodology within the social sciences. Rather, we are seeking to illustrate the role of methodology in research design and, by extension, its influence on conservation practice.

### 2.5.1 Experimental Methodologies

Experimental methodologies seek to test whether a specific treatment affects an outcome by applying a treatment to one group (experimental) but not to another (control) (Creswell & Creswell 2017). Ideally, assignment to these groups is random, which is meant to ensure there are no systematic differences between groups. Such randomized control trials (RCTs) allow the analyst to assume that any effects seen in the treatment group are due to the treatment rather than some other confounding factor. This logic is also used in "natural experiments" and other research designs that seek to construct treatment and controls to account for potential confounders (Ferraro 2009; Gertler et al. 2016). Such quasi-experimental approaches are often more feasible than conducting an experiment for reasons of cost or ethics, among others (Pynegar et al. 2021). The process of randomization involves withholding the intervention from the control group, which means randomly assigning it is not morally neutral. For example, randomly allocating areas to be deforested or not, so as to examine biodiversity impacts, would raise major ethical concerns. However, there are many examples of cases where use of RCTs would be appropriate and could help build knowledge of conservation mechanisms and outcomes (see Box 2.5).

---

**Box 2.5  Methods: Experimental methodologies**

Experimental methodologies are increasingly used in several different social science disciplines, notably economics, political science, and psychology. For example, Ferraro and Pattanayak (2006) have helped spur use of such methodologies to understand what policies and programs work, in which contexts, and why. A large pool of quasi-experimental impact evaluation studies now exists on a wide range of conservation interventions, particularly in relation to forest ecosystems (Börner et al. 2020). Until recently, RCTs had not been used in conservation impact evaluation, but this approach is increasingly employed. The Watershared program in the Bolivian Andes has been particularly well studied (Asquith 2020). Established in 2003, Watershared is a conservation program begun in Bolivia that seeks to use in-kind incentives to protect forests in riparian zones and improve downstream water quality. By 2016, Watershared had 210,000 ha (4500 households) under conservation agreements and had enough experience, technical knowledge, and resources to undertake an RCT to assess program impacts (Asquith 2020). Results were mixed, with some surprises. For example, Watershared did not appear to slow deforestation or improve water quality at the landscape scale in its first five years. However, it was found to have changed land-use practices, such as increasing the area of improved grazing for cattle and significantly boosting fruit tree production, and positively affected household perceptions of environmental quality measures such as water quantity and forest condition (Wiik et al. 2020).

---

### 2.5.2 Nonexperimental Methodologies

Nonexperimental, or survey, methodologies aim to provide a numeric description of attitudes, opinions, preferences, or trends of a population by studying a sample of that population (Creswell & Creswell 2017). A common assumption of nonexperimental methodologies is that when sampled appropriately, the responses from the group of people within the sample can be generalized to the population of that sample (Babbie 2013). Methodologies include longitudinal and cross-sectional research, commonly using questionnaires or structured interviews. Using the tropical deforestation example, researchers might ask how use of and attitudes toward forest resources change over time as a result of legislative change (e.g. a law banning forest clearance). They might survey a sample of the population before and after the legislative change. Nonexperimental survey research is appropriate here because the aim of the research is to measure quantitative change over time.

### 2.5.3 Ethnography

Ethnography is used to provide "a social scientific description of a people and the cultural basis of their peoplehood" (Vidich & Standford 2000, p. 38). The central guiding assumption of ethnographic inquiry is that any "group of people interacting together for a period of time will evolve a culture" (Patton 2002, p. 81). Culture is defined as a collection of behavior patterns and beliefs that constitute standards for deciding what is, what can be, how one feels about it, what to do about it, and how to go about doing it (Goodenough 1971). Ethnographers primarily use the method of **participant observation**, following the tradition of anthropology (see Chapter 3 for more on participant observation). In ethnographic research, the researcher deems it necessary to embed themselves within the culture to engender trust and thus generate meaningful and "truer" insights. Using the tropical deforestation example, a researcher might be interested in exploring the culture of both legal and illegal forest practices, including examining the roles of power, corruption, and violence (e.g. Boekhout van Solinge 2014).

### 2.5.4 Grounded Theory

Grounded theory is used to develop a general theory about a topic through multiple phases of data collection and analysis to explain what is observed. Unlike the methodologies described above that "direct us to particular aspects of human experience as especially deserving of our attention in our attempt to make sense of the social world ... grounded theory focuses on the process of generating theory, rather than a particular theoretical content" (Patton 2002, p. 125).

Grounded theorists typically study a phenomenon (a way an individual or group experiences the world) (McCaslin & Scott 2003; Creswell & Creswell 2017), so the methodology has strong links to phenomenology (see Section 2.5.7). The method of data analysis in grounded theory is systematic and procedural, where the researcher is seeking to describe a version, or versions, of reality through organized analysis of data about a phenomenon. For example, a researcher might want to explore the ethical, philosophical, and practice-based journeys of landholders who see property rights over forest resources as owing a responsibility to country and community, rather than an exclusive right of private possession (e.g. Moon et al. 2020). Grounded theory is a suitable methodology in this example, because the researcher is seeking to develop a specific theory of responsibility-oriented land stewardship.

### 2.5.5   Narrative Analysis

Narrative analysis is used to study people's lives to elicit and then interpret stories to understand life and culture. Stories are at the heart of narrative analysis, where they are considered to "offer especially translucent windows into cultural and social meanings" (Patton 2002, p. 116). With narrative analysis, the researcher asks one or more individuals to provide stories about (an aspect of) their lives, which are often then retold by the researcher. This approach typically results in a combination of views of the participant and the researcher in a collaborative narrative, such that the methodological focus of narrative analysis is the process of interpretation, aligning this methodology with interpretivist philosophies (e.g. Denzin 1989; Clandinin & Connelly 2000). Using the tropical deforestation example again, a researcher might be interested in rich, descriptive stories of different community members and how they tie together to tell a nuanced story of social and ecological change over time (Rissman & Gillon 2017). Narrative analysis is a suitable methodology here because the researcher is seeking rich descriptions that, in this example, could be used to ascribe intentionality to behavior on the basis of specific, embedded, contextual stories (Creswell & Miller 2000).

### 2.5.6   Orientational Research

Orientational qualitative research "begins with an explicit theoretical or ideological perspective that determines what conceptual framework will direct fieldwork and the interpretation of findings" (Patton 2002, p. 129). Unlike grounded theory, social scientists using orientational research choose categories and collect data that explicitly reflect their political/ideological views on issues such as feminism, racism, and socioeconomic class. This methodology therefore aligns well with the critical theory research paradigm, which aims to reveal oppression and bring about change. Using orientational research, one could ask, for example, how the lens of gender shapes and affects how we understand the social world and act within it. Researchers must be clear about how any theoretical framework is used in their research and how it influences the research design, including research questions, data collection, and analysis (Patton 2002). Researchers who adopt this methodology are often seeking not just to understand society, but to change it, through change-oriented forms of engagement (Patton 2002). Other methodologies can be considered as orientational, depending on the degree to which they embed theory into the design. In the tropical deforestation example, a researcher using a postcolonial orientational approach might analyze the drivers of land use and forest loss as products of the history and legacies of colonialism. They might focus on how colonial governments changed land tenure and land-use practices, for example, by claiming all forested land on behalf of the state and dispossessing indigenous communities of their communally managed ancestral forests. Orientational research is a suitable methodology here because the researcher is seeking to determine how social constructs (e.g. of sovereignty and property rights) interact with power and consider the consequences for forest resources, with an agenda to change those power structures (for example, by returning forests to indigenous tenure and management).

### 2.5.7 Phenomenological Analysis

Phenomenologists explore "how human beings make sense of experience and transform experience into consciousness, both individually and as shared meaning [thereby] capturing and describing how people experience some phenomenon—how they perceive it, describe it, feel about it, judge it, remember it, make sense of it, and talk about it with others" (Patton 2002, p. 104). The intention is to "describe *phenomena*, in the broadest sense as whatever appears in the manner in which it appears, that is as it manifests itself to consciousness, to the experiencer. ... Explanations are not to be imposed before the phenomena have been understood from within" (Moran 2002, p. 4).

What distinguishes phenomenological analysis from other methodologies is that research participants are those with direct experience of the described phenomenon. As such, the researchers put their own systems of meaning aside, enabling a focus on the lived experience of participants. "Phenomenologists focus on how we put together the phenomena we experience to make sense of the world and, in doing so, develop a worldview. There is no separate (or objective) reality for people. There is only what they know their experience is and means" (Patton 2002, p. 106). In the context of tropical deforestation, a researcher might be interested in knowing what the experience of deforestation feels like and means to different people who engage in or are affected by deforestation practices. Phenomenology is a suitable methodology because the researcher is seeking to understand the intimate relationships between forest resources and livelihoods as experienced by these people as the system undergoes change.

### 2.5.8 Participatory Research

Participatory approaches are based on a diverse set of techniques drawn from activist participatory research, applied anthropology, and work on farming systems. Participatory approaches owe much to the work of Freire (1968) and Chambers (1992), who have argued that rather than being studied by experts in the field of development, the poor and oppressed should be empowered to conduct their own analyses of their realities and challenges, and thereby shape solutions. There are three core principles to participatory methods (Chambers 1992): (i) poor people are creative and capable, and can and should do much of their own investigation, analysis, and planning; (ii) outsiders have a role as convenors, catalysts, and facilitators; (iii) the un- or dis-empowered should be empowered. Participatory methods are not meant to offer a prescriptive set of methods. Instead, they are meant to provide a loose and flexible toolkit. Common methods include informal mapping (e.g. sketch maps of villages with key features and points of interest/concern), semi-structured interviews (i.e. interviews wherein the list of questions does not overly constrain the interviewer or interviewee), diagramming (e.g. compiling seasonal calendars of household activities to understand critical periods and activities), ranking exercises (e.g. ranking preferences and priorities or using wealth ranking to understand local notions of poverty), and transect walks (e.g. villagers walk with key informants to discuss problems and priorities). By embracing mixed methods, participatory approaches are firmly grounded in the notion of triangulation (Box 2.6).

---

**Box 2.6   Methods: Triangulation**

Triangulation is defined as "the mixing of data or methods so that diverse viewpoints or standpoints cast light upon a topic" (Olsen 2004, p. 104). Triangulation is based on the principle that (i) different methods provide different types of data on different sorts of social phenomena (i.e. different methods can *complement* each other) and (ii) data from different methods can be used to check whether the same patterns and processes are found across methods (i.e. different methods can be used to *validate* each other). The concept of triangulation has influenced many branches of the social sciences, which often draw on a mix of methods that generate both quantitative and qualitative data.

---

## 2.6   Conclusion

Understanding the philosophical basis of research is critical to ensuring that social research outcomes are appropriately and meaningfully interpreted. An examination of the points of difference and the intersection of the philosophical approaches adopted in the social sciences and those of the natural sciences can generate critical reflection and debate about what we can know, what we can learn, and how this knowledge can affect the way conservation research is conducted. A deeper understanding of the philosophical basis of any discipline used when conducting and interpreting research outcomes relating to conservation can ensure the research is clear, well-articulated as a coherent research design, and defensible in terms of the knowledge developed.

## For Further Reading

1 Evaluating the best available social science for natural resource management decision-making (Charnley et al. 2017, *Environmental Science & Policy* 73: 80–88).
2 A basic guide for empirical environmental social science (Cox 2015, *Ecology and Society* 20).
3 *Research Design: Qualitative, Quantitative, and Mixed Methods Approaches* (Creswell & Creswell 2017, SAGE, Thousand Oaks, CA).
4 *The Foundations of Social Research: Meaning and Perspectives in the Research Process* (Crotty 1998, SAGE, London).
5 The influence of philosophical perspectives in integrative research: a conservation case study in the Cairngorms National Park (Evely et al. 2008, *Ecology and Society* 13: 52).
6 *Applied Social Science Methodology: An Introductory Guide* (Gerring & Christenson 2017, Cambridge University Press, Cambridge, UK).
7 A Guide to understanding social science research for natural scientists (Moon & Blackman 2014, *Conservation Biology* 28: 1167–1177).

**8** Expanding the role of social science in conservation through an engagement with philosophy, methodology, and methods (Moon et al. 2019, *Methods in Ecology and Evolution* 10 (3): 294–302).

**9** *Conducting Research in Conservation: Social Science Methods and Practice* (Newing 2010, Routledge, Milton Park, Oxon, UK).

**10** *Social Science Theory for Environmental Sustainability: A Practical Guide* (Stern 2018, Oxford University Press, Oxford, UK).

# References

Agrawal, A. (1995) Dismantling the divide between indigenous and scientific knowledge. *Development and Change* 26: 413–439.

Althaus, C. (2020) Different paradigms of evidence and knowledge: recognising, honouring, and celebrating Indigenous ways of knowing and being. *Australian Journal of Public Administration* 79: 187–207.

Asquith, N. (2020) Large-scale randomized control trials of incentive-based conservation: what have we learned? *World Development* 127: 104785.

Audi, R. (1998) *Epistemology: A Contemporary Introduction to the Theory of Knowledge.* London: Routledge.

Babbie, E.R. (2013) *The Basics of Social Research*, 6th ed. Belmont, CA: Wadsworth Cengage Learning.

Belcher, B.M., Rasmussen, K.E., Kemshaw, M.R. et al. (2015) Defining and assessing research quality in a transdisciplinary context. *Research Evaluation* 25: 1–17.

Berger, R. (2015) Now I see it, now I don't: researcher's position and reflexivity in qualitative research. *Qualitative Research* 15: 219–234.

Berkes, F., Colding, J. & Folke, C. (2000) Rediscovery of traditional ecological knowledge as adaptive management. *Ecological Applications* 10: 1251–1262.

Bernard, H.R. (2013) *Social Research Methods: Qualitative and Quantitative Approaches.* Thousand Oaks, CA: SAGE.

Bietsa, G. (2010) Pragmatism and the philosophical foundations of mixed methods research. In: *SAGE Handbook of Mixed Methods in Social & Behavioral Research*, 2nd ed. (ed. A. Tashakkori & C. Teddlie), 95–118. Thousand Oaks, CA: SAGE.

Biggs, D., Courchamp, F., Martin, R. et al. (2013) Legal trade of Africa's rhino horns. *Science* 339: 1038–1039.

Boekhout van Solinge, T. (2014) Researching illegal logging and deforestation. *International Journal for Crime, Justice and Social Democracy* 3: 35–48.

Bohensky, E.L. & Maru, Y. (2011) Indigenous knowledge, science, and resilience: what have we learned from a decade of international literature on "integration?" *Ecology and Society* 16 (4): art. 6. http://doi.org/10.5751/ES-04342-160406.

Bohman, J. (2016) Critical theory. In: *The Stanford Encyclopedia of Philosophy* (ed. E.N. Zalta). Stanford, CA: Metaphysics Research Lab, Center for the Study of Language and Information, Stanford University. https://plato.stanford.edu/entries/critical-theory.

Börner, J., Schulz, D., Wunder, S. et al. (2020) The effectiveness of forest conservation policies and programs. *Annual Review of Resource Economics* 12: 45–64.

Brewer, T.D. (2013) Dominant discourses, among fishers and middlemen, of the factors affecting coral reef fish distributions in Solomon Islands. *Marine Policy* 37: 245–253.

Brown, K. (2003) Three challenges for a real people-centred conservation. *Global Ecology & Biogeography* 12: 89–92.

Bruner, J. (1986) *Actual Minds, Possible Worlds*. Cambridge, MA: Harvard University Press.

Bryman, A. (2016) *Social Research Methods*. Oxford, UK: Oxford University Press.

Burgman, M. (2005) *Risks and Decisions for Conservation and Environmental Management*. Cambridge, UK: Cambridge University Press.

Castleden, H., Hart, C., Cunsolo, A. et al. (2017) Reconciliation and relationality in water research and management in Canada: implementing indigenous ontologies, epistemologies, and methodologies. In: *Water Policy and Governance in Canada* (ed. S. Renzetti & D.P. Dupont), 69–95. Basel: Springer International.

Celermajer, D., Lyster, R., Wardle, G.M., Walmsley, R. et al. (2021) The Australian bushfire disaster: how to avoid repeating this catastrophe for biodiversity. *WIREs Climate Change* 12 (3): e704.

Chambers, R. (1992) *Rural apprasial: rapid, relaxed and participatory*. IDS Discussion Paper 311. Brighton: IDS.

Chilisa, B. (2020) *Indigenous Research Methodologies*. Thousand Oaks, CA: SAGE.

Cinner, J.E., Daw, T. & McClanahan, T.R. (2009) Socioeconomic factors that affect artisanal fishers' readiness to exit a declining fishery. *Conservation Biology* 23: 124–130.

Clandinin, D.J. & Connelly, F.M. (2000) *Narrative Inquiry: Experience and Story in Qualitative Research*. San Francisco, CA: Jossey-Bass.

Creswell, J.W. & Creswell, J.D. (2017) *Research Design: Qualitative, Quantitative, and Mixed Methods Approaches*. Thousand Oaks, CA: SAGE.

Creswell, J.W. & Miller, D.L. (2000) Determining validity in qualitative inquiry. *Theory Into Practice* 39: 124–130.

Cronon, W. (1996) The trouble with wilderness: or, getting back to the wrong nature. *Environmental History* 1: 7–28.

Crotty, M. (1998) *The Foundations of Social Research: Meaning and Perspectives in the Research Process*. London: SAGE.

D'Cruz, H., Gillingham, P. & Melendez, S. (2007) Reflexivity, its meanings and relevance for social work: a critical review of the literature. *British Journal of Social Work* 37: 73–90.

Denzin, N.K. (1989) *Interpretive Biography*. Thousand Oaks, CA: SAGE.

Denzin, N.K. & Lincoln, Y.S. eds. (2011) *The SAGE Handbook of Qualitative Research*, 4th ed. Thousand Oaks, CA: SAGE.

Evely, A.C., Fazey, I.R.A., Pinard, M. et al. (2008) The influence of philosophical perspectives in integrative research: a conservation case study in the Cairngorms National Park. *Ecology and Society* 13: art. 52. http://www.ecologyandsociety.org/vol13/iss2/art52.

Ferraro, P.J. (2009) Counterfactual thinking and impact evaluation in environmental policy. *New Directions for Evaluation* 2009: 75–84.

Ferraro, P.J. & Pattanayak, S.K. (2006) Money for nothing? A call for empirical evaluation of biodiversity conservation investments. *PLoS Biology* 4 (4): e105.

Feyerabend, P.K. (1981) *Problems of Empiricism: Philosophical Papers*. Cambridge, UK: Cambridge University Press.

Fishbein, M. & Ajzen, I. (1975) *Belief, Attitude, Intention, and Behavior*. Reading, MA: Addison-Wesley.

Freire, P. (1968) *Pedagogy of the Oppressed*. New York: The Seabury Press.

Gertler, P.J., Martinez, S., Premand, P. et al. (2016) *Impact Evaluation in Practice*. Washington, DC: The World Bank.

Goodenough, W. (1971) *Culture, Language and Society*. Reading, MA: Addison-Wesley.

Graham-Rowe, D. (2011) Biodiversity: endangered and in demand. *Nature* 480: S101–S103.

Guba, E.G. (1981) Criteria for assessing the trustworthiness of naturalistic inquiries. *Educational Technology Research and Development* 29: 75–91.

Guba, E.G. (1990) *The Paradigm Dialog*. Newbbury Park, CA: SAGE.

Guba, E.G. & Lincoln, Y.S. (1994) Competing paradigms in qualitative research. In: *Handbook of Qualitative Research* (ed. N.K. Denzin & Y.S. Lincoln), 105–117. Thousand Oaks, CA: SAGE.

Hammersley, M. & Atkinson, P. (2007) *Ethnography: Principles in Practice*. London: Routledge.

Horsburgh, D. (2003) Evaluation of qualitative research. *Journal of Clinical Nursing* 12: 307–312.

Jackson, S.E., Douglas, M.M., Kennard, M.J. et al. (2014) "We like to listen to stories about fish": integrating indigenous ecological and scientific knowledge to inform environmental flow assessments. *Ecology and Society* 19: 43.

Johnson, B. & Gray, R. (2010) A history of philosophical and theoretical issues for mixed methods research. In: *SAGE Handbook of Mixed Methods in Social & Behavioral Research*, 2nd ed. (ed. A. Tashakkori & C. Teddlie), 69–94. Thousand Oaks, CA: SAGE.

Kinnane, S. (2005) Indigenous sustainability: rights, obligations, and a collective commitment to country. In: *International Law and Indigenous Peoples* (ed. J. Castellino & N. Walsh), 159–194. Leiden, the Netherlands: Martinus Nijhoff.

Kirkness, V.J. & Barnhardt, R. (1991) First nations and higher education: the four R's – respect, relevance, reciprocity, responsibility. *Journal of American Indian Education* 30: 1–15.

Koch, T. & Harrington, A. (1998) Reconceptualizing rigour: the case for reflexivity. *Journal of Advanced Nursing* 28: 882–890.

Kolkman, M.J., Veen, A.V.D. & Geurts, P.A.T.M. (2007) Controversies in water management: frames and mental models. *Environmental Impact Assessment Review* 27: 685–706.

Lincoln, Y.S. & Guba, E.G. (1985) *Naturalistic Inquiry*. Newbury Park, CA: SAGE.

Marx, K. (1845[1992]) *Early Writings*. London: Penguin Books.

Maynard, M. (1994) Methods, practices and epistemology: the debate about feminism and research. In: *Researching Women's Lives from a Feminist Perspective* (ed. M. Maynard & J. Purvis), 10–26. London: Taylor & Francis.

McCaslin, M.L. & Scott, K.W. (2003) The five-question method for framing a qualitative research study. *The Qualitative Report* 8: 447–461.

Moon, K. (2020) Understanding the experience of an extreme event: a personal reflection. *One Earth* 2: 493–496.

Moon, K., Adams, V.M. & Cooke, B. (2019) Shared personal reflections on the need to broaden the scope of conservation social science. *People and Nature* 1: 426–434.

Moon, K. & Blackman, D. (2014) A guide to understanding social science research for natural scientists. *Conservation Biology* 28: 1167–1177.

Moon, K., Blackman, D.A., Adams, V.M. et al. (2019) Expanding the role of social science in conservation through an engagement with philosophy, methodology, and methods. *Methods in Ecology and Evolution* 10: 294–302.

Moon, K., Blackman, D.A. & Brewer, T.D. (2015) Understanding and integrating knowledge to improve invasive species management. *Biological Invasions* 17: 2675–2689.

Moon, K. & Browne, N.K. (2021) Developing shared qualitative models for complex systems. *Conservation Biology* 35: 1039–1050.

Moon, K., Cvitanovic, C., Blackman, D.A. et al. (2021) Five questions to understand epistemology and its influence on integrative marine research. *Frontiers in Marine Science* 8: 173. https://doi.org/10.3389/fmars.2021.574158.

Moon, K., Guerrero, A.M., Adams, V.M. et al. (2019) Mental models for conservation research and practice. *Conservation Letters* 12: e12642. https://doi.org/10.1111/conl.12642.

Moon, K., Marsh, D. & Cvitanovic, C. (2020) Coupling property rights with responsibilities to improve conservation outcomes across land and seascapes. *Conservation Letters* 13: e12767. https://doi.org/10.1111/conl.12767.

Moran, D. (2002) *Introduction to Phenomenology*. London: Routledge.

Morgan, D.L. (2007) Paradigms lost and pragmatism regained: methodological implications of combining qualitative and quantitative methods. *Journal of Mixed Methods Research* 1: 48–76.

Morton, A. (1996) *Philosophy in Practice: An Introduction to the Main Questions*. Oxford, UK: Blackwell Publishers.

Moses, J.W. & Knutsen, T.L. (2012) *Ways of Knowing: Competing Methodologies in Social and Political Research*, 2nd ed. Hampshire, UK: Palgrave Macmillan.

Nadasdy, P. (1999) The politics of TEK: power and the 'integration' of knowledge. *Arctic Anthropology* 36: 1–18.

Olsen, W. (2004) Triangulation in social research: qualitative and quantitative methods: can they really be mixed. In: *Developments in Sociology* (ed. M. Holborn), 103–121. Ormskirk: Causeway Press.

Parsons, M. & Fisher, K. (2020) Indigenous peoples and transformations in freshwater governance and management. *Current Opinion in Environmental Sustainability* 44: 124–139.

Patton, M.Q. (2002) *Qualitative Research & Evaluation Methods*. Thousand Oaks, CA: SAGE.

Popper, K.R. (1963) *Conjectures and Refutations: The Growth of Scientific Knowledge*. Abingdon, UK: Routledge & Kegan Paul.

Pratt, D.D. (1998) *Five Perspectives on Teaching in Adult and Higher Education*. Malabar, FL: Krieger.

Pynegar, E.L., Gibbons, J.M., Asquith, N.M. et al. (2021) What role should randomized control trials play in providing the evidence base for conservation? *Oryx* 55 (2): 235–244.

Rissman, A.R. & Gillon, S. (2017) Where are ecology and biodiversity in social–ecological systems research? A review of research methods and applied recommendations. *Conservation Letters* 10: 86–93.

Roebuck, P. & Phifer, P. (1999) The persistence of positivism in conservation biology. *Conservation Biology* 13: 444–446.

Scales, I.R. (2012) Lost in translation: conflicting views of deforestation, land use and identity in western Madagascar. *The Geographical Journal* 178: 67–79.

Schwandt, T.A. (1994) Constructivist, interpretivist approaches to human inquiry. In: *Handbook of Qualitative Research* (ed. N.K. Denzin & Y.S. Lincoln), 118–137. Thousand Oaks, CA: SAGE.

Scott, J. & Marshall, G. eds. (2009) *A Dictionary of Sociology*. Oxford, UK: Oxford University Press.

Shackeroff, J. & Campbell, L. (2007) Traditional ecological knowledge in conservation research: problems and prospects for their constructive engagement. *Conservation and Society* 5: 343–360.

Slovic, P. (2000) *The Perception of Risk*. London: Earthscan.

Soulé, M.E. (1985) What is conservation biology? *BioScience* 35: 727–734.

Stocker, L., Collard, L. & Rooney, A. (2016) Aboriginal world views and colonisation: implications for coastal sustainability. *Local Environment* 21: 844–865.

Stokes, G. (1998) *Popper: Philosophy, Politics and Scientific Method*. Oxford, UK: Blackwell.

Tashakkori, A. & Teddlie, C. eds. (2010) *SAGE Handbook of Mixed Methods in Social & Behavioral Research*. Thousand Oaks, CA: SAGE.

Tuhiwai, L. (2021) *Decolonizing Methodologies: Research and Indigenous Peoples*. London: Bloomsbury.

Vidich, A.J. & Standford, M.L. eds. (2000) *Qualitative Methods: Their History in Sociology and Anthropology*. Thousand Oaks, CA: SAGE.

Waltz, K.N. (2010) *Theory of International Politics*. Long Grove, IL: Waveland Press.

Waylen, K.A., Fischer, A., McGowan, P.J.K. et al. (2010) Effect of local cultural context on the success of community-based conservation interventions. *Conservation Biology* 24: 1119–1129.

Weeks, P. & Packard, J.M. (1997) Acceptance of scientific management by natural resource dependent communities. *Conservation Biology* 11: 236–245.

Wiik, E., Jones, J.P., Pynegar, E. et al. (2020) Mechanisms and impacts of an incentive-based conservation program with evidence from a randomized control trial. *Conservation Biology* 34 (5): 1076–1088.

Wilson, S. (2008) *Research Is Ceremony: Indigenous Research Methods*. Winnipeg: Fernwood Publishing.

Wolcott, H.F. (2002) Writing up qualitative research ... better. *Qualitative Health Research* 12: 91–103.

Woodward, E., Hill, R., Harkness, P. et al. (2020) *Our Knowledge Our Way in Caring for Country: Indigenous-led Approaches to Strengthening and Sharing Our Knowledge for Land and Sea Management. Best Practice Guidelines from Australian Experiences*. Cairns: North Australian Indigenous Land and Sea Management Alliance and Commonwealth Scientific and Industrial Research Organisation.

# 3

# Anthropology and Conservation

*Diane Russell and C. Anne Claus*

## 3.1 Defining Anthropology

Anthropology is the holistic study of the human species: humankind's past and present biological, linguistic, social, and cultural variations. Archaeology, biological anthropology, cultural anthropology, and linguistics make up the major four fields of the discipline. This chapter focuses on cultural anthropology and its subdiscipline environmental anthropology, noting relevant concepts and contributions from the other three fields (Table 3.1). There is a great deal of cross-fertilization between academic and applied approaches to anthropology. The former pursues scientific knowledge through research and theory building, while the latter deploys this body of knowledge and theory to address social and environmental problems.

Anthropologists also work across a wide range of scales—from the intimate to the global. Cultural anthropological explanations are often grounded in and documented by rich detailed case studies of people, places, and issues, known as ethnographies. Ethnology (cross-cultural analysis of ethnographic data) is used to describe patterns of cultural diversity and similarity. **Ethnographic** knowledge emerges from rigorous **qualitative** data collection and analysis.

### 3.1.1 Major Themes and Questions in Anthropology

In examining the history and diversity of the human species, anthropology delves into virtually every aspect of human existence. Four themes have consistently held the attention of the discipline: (i) human evolution; (ii) culture; (iii) behavior; and (iv) social structure.

#### 3.1.1.1 Human Evolution and Variability

Anthropology describes how *Homo sapiens* emerged from within the genus *Hominidae*, and how human groups radiated out over the Earth, adapting to and changing terrains and climates. In tracing human migration, another key anthropological theme is the nature of variability within the human species: how and why racial and ethnic varieties emerged.

Anthropology as a discipline has confronted racism and ethnocentrism, albeit incompletely as described below. Ethnographers historically demonstrated that people deemed "primitive" were skillful and innovative, contrary to neo-Darwinian stereotypes.

*Conservation Social Science: Understanding People, Conserving Biodiversity*, First Edition.
Edited by Daniel C. Miller, Ivan R. Scales, and Michael B. Mascia.
© 2023 John Wiley & Sons Ltd. Published 2023 by John Wiley & Sons Ltd.

**Table 3.1** Anthropological subfields and disciplines relevant to conservation.

| Subfield/subdiscipline | Disciplinary overlap | Relevance to conservation |
| --- | --- | --- |
| Archaeology | | |
| Classical | History/classics<br>Area studies<br>Human geography | Place-specific natural resource management |
| Prehistoric | Paleoecology<br>Human geography | Evolution of human groups within biomes |
| Historical ecology | Paleoecology<br>Human geography | How humans transformed landscapes |
| Biological anthropology | | |
| Primatology | Zoology | Great ape and primate conservation; insights into human behavior |
| Human evolution | Human biology | Radiation into different areas; technology development; natural resource management |
| Human populations and diversity | Human biology | Diversity of human societies in relation to place |
| Anthropological linguistics | | |
| Language area studies | Area studies<br>Human geography | Ethnotaxonomies |
| Comparative linguistics | Modern languages<br>Semiotics | Comparative ethnobiology |
| Ethnobiology and related subfields | Botany and ecology<br>Cultural anthropology | Local ecological knowledge |
| Cultural/social anthropology | | |
| Social anthropology | Sociology<br>Rural sociology | Social systems, institutions, methodologies |
| Ecological anthropology | Ecology<br>Conservation biology | Human ecology |
| Environmental anthropology | Environmental studies | Environmental justice |
| Economic anthropology | Economics<br>World-systems theory<br>Political economy | Markets<br>Class<br>Globalization and commoditization |
| Legal anthropology | Legal studies | Bylaws and natural resource management regulations |
| Political anthropology | Political science | Patron–client relations<br>Elite capture |

**Table 3.1** (Continued)

| Subfield/subdiscipline | Disciplinary overlap | Relevance to conservation |
|---|---|---|
| Psychological anthropology | Psychology | Worldviews—the way people see the world, their beliefs, and knowledge systems |
| | | Concepts of nature |
| Anthropology of consciousness | Religious studies | Worldviews |
| | | Sacred and spiritual conservation |
| Cross-disciplines | | |
| Agrarian studies | Political science | History of rural populations; state–society struggles; globalization and neoliberalism |
| | Economics | |
| | History | |
| Study of the commons | Political science | Common-property natural resource management institutions and models |
| | Economics | |
| | Human geography | |
| Historical ecology | Geography | Land histories; anthropogenic landscapes |
| | Ecology | |
| World systems | Political economy | Market systems and commoditization; neoliberalism |
| Political ecology | Geography | Neoliberalism; environmental justice; rights-based approaches |
| | Political science | |
| | Sociology | |
| | Ecology | |

Anthropological research reveals that assumptions about human diversity often reflect power and status hierarchy more than actual biological differences, and concepts of race are thickly intertwined with colonial modes of classification. For instance, the American Anthropological Association finds that "from its inception, this modern concept of 'race' was modeled after an ancient theorem of the Great Chain of Being, which posited natural categories on a hierarchy established by God or nature. Thus 'race' was a mode of classification linked specifically to peoples in the colonial situation" (AAA 1998).

Following this thread, anthropologists have been in the forefront of democratizing conservation, highlighting the need for local and indigenous communities to lead conservation efforts based in their knowledge and within their social systems. Fighting racism is part of the wider anthropological preoccupation with how power and hierarchy impact human decisions and institutions, and many anthropologists are committed to using their privilege and access to create space for these initiatives.

### 3.1.1.2 Culture

To help describe modern human variability, anthropologists use the concept of **culture**. Material culture refers to the physical artifacts and structures that people create. But the essence of culture is patterns of thinking and acting that are passed on via learning rather than or alongside genetic inheritance. A culture is thus a constellation of learned behavior patterns. Culture, in its full complexity, has been described as a uniquely human characteristic even as nonhuman primates have been found to employ cultural transmission of knowledge such as hunting and gathering techniques. While culture and identity—your self-concept of who you are in a group and in the world—are closely intertwined, identities transcend cultures and elements of culture diffuse across groups. A society, or large social group sharing the same geographic or social territory, may comprise one or many cultures.

Concepts such as time, space, mortality, spirituality, and causality are defined and delineated by cultural understandings; thus culture prescribes how individuals relate to the world and to each other. Traditions and customs distill and communicate meaning within and between cultures. With their deep and broad viewpoints, anthropologists do not see traditions, cultures, or human societies as static and immutable, but as fluid and ever-changing. From early studies of acculturation (the process of acquiring culture) to recent research on the impact of social media, anthropologists have recorded cultural evolution, cultural transmission, and how individuals within groups experience the differential impacts of change depending on social role, status, gender, and other variables. Perceptions about nature, and conservation, are being constantly created and transmitted through the cultural media of language, art, music—and science.

### 3.1.1.3 Behavior

Culture shapes not only how people perceive other humans and the natural world; it structures how people act and how they think they *should* act. The core anthropological research method of **participant observation** involves living and working in an area for extended periods of time, getting to know a group of people and particular topics holistically (see Chapter 2 for more on methods). This methodology can reveal how actual behavior (what people do) deviates from **normative** behavior (what they say they should do) as anthropologists carry out structured long-term data collection, and triangulate information sources (see Box 2.6, Chapter 2 on Social Science Foundations for more on triangulation). This approach to understanding behavior is critical to crafting effective conservation strategies, as it can not only reveal disconnects between actual and normative behaviors, it can illuminate *why* people behave in a certain way—their motivations, beliefs, and constraints—through a holistic grasp of the cultural context.

While understanding cultural context is critical, anthropologists show how political–economic forces within and outside a group shape behavior and how behavior is perceived. For instance, when a protected area (PA) is created within an ethnic group's ancestral territory, hunting may be perceived as illegal and even immoral by state authorities and conservation organizations while the group sees it as legitimate and culturally appropriate. Holistic understandings of behavior challenge notions about what might be entailed in "behavior change" for conservation objectives and indicate where such efforts may be unfeasible or unethical.

### 3.1.1.4 Social Structure

Humans have crafted a wide variety of social structures in order to live and make a living together within a multitude of **ecosystems** and within increasingly large and complex human settlements and networks. Social structure encompasses abstract rules and **norms** as well as physical groups of people. Anthropologists have documented this diversity in social structures over time and space, including a vast array of marriage and kinship or familial structures, as well as social structures for natural resource management (NRM), trade, and governance, among others. There are important intersections among social structures; for instance, forms of kinship and inheritance can shape access to, and management of land and natural resources.

As elaborated in this chapter, social structures are fluid and dynamic, imbued with relations of power both internally and in interaction with external actors and institutions. For example, the structure of the family or household unit in a given culture can change dramatically with migration, economic circumstance, and gender dynamics. Conservation organizations and programs are themselves social structures, and similarly their approaches and ultimately their impacts are shaped by relations of power, conflict, and cooperation in interaction with other social structures.

### 3.1.2 Anthropology and Conservation

This chapter explores the multiple ways that anthropological research intersects with and contributes to understanding conservation. Conservation as a field of research and practice is an exciting and crucially important arena for anthropological involvement. It has opened up opportunities for engaging in conservation initiatives and for cross-disciplinary collaboration (Table 3.1). Many anthropologists are passionate about conservation and, like the authors of this chapter, study ecology, forestry, conservation biology, and other subjects to be able to integrate their knowledge with these other sciences. Anthropologists also teach in interdisciplinary programs and mentor conservation practitioners. Thus, this chapter is relevant to both conservation science (e.g. examining and proposing research methodologies) and conservation practice (analyzing a range of conservation approaches).

To anthropologists, the kind of profound understanding derived from participant observation is needed before even posing questions and developing hypotheses with respect to conservation. For instance, research questions need to be framed within an **epistemologically** appropriate context. Participant observation allows anthropologists to make systematic connections that might otherwise be missed, to observe trends, and to understand influential local political economies and power dynamics. Many anthropologists have illustrated how ignoring this in-depth knowledge has serious consequences for peoples and environments.

As we see from this brief overview, anthropological knowledge and experience relevant to conservation practice is vast. There are many pertinent avenues of inquiry within the field in addition to research on conservation itself. This chapter focuses on four critical areas of inquiry distilled from this body of knowledge that are especially relevant for conservation: (i) culture and identity; (ii) communities and institutions; (iii) **political economy**; and (iv) diversity and variability.

#### 3.1.2.1 Culture and Identity

In what ways do biological and cultural diversity converge? How do cultures shape values, knowledge, and behavior? What is the relationship between environments and particular identities, like indigeneity? How is ecological knowledge shaped by factors like gender and class? How can conservation initiatives be culturally literate and grounded in what matters to people who implement them and are impacted by them? In short, how is an understanding of culture and identity important for conservation?

### 3.1.2.2 Communities and Institutions

How do formal and informal institutions like land tenure, property rights, and kinship shape the ways that people manage their environments? In what ways do community complexity and diversity shape conservation? How do the structures of conservation organizations impact and reshape socioenvironmental dynamics? How can moving beyond the overused term "community" reveal local institutions that may be overlooked but are critically important for conservation? In short, how do anthropologists define and study communities and institutions in conservation?

### 3.1.2.3 Political Economy, Power, and Conservation

How do power relations at all levels and across time and space structure societies' management of natural resources? How do they interface with conservation initiatives and the voices people have in making decisions about their environments? In what ways do conservation projects build on problematic colonial and neocolonial relationships? How is conservation shaped by power derived through political, material, and discursive means? What is the role of markets and commodities in **neoliberal** conservation? How are anthropologists involved in challenging dominant conservation paradigms through their research? In short, how do relations of power and economic forces shape conservation practice and outcomes?

### 3.1.2.4 Diversity and Variability in Environmental Management

How does **biodiversity conservation** intersect with NRM systems such as agriculture? How can the diversity and innovation in these systems be harnessed for holistic and culturally relevant conservation? In what ways do NRM systems adapt and evolve in relation to the environment, and does overexploitation lead to societal collapse? How does losing traditional technologies impact knowledge about natural resources? How do cultural changes increase vulnerability and economic insecurity that may undermine conservation efforts? In short, what have anthropologists learned about how societies manage environments?

While these questions address variable scopes from individual to global systems, there are important connections among them. Identity, culture, and institutions drive the environmental shifts that have catapulted us into a new geological epoch, the **Anthropocene**, characterized by the depth of human impact on the Earth's ecosystems.

## 3.2 A Brief History of Anthropology

Anthropology emerged from the study of non-Western peoples as they were "discovered" through increased trade, travel, and Western drives for colonization during the 1800s. Scholars and philanthropists in this period attempted to work against racist paradigms such as the depiction of some peoples as less than human, although their analyses still placed humankind on an evolutionary scale with Western societies as the most enlightened civilization. This "armchair anthropology" used the observations of missionaries and explorers, as well as legends and myths such as *The Golden Bough* by James Frazier (1993), for analysis from afar. By the early 1900s, however, Franz Boas and Bronislaw Malinowski, among others, fashioned an anthropology that questioned racist and simplistic hierarchies of civilizations. These disciplinary forebears developed two components of the modern anthropological approach: culture as an explanatory factor in human variability and extended fieldwork as a methodological approach to enable the researcher to understand people on their own terms and not through externally derived questions that could be easily misunderstood or misinterpreted.

Through the students of Boas and Malinowski, notably Ruth Benedict, Margaret Mead, and Alfred Kroeber in the USA, anthropology became institutionalized in universities. In the USA, anthropology departments proliferated after World War II as returning war veterans pursued the discipline on government scholarships. The road to academic institutionalization was more convoluted in Britain, where anthropology was concentrated around particular schools and strong personalities, like Max Gluckman of the University of Manchester. Though academic anthropology was largely contained within national borders in this period, there was always some cosmopolitan crossover, especially with French and German scholars.

Due to the dubious evidence base of armchair anthropology, scholars took pains to illustrate the scientific rigor of anthropological methods. For Boas and Malinowski, this rigor stemmed from long-term fieldwork to garner an understanding of all components of societies (e.g. material culture, NRM systems, kinship, language, religion) and how they are connected. Anthropological research has long been comparative as well: some anthropologists study how social, economic, and ecological systems adapt and change in space and over time (e.g. Orlove & Guillet 1985 on mountain cultures; Netting 1993 on small-scale agricultural societies) and develop formal comparative models such as ethnobiological taxonomies to understand global patterns of classification (Berlin 1992). Anthropological collaborations with other disciplines such as botany, ecology, geography, and history are also common. Differing **epistemologies** and **methodologies** can both challenge and enrich these interdisciplinary ventures, as seen in Box 3.1.

As in other sciences, substantiating or refuting theories is important in anthropology. The history of the discipline reveals broad theoretical shifts. After the dominance of theories that centered around inventorying and describing the components of particular cultures (e.g. their kinship systems, modes of subsistence), British structural functionalism focused on the relationships among these components or structures. For instance, later in the chapter we talk about how marriage and kinship systems shape land tenure and land management.

The 1960s gave rise to the structuralism presented by French philosopher–anthropologist Claude Lévi-Strauss and British anthropologist Mary Douglas, among others. Lévi-Strauss was also interested in the elements of culture but sought to reduce those elements to the essentials common to all humans. He argued that symbols and themes found in myth and language such as duality (man/woman, hot/cold) reflect human mental structures and thus provide insights into human cognition and, ultimately, behavior. While the earlier theories were influenced by natural science, structuralism was influenced by and influenced linguistics, philosophy, art, and literature. Around the same time, anthropologists like Clifford Geertz argued that anthropology is an interpretive science that seeks to understand how people derive meaning from the world around them. Both of these approaches celebrated the human capacity for symbolic thought and attempted, in very different ways, to understand how symbolism operated within specific cultures.

In contrast to structuralism and interpretivism, materialist approaches in anthropology that emerged from Marxist analyses explained social relations as derived from the struggle for control over the means of production (see also Chapter 5, Human Geography and Conservation, and Chapter 8, Sociology and Conservation). Such analyses focused, for

**Box 3.1   Crossing boundaries: interdisciplinary collaboration and primate conservation**

Primatology is closely linked to zoology and conservation biology. It studies nonhuman primates (NHPs) on their own terms as well as the evolutionary and behavioral links among all primates, including humans. Perhaps the best-known anthropologist in the world, Jane Goodall, is a primatologist. Her work depicts primate conservation as a moral imperative, given the close relations between humans and great apes: we share 96% of our genes. Primates, especially the Great Apes—gorillas, bonobos (Figure 3.1), chimpanzees, and orangutans—are among the most endangered of species, predicted to go extinct rapidly unless conservation is more successful.

NHP reproductive strategies involve long gestation periods and significant investments in infants, making them more vulnerable than species with quick reproductive cycles. Although NHPs have been known to live compatibly with human populations, this is increasingly not the case. First, NHPs share many food preferences with humans and thus come into conflict with farmers due to "crop raiding." Second, farming and other land uses cause deforestation that reduces their habitat. Third, NHPs become collateral damage in wars and conflicts such as the slaughter of gorillas in the Virungas region of the Democratic Republic of the Congo as a result of militia action. Fourth, apes and humans are so closely related that they can transmit diseases such as Human Immunodeficiency Virus (HIV) and Ebola—placing both at risk. Finally, in many parts of the world, but particularly in West and Central Africa, NHPs are hunted for food and the pet trade. In these locations, wild meat—known as bushmeat—is an important foodstuff that contributes to nutrition as well as livelihoods. Increasingly, the bushmeat trade is big business with bushmeat from Africa found in markets and dinner tables in Europe and the USA.

The collaboration of Rebecca Hardin, a cultural anthropologist, and Melissa Remis, a primatologist, was designed to enhance their respective research projects on the consequences of displacement and the transformation of hunting and gathering to new forms of forest use in the Dzangha-Sangha Reserve of the Central African Republic, and to inform approaches to gorilla conservation (Hardin & Remis 2006). Hardin and Remis note that their differing expertise in cultural anthropology and primatology allowed them to uncover changes in forest use linked to gender and species-specific trends, such as ways that men and women

**Figure 3.1**   Bonobo at Lola Ya Bonobo Reserve, Democratic Republic of Congo, 2019. The bonobo, humanity's closest relative, has been observed using tools. Photo by Diane Russell.

| Box 3.1 (Continued) |
| --- |

differentially collect and use species and the impacts on those species. Their research ultimately found that an expanding logging frontier is outpacing creative human–animal adaptations to forest disturbance (Hardin & Remis 2006, p. 275). Collaborating led them to question their own disciplinary theories and methodologies in productive ways; they argue that "anthropology remains one of the few long-term intellectual 'homes' for widely divergent ways to study both human and nonhuman life. It is thus well positioned for innovation in the changing worlds of critical and scientific inquiry" (Hardin & Remis 2006, p. 284).

This type of collaboration has significantly advanced since Hardin and Remis worked together. Emerging research in anthropology questions even very basic categories that we tend to take for granted, such as the categories of human and nonhuman. It examines what it means to be social by looking, for example, at communication networks among trees or how bees collaborate. Thus anthropologists in some ways are paying new attention to the work of ecologists, but through an anthropological perspective that considers the ways in which history and humanity contribute to the lives of species-in-formation. This new field of research draws heavily on indigenous philosophies to further ask how a decentering of human exceptionalism might engender new forms of collaboration within conservation programs. Parathian et al. (2018) provide an overview of human–wildlife interactions seen through the lens of multispecies ethnography while Riley (2018) describes the maturation of the field of ethnoprimatology to account for these new philosophies.

example, on ways that rural communities under colonial or postcolonial regimes struggled to gain, retain, or control access to land and natural resources needed for their livelihoods through aligning with powerful actors or conversely finding common cause to fight actors inhibiting access.

These concepts were followed in the 1970s by two main theoretical paradigms: **postmodernism** and **post-structuralism**. The influence of these theories reverberated not only in the products but also the practice of anthropological research (Box 3.2). Anthropologists continue to study themes that originally enticed the discipline's early researchers, but theories and methods of research have transformed considerably. Since its inception, anthropology has grown into a rich suite of subdisciplines and specializations that intersect with conservation, as depicted in Table 3.1.

Ecological anthropology emerged largely in the USA fifty years after the professionalization of the discipline when Julian Steward and Leslie White, considered founders of this subfield, began to research environmental impacts on culture. By the 1960s, ecological anthropologists were conducting innovative cross-disciplinary work that viewed human beings not as an exceptional species, but as one component of a broader ecosystem. Rappaport (1968) pioneered the use of ecological concepts in ethnography in *Pigs for the Ancestors*, devising functional explanations for the husbanding of pigs that seemingly consumed more energy than they provided for the surrounding human populations (Box 3.3). Bateson (1972) turned to systems theory and cybernetics to craft a framework for understanding how components of society interact and shift systems. He

**Box 3.2    Debate: crises of representation**

Long concerned with how experts and expertise are crafted in the societies that they study, anthropologists have recently begun to ask similar questions of their own discipline. Whose work is viewed as authoritative within the halls of anthropology departments? Whose work is regularly cited for its ability to generate theoretical insights? In efforts to create research that responds to widely known scholarly debates and conversations, are anthropologists sidelining the work of other compelling and intellectually rich theoreticians based on their gender, school, educational background, or ethnicity? These questions about representativeness and authoritativeness have historical precedents in the discipline. They resonate with similar questions within conservation about whether mainstream conservation approaches such as PAs are grounded in Western notions of protectionism and lack the perspectives of indigenous people and people of color.

In the mid-1980s, a palpable sense of crisis permeated anthropology. Margaret Mead, the public figurehead of anthropology in the United States, was accused of misinterpreting Samoan adolescents' accounts of their sexual freedom fifty years earlier, when she reported that girls deferred marriage for many years and indulged in casual sex before choosing a husband. Her accuser claimed that what Mead had described were inventive tales rather than fact and that later research did not corroborate those accounts. This controversy reverberated widely because the objectivity of research and the ability of researchers to adequately represent their findings—assumptions on which the field had been founded—were called into question. This questioning led to what is now called the "crisis of representation" within anthropology.

Feminist and critical anthropologists laid the foundation for this self-critique by questioning the anthropological project itself (e.g. Hymes 1972; Asad 1973; Moore 1988). Charges levied against the field were serious. Anthropologists were accused of complicity with colonial rule by using colonial structures to gain access to field sites, and in some cases even participating in colonial governments, as well as failing to acknowledge the institutional and personal limitations under which anthropological knowledge was produced as a result of power differentials. Reactions to these accusations were wide-ranging: some viewed the debate as lacking salience for the discipline and, in the case of Mead, pointed out differences that could account for discrepancies in researcher accounts of adolescent Samoan life. For example, while Mead's accuser went to the region where she had done her fieldwork, he was a male, arrived thirty years later, and interviewed not young girls but male chiefs. Others, arguing that the inability to be objective altered the integrity of the discipline's primary approach, called for the abandonment of ethnographic research altogether.

Ultimately, the crisis of representation engendered changes in the way anthropologists conduct their research. In the USA, more robust ethical codes were put in place by the discipline's professional body, the American Anthropological Association (AAA 2012). Additional impacts were felt in the production of anthropological texts. Ethnographies became more self-aware with the growing disciplinary concern about the privileged position of the ethnographer in describing "reality." Presenting findings with large sections of dialogue or in testimony form was a response of some researchers who were concerned with presenting the "voice" of their informants with as little researcher interference as possible.

**Box 3.2 (Continued)**

**Figure 3.2** Interviewing staff of a sustainable charcoal project, Zambia, 2015. Photo by Diane Russell.

Anthropologists represent complex issues and diverse viewpoints by using multiple data collection and analysis methods (Figure 3.2). As such, some experimental works placed field notes, fiction, and academic texts alongside one another to provide readers with tools to make their own conclusions about the subjects studied. Finally, this broader discussion generated debate about how complexity, change, and diverse social identities are depicted. This concern is addressed in feminist anthropology, for example in showing that men's and women's experiences are often very different and that the gender and status of the researcher also shape research outcomes.

The legacy of these crises of representation lingers in the discipline. For anthropologists, their own **positionality** is also data that helps readers evaluate the conditions of knowledge production that underlie the research they produce, and for this reason, reflexive assessments are often part of ethnographic texts (see Chapter 2). Such reflexive thinking is highly valuable for conservation as planners consider how their culture, status, and disciplinary frameworks structure their thinking not only about conservation approaches but about conservation priorities, even about nature itself, and how such thinking may affect potential partnerships and conservation outcomes.

---

**Box 3.3 Methods: non/equilibrium models**

---

Interdisciplinary borrowing between the natural and social sciences extends to the way research questions are framed, as understandings of ecology impact how anthropologists view social changes and vice versa. An example is the ecological concept of equilibrium. The equilibrium framework characterizes change as a perturbation. It sees an ecosystem, after having faced an exogenous disturbance, as progressing through various predictable stages to reach a stable "climax" community.

This approach shaped thinking in both natural and social sciences for much of the twentieth century (see also section 5.3.2.3, Chapter 5, Human Geography and Conservation, for a discussion of equilibrium and nonequilibrium models). Rappaport (1968) epitomized an equilibrium approach. He presented his research population, the Tsembaga of Papua New Guinea, as being in balance with their surroundings, as one part of the ecosystem. This view assumed that the Tsembaga and their environment together comprised a bounded system constantly working toward homeostasis, a process that balances out disturbances to the system so that it returns to its original or "natural" state. Equilibrium approaches within anthropology tended to be applied to two types of communities: isolated hunter-gatherers and self-sufficient "traditional" farming communities, both enforcing ideas of bounded social systems.

Yet questions about where to draw the boundary of an ecosocial system, along with evidence that even dramatic shifts in social and natural systems are common, have largely overturned the assumptions that underlie equilibrium theories. Over the past decades, both the natural and social sciences have shifted from equilibrium to nonequilibrium approaches (Table 3.2) as the paradigm that sees nature as "balance" has been questioned (Biersack 1999). Rather than climax, stasis, and adaptation, nonequilibrium ecologies emphasize disturbance, flux, and instability. In environmental anthropology, this has translated into a focus on historical perspectives, how complexity and uncertainty shape environmental behaviors, and interest in the spatial and temporal dynamics of change. One implication for conservation from this body of work is that a history of complex social interactions shapes the way people behave and transform natural systems; delving into this history promotes richer engagement with people and more nuanced understanding of environmental change.

Still, the extent to which equilibrium views have been extinguished is questionable. Within conservation, approaches that do not consider the realities of rapid

**Table 3.2** Equilibrium versus nonequilibrium models.

| | Equilibrium | Nonequilibrium |
|---|---|---|
| *Scale* | Local, bounded | Links to wider political–economic systems |
| *Representation* | Stability/homeostasis/order | Instability/disturbance/complexity |
| *Metaphor* | Balance | Chaos |

---

**Box 3.3 (Continued)**

sociopolitical and climatic changes, and the diverse responses to them, remain rooted in the philosophies of equilibrium science. Resilience studies—the analysis of how systems weather shocks and avoid degradation—may fall prey to this mode of thinking. One review of the field posits that "resilient systems are those wherein both processes of adaptation and modification exist that positively and mutually support [social and environmental] goals for the overall betterment of the collectivity, as a whole" (Stokols et al. 2013, p. 7). Such a definition of resilience may fail to consider the power relations and complexity underlying the concept of maintaining societal or collective interests.

---

saw the world as a series of linked systems of individuals, societies, and ecosystems where feedback loops controlled adaptation and change by shifting multiple variables within those systems.

In addition to the expansion of systems theory, ecological anthropology of the 1960s and 1970s provided novel insights into the convergence of cultural practices, human adaptation, and ecological maintenance. Research found that human *adaptation* to an environment is not determined by ecological conditions alone; societies may exploit the same resources or terrain in very different ways. For instance, colonial settlers introduced technologies that transformed the ecosystems of their new homes.

Environmental anthropology has changed considerably since the 1960s. Thematically, environmental anthropologists now address a vast range of topics beyond human adaptation to ecosystems, including economies, extraction, disasters, landscapes/seascapes (Box 3.4), **globalization**, and **development**. Take "disasters," for example: anthropologists assess popular and scientific ideas about what causes disasters, how disasters unfold, their differential impacts on populations, the way responsibility is taken or avoided for them, political implications of their occurrence, their long-term societal impacts, and how preventative measures are interpreted by citizens, among other inquiries. Amidst their wide-ranging thematic interests, anthropologists are united by their desire to challenge common assumptions through careful, long-term fieldwork that seeks to question and produce theories in light of political awareness and oftentimes with policy concerns in mind.

The differences between ecological and environmental anthropology, **political ecology** (see Chapter 5), and human and social ecology are not clearly delineated; in fact, there is substantial overlap and collaboration within these fields (Table 3.1). Some ways that specialists in these fields collaborate is through global communities of practice such as the Political Ecology Network, supporting initiatives such as Indigenous and Community Conserved Areas, and research into policy regimes such as Reducing Emissions from Deforestation and Forest Degradation (REDD+). In this chapter, the various aforementioned subfields are collectively referred to as environmental anthropology.

---

**Box 3.4   Applications: seascapes**

Anthropological accounts of conservation tend to be terracentric, though some recent ethnographies attempt to counter this trend. The newer focus on ocean conservation arises out of a long-standing interest in marine common-property systems (e.g. Akimichi & Ruddle 1984); ocean navigation and exchange (e.g. Gladwin 1970); and fisheries policy (e.g. McGoodwin 1990). Emerging work on seascapes addresses a few central concerns. First, attention has been given to the social processes surrounding resource rights reallocation, work that builds on and contributes to similar research agendas in terrestrial PAs. In many marine reserves, tourism displaces fishing, dispro-portionately impacting populations living nearby, thus the impacts of marine reserves differ for local residents, tourists, and distant resource users (Walley 2004; Mascia & Claus 2009). Second, anthropologists have studied the influence of new technol-ogies on local cultural institutions and human health, as well as on marine resources. Technological advances in cyanide fishing and scuba diving may increase marine mobility but they can also have significant detrimental effects on both the health of human beings (Lowe 2006) and the prevalence of marine resources (Berkes et al. 2006).

A third focus of seascape research is its application to conservation problems. Ethnographic accounts of marine customary management provide valuable information on diverse methods for managing resource use. Similar to techniques promoted by international conservation organizations based in the USA and Western Europe, cus-tomary management places restrictions on space, temporality, gear, and catch to limit the harvesting of marine plants and animals. Though there are similarities across many systems of marine management, they may differ in their scale, underlying phi-losophy, and intent. This has led anthropologists to advocate for hybridized customary and marine conservation management systems that have proven to be socially suc-cessful in some island states (e.g. Cinner & Aswani 2007).

## 3.3   Concepts, Approaches, and Areas of Inquiry in Anthropology

Anthropology contributes in multiple ways to our understanding of both the human and ecological dimensions of conservation. Here we look more closely at the four themes intro-duced in Section 3.1.2: culture and identity, communities and institutions, power and political economy, and how these and other factors shape the diverse systems that humans have devised to manage their environments. These themes are closely connected within environmental anthropology not only because humans have an impact on the environ-ment, but physical modifications to the environment such as commodity production and land-use change also directly impact social organization and governance.

### 3.3.1   Culture and Identity

Culture is perhaps *the* defining concept in anthropology. Culture refers to a dynamic system of shared symbols, meanings, and norms. It informs how people store, transmit, and understand knowledge, such as understandings of nature and the environment. Human behavior is situated in this complex and ever-changing social and political context. Culture is important to conservation since societies act in and on their environments based on their perceptions. Indeed, culture is the *foundation* for conservation, as it gives rise to environmental values. Culture also shapes knowledge about the environment, resulting in the rich complexity of local ecological knowledge (LEK), which integrates theory and observations about the natural world into practices, beliefs, and customs that govern the management of natural resources and natural systems.

Identity refers to the way human beings perceive themselves and are perceived by others as individuals and as members of groups that may share the same culture. Categories of identity, such as gender, ethnicity, race, class, religion, age, and nationality, give order to the social world. Rather than indicating a core or defining essence however, identities are *relational*; they are a mix of self-determined and socially imposed notions. Anthropologists tend to approach identity in three distinct ways: by focusing on politics, **ideology**, and experiences. "Identity politics" refers to the negotiation of identities, the way identities shape behavior, and the implication of identities in the way groups interact with one another in larger political formations such as nations (see Chapter 6). The relationship between identity politics and resource access, use, and management is an important area of inquiry for anthropologists in conservation.

### 3.3.1.1 The Convergence of Biological and Cultural Diversity

**Biodiversity** is richer where there are numerous microclimates due to variable topography and distinctive landforms. Intriguingly, in such areas, significant cultural diversity is also found. What variables explain this co-constituted biological and cultural diversity? Research into biocultural diversity brings together anthropologists, ethnobiologists, linguists, archaeologists, and cultural historians to trace out the numerous connections between human and biological diversities.

Linguistic analysis plays a significant role in this scholarship. Language is a defining feature of culture and reflects relations with the natural world. The extent to which a language itself shapes perception of the world has long been debated, but it is clear that terms, classifications, and taxonomies used to describe the natural world are a rich source of information on how a language group understands and uses natural resources. Language plays a large part in biocultural diversity studies. As Hunn (1996, p. 5) notes, "variation among ethnobiological systems reflect[s] not only cultural but also biological diversity, since nomenclatural recognition first of all requires that a species exist locally. Ethnobiological diversity will 'capture' biological diversity in proportion to the cultural and/or linguistic endemism within a region." For instance, Papua New Guinea, a megadiverse country with many ecotypes, has 823 distinct language groups. Studies of biocultural diversity focus on specific *biomes*, such as mountains (Stepp et al. 2005), or on resource *management* in "biocultural hotspots" (Maffi 2001). Innovatively mapping these "hotspots" using geographic information systems (GIS) increases the ability of nonspecialists to interpret and understand overlapping biocultural diversities (Figure 3.3).

In recognizing that similar forces threaten linguistic, cultural, and biological diversities, scholars argue that failure to pursue integrated solutions will have dramatic consequences for human and nonhuman life. While the correlation between biological and cultural diversity reinforces the significance of local knowledge and socioeconomic institutions for

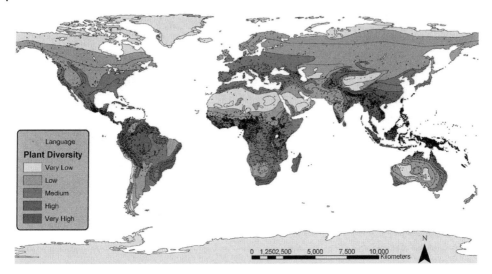

**Figure 3.3** Global plant diversity and language distribution. *Source:* Stepp et al. (2004), Figure 1, p. 271, with permission.

conservation in particular terrains, questions emerge for conservationists about how to define cultural diversity and the role it can play in overall conservation policy and strategy. Cultural diversity coevolves with biodiversity: human use impacts species and natural resources (e.g. wild plants and animals, water, soil), while ecosystem and species changes impact farming, hunting, and other human activities. Knowledge about practices and ecosystem changes is transmitted culturally, thus different cultures experience and shape their environments differently. While this diversity of knowledge and practice is a rich resource, it could intensify ecological fragmentation unless efforts are made to manage across cultural boundaries. When traditional means of communicating and sharing across natural and cultural boundaries break down, conflicts can arise between neighbors and ethnic groups that use the same resources (e.g. farmers and pastoralists). There are also questions about what kind of cultural diversity promotes biodiversity and if it is all similarly valued. Can recent immigrants be seen to expand diversity in knowledge and practice? What about gender and status diversity *within* cultures? Despite the challenges of managing cultural diversity, is it not better (for people and biodiversity) to make that effort than to blanket the landscape with monocultures? Such questions continue to drive research in this field.

### 3.3.1.2 Defining Indigeneity and Being Indigenous
Conservation is embedded in place and territory (see Chapter 5 for more on the importance of place). These places and territories are often the homes of indigenous people. Research in conservation and indigeneity is driven by three overarching questions.

#### 3.3.1.2.1 What Is Indigeneity?
The International Labor Organization (ILO 1989, Article 1) defines indigenous peoples as "tribal peoples in independent countries whose social, cultural and economic conditions distinguish them from other sections of the national community, and whose status is

regulated wholly or partially by their own customs or traditions or by special laws or regulations." Anthropologists however are often less concerned with defining particular groups of people as indigenous (preferring self-determination as indigenous by the peoples themselves) than with researching how indigeneity is articulated, problematized, and formed. Indigeneity is an identity that may or may not conform to the above legal definition in the eyes of those who determine themselves as indigenous. Yet the stakes of indigenous identity are often high. As long-standing residents and users of many lands/seas that conservationists seek to preserve, indigenous peoples have been thrust into debates over property rights. Oftentimes these debates rest on questions of who was there "first."

In reality, complex relationships among ethnic groups confound easy declarations about indigenousness as some groups are able to position themselves as indigenous, thereby securing property rights, while others with similar histories are not as successful (Li 2000). In the Amazon, some question why Brazilian *mestizos*—individuals of non-indigenous or mixed origin who are long-standing rural residents—are not deserving of the recognition afforded to indigenous Amazonians (Penna-Firme & Brondizio 2007). In other cases, as in Japan, it is disadvantageous to be labeled indigenous because of popular associations of indigeneity with backwardness and traditionalism (Tsunemoto 2001).

Whether populations are considered indigenous or not, therefore, impacts not only the ability of residents to access natural resources that they previously used, but also whether those land/seascapes become available for conservation projects. In many cases, unequal power relations and limited economic resources prevent full participation of indigenous community members in decision-making for conservation (Paulson et al. 2012).

### 3.3.1.2.2 How Is Indigeneity Defined, and by Whom?

How is evidence of indigeneity procured, and who is in the powerful position of determining whether groups are indigenous or not? An indigenous identity can be forged by, attributed to, and imposed upon particular populations over others in both local and global spaces. In China, a place rarely associated with indigeneity, Chinese public intellectuals have successfully defined certain populations as indigenous and have garnered attention from the international environmental community in the process (Hathaway 2010). Given the high stakes of claiming indigeneity in some places, the authenticity of these claims is often debated to political ends. Understanding the intricate power relationships underlying identity presentation and the acceptance or contestation of indigenous identities is important for settling conservation and development rights claims.

### 3.3.1.2.3 Are Indigenous People Conservationists?

Other conservation-relevant debates about indigeneity argue about the degree to which indigenous peoples are "ecologically noble savages," or conversely, "ignoble savages." The myth of the ecologically noble savage asserts that indigenous people naturally live in benevolent harmony with their environments. Due to their long-standing interactions in their ecosystems, the argument goes, indigenous people have developed superior NRM systems that conservationists should adopt. The ignoble savage, on the other hand, is depicted as a reckless and inefficient killer of wild animals. Photos of indigenous people are often present in conservation advertising, a phenomenon that can be traced to the mid-1980s when human rights groups "went green" and environmental nongovernment organizations

(NGOs) "went native" to broaden the appeal of both groups. Yet, the tensions that arise when actual indigenous lives meet conflicting ideas of what behaviors qualify as indigenous in the public's eye complicate matters and can result in political consequences for all involved (e.g. B. Conklin & Graham 1995). This research illustrates that assessments of the ecological benevolence of particular resource users are often based in popular imaginations rather than demonstrated practices. Indigenous groups, like other communities, at times exhibit conservation ethics and at other times do not.

### 3.3.1.3  Valuation, Values, and Behavior

In the context of conservation, values refer to both the economic sense of the worth of an object and the sociological sense of standards for behavior. Uncovering values in the economic sense is a crucial step for protecting **ecosystem services**, initiatives that aim to quantify the environment in terms of its provisioning, regulating, aesthetic, and spiritual worth (see also Chapter 4). Anthropologists involved in determining the worth of objects and activities for these kinds of projects have rarely found evidence of the economic theory of self-interest, whereby individuals seek to minimize their effort in order to get maximum value (see Chapter 4 for more on humans as rational actors). Even Malinowski, writing in 1922, found that the concepts of the inhabitants of New Guinea's Trobriand Islands of what is valuable in a garden were counter to this principle—their yam gardens were considered more valuable the more effort gardeners invested in them, but the resulting vegetables were counter-intuitively often left to rot because it was the effort rather than the product that garnered prestige. Similarly, quantifying the worth of some ecosystem services, such as the availability of freshwater, can be easier than others, because researchers themselves often value certain environmental services over others. Fine-grained ethnographic research is often necessary to uncover the cultural and spiritual values of an environment, which is critical for indigenous and local conservation efforts that do not lend themselves to state or market-driven approaches.

Investigations into values in the sociological sense look at value orientations. At a very large scale, this research on values might identify concepts, like "balance" or "purity," and situate them within broader cultural frameworks of meaning and action. Within North America, for example, there is a popular tendency to value natural areas as "wildernesses," whereas in Japan wilderness has less widespread appeal than the domesticated nature of gardens (see Chapter 5 for more on the cultural values surrounding wilderness). Indeed, in many cultures a forest or designated wilderness is the domain of specialized groups, such as hunters, and few people wish to venture into these areas. Values regarding the treatment of animals radically differ as well, with some cultures expressing disgust at the consumption of particular animals (like dogs), while they freely consume other animals (like cows). Ethnographers are also attentive to how value orientations change, and have probed into the shift, for example, from aggressive whale hunting to whale conservation in much of Europe and North America over the past century (Kalland 2009). These values play out in conservation policy as groups pressure governments to ban or support hunting, or craft campaigns against bushmeat consumption in Africa, which are themselves judgments about which animals are considered appropriate for consumption and by whom.

Value orientations within spiritual traditions are clearly more than just philosophies—they orient behavior. Sacred groves are a form of resource management that draws on spiritual values. Sacred groves often have positive ecological values, but a primary rationale for their conservation derives from their salience to religion and community. Because sacred groves are viewed as arising from societal values, they are often perceived as more appropriate than protectionist-oriented PAs in certain regions (Apffel-Marglin 1998). They have been part of conservation organization and government planning in India, where World Wildlife Fund India (now World Wild Fund for Nature India) featured sacred groves prominently in its strategies, creating confluence between religious and conservation activities. In some cases, sacred groves have been converted to large-scale PAs. Silent Valley National Park in India was established because it was deemed sacred by local people. Vana Degula, a garden of sacred plants, was established in Sri Lanka after forests became threatened by eucalypts. Defining community values was the first step in designing those conservation initiatives that have had widespread local appeal.

### 3.3.1.4   Local Environmental Knowledge

LEK refers to knowledge, practices, and beliefs maintained by people with histories of ecological interaction. Due to environmental and cultural variations, these knowledge systems are highly variable worldwide. Traditional ecological knowledge (TEK), indigenous knowledge (IK), folk science, and ethnoecology (Box 3.5) are conceptually similar to LEK, which is the broadest of these terms as it encompasses knowledge systems of urban and rural, indigenous and long-standing populations. Common to all of these terms is a focus on history, lived experience, and place. Two basic assumptions underlie the concept of LEK: first, long-resident people have developed sophisticated systems of resource management that are sensitive to local conditions; second, these systems of knowledge are critical to understanding the relationship of humans with their non-human surroundings.

Anthropological research on LEK has shifted over time. One strand of LEK studies presents fine-grained descriptions of local residents' everyday practices like environmental taxonomies, technologies, management practices, and underlying cultural beliefs. Harold Conklin's (1957) pioneering study, for example, systematically laid out the technical specifications of the Hanunoo's swidden (or shifting cultivation) agricultural system on the island of Mindoro in the Philippines. Alongside details of tropical cultivation, Conklin presented Hanunoo rituals and traditions, illustrating their importance to the overall agricultural system. These *content* studies aim to understand how cultural representations relate to environmental decision-making and behavior.

Detailed information about local resource management systems can aid in the development of natural resource policy. Conklin's study in the Philippines illustrated how, contrary to popular opinion, Hanunoo's fallow land was intentionally left in a state of regeneration rather than abandoned. This method minimized risk and maximized use of the environment and was economically more efficient for the region than the modern intensive agriculture that the government was implementing at the time. Similarly, a study of the Nuaulu of Indonesia's taxonomy of tree species demonstrates that their forest categories were, unlike official ones, dynamic and multidimensional. For instance, their classification of tree species shifted based on whether the trees were categorized as female or male, fruiting or non-fruiting (Ellen 2007). These taxonomies anticipated scientific ecology's modeling of rainforest patchiness.

**Box 3.5   Crossing boundaries: ethnoecological approaches**

Ethnobiology—the study of how people identify, classify, and use natural resources—was an early area of intersection between anthropology and conservation. Indigenous and local people together with anthropologists as translators have often been guides for conservation initiatives, aiding in identification of local flora and fauna and describing their uses. In addition, anthropologists have contributed extensively to plant and animal collections as they were often the only scientists to visit remote areas. This practice has, however, since been denounced by many indigenous groups and nations as theft of their intellectual property. In response, guidelines and regulations now govern such collections. For example, the advocacy group for indigenous peoples, Cultural Survival, details several tools and approaches (Cultural Survival 2000).

Ethnoecology is the study of how people understand their environments and their relationships to them. Ethnoecologists broadly take one of three approaches to their work: economic, ecological, or cognitive. Economic approaches seek to compile basic inventories of environmental information, such as economic uses of natural resources, while ecological approaches seek to answer how human societies interact with plants, animals, and the biophysical environment.

Cognitive studies are the most prominent approach in ethnoecology today. The driving question here is "how and in what ways do human societies view and think about the environment?" To answer this question, ethnoecologists undertake various types of research. Basic qualitative research documents botanical knowledge such as the common or vernacular names of plants, how they are harvested and stored, their medicinal, industrial, or food uses, frequency of collection, economic and cultural importance, and management status. **Quantitative** research utilizes structured interviews and questionnaires to systematically evaluate factors such as use-value and the relative economic importance of a given species. Experimental methods are used to assess the benefits of plants for subsistence and seek to maximize the value that local people attain from their natural resources. Often a mixture of these research types is deployed to uncover taxonomies. Taxonomy is the practice and science of classification. Ethnoecologists look at how knowledge of the environment is classified, since this pattern illuminates the importance of plants in and among societal groups. Horticulturalists, for example, tend to know more taxa than foragers, illustrating the importance of particular types of environmental knowledge for cultivation (Balée 1999). Refined or detailed naming tends to signal more reliance on a plant or animal for subsistence or cultural factors (the often quoted but apocryphal argument about the myriad names that Eskimos have for snow illustrates this point). Two features of ethnobiological classification are its perceptual basis and the organization of categories into hierarchical structures.

Perhaps more than other research in environmental anthropology, ethnoecology remains comparative, in that it focuses not just on one local taxonomic system but on comparing systems. Comparative taxonomic analyses have led some authors to hypothesize that there are underlying principles of ethnobiological classification worldwide (Berlin 1992). For example, there are no less than three and no more than six ethnoecological ranks into which all recognized plant and animal taxa can be accommodated: kingdom, life-form, intermediate, generic, specific, and varietal (Figure 3.4). The comparative nature of ethnoecology, along with its melding of the natural and social sciences, ensures its continued relevance to conservation.

**Box 3.5   (Continued)**

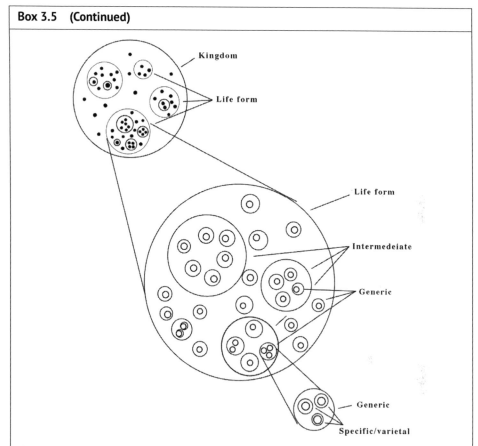

**Figure 3.4**   A schematic representation of ethnobiological taxonomic ranks. Relative positions convey how closely related organisms are—the closer the circles, the more closely related the organisms. *Source:* Berlin (1992) Figure 1.2, p. 23.

Application of LEK to conservation problems similarly seeks to integrate this environmental knowledge into natural resource policy. LEK research has yielded insights into behavioral ecology, environmental fluctuations, and linkages among environmental processes. Social applications of LEK include alternative methods for resource management and conservation that can be implemented in diverse environments like multiple species management, resource rotation, and succession management (e.g. Berkes et al. 2000).

Another strand of LEK research focuses on the *context* of LEK. As anthropologists began to question the generation of anthropological knowledge in the 1990s, the context of LEK generation was also questioned. Researchers noted diversity in LEK arising from informants' gender, occupation, and age (Robbins 2003). The simple association of LEK with traditionalism and community is questioned in these studies that highlight the many facets of how and by whom LEK is generated (e.g. Nazarea 2006; Forsyth & Walker 2008).

Attention to the broader milieu of LEK led to a third strand of LEK studies, regarding its *production*. Though LEK is set up in contrast to Western science, this boundary is fuzzy given the long interaction of these knowledge systems since the fifteenth century (Agrawal

1995). Some European plant classification systems, for example, were based in the LEK of low-caste Indian forest dwellers (Grove 1996). Research of this type tends to see LEK as everyday understanding and practice with high variability (e.g. Sillitoe 2007).

Natural scientists continue to learn from local communities' taxonomic classification of species previously unknown to Western science. However, some have argued that the decontexualization of LEK that occurs when aspects are extracted for conservation applications results in reduced effectiveness of LEK. In particular, an instrumental orientation focused on practical use of LEK may neglect underlying belief systems and can result in contradictory philosophical bases for conservation action. For example, ethnobotanists have voiced concern about use of LEK for pharmaceutical research. And they note that species identification is only one part of the complex system of management of the species and habitat, treatment of disease, as well as wider spiritual and cultural considerations (Martin 2004).

Others have argued that it is not the underlying value systems that distinguish LEK and Western science, since all science is culturally embedded. Rather, what counts as "knowledge" depends on who is determining its value. As such, anthropologists have urged critical reflection before applying LEK to conservation problems (Dove 2006). Some researchers have suggested that using the term *situated knowledges* rather than enforcing a dichotomy between "LEK" and "science" is one way to promote critical reflection of how all knowledge systems are contingent and shaped by culture.

LEK remains a significant concept for conservation, however. The Convention on Biological Diversity recognized the importance of LEK to conservation, stating that

> each contracting Party shall, as far as possible and as appropriate, subject to national legislation, respect, preserve and maintain knowledge, innovations and practices of indigenous and local communities embodying traditional lifestyles relevant for the conservation and sustainable use of biological diversity and promote their wider application with the approval and involvement of the holders of such knowledge, innovations and practices and encourage the equitable sharing of the benefits arising from the utilization of such knowledge innovations and practices.
>
> *(CBD 1992)*

### 3.3.2 Communities and Institutions: The Foundation for Action

Anthropologists study how community and individual identities are linked, through concepts such as kinship and ethnicity. Communities are typically distinguished from societies, which are larger entities with a common identity (e.g. Javanese society) or structure (e.g. industrial societies). While we continue to employ the term *community* in this chapter, it is important to foreground that its vagueness hampers conservation efforts that require fine-grained knowledge of people and localities (Table 3.3).

Institutions are not just formal groups, but also the rules and regulations that groups abide by, and anthropologists have sought to make visible the importance and power of "informal" institutions such as customary tenure systems and bylaws created by communities for common-property NRM. Anthropologists also study local institutions within the context of community-based natural resource management (CBNRM) (e.g. Schnegg 2018), indigenous and community-based conservation (CBC) (e.g. Stevens 2014), and integrated conservation and development projects (ICDPs) (e.g. Gezon 1997).

**Table 3.3** CAUTION: Use with care!

| Term | In theory | In practice |
| --- | --- | --- |
| Community | A group of people living in the same place or having characteristics in common | Community is an umbrella term that can obscure social relationships that relate to conservation projects. Use more specific terms such as village, ethnic group, or population living within 20 kilometers of a protected area to describe populations and groups. |
| Participation | The action of taking part in something | Participation is a slippery term because it encompasses a wide range of activities, from accepting information to giving consent, to determining actions. Describe what it involves in any given context by carefully defining who is involved and the nature of their involvement. |
| Stakeholder | A person or group with an interest or concern in something | Stakeholders have different stakes in conservation projects; they are not easily collapsed into this overarching term. Clearly describe and differentiate the rights, roles and responsibilities of relevant groups. |
| Livelihood | A means of securing the necessities of life | This term downplays the mediation of markets and the multiple strategies used by people to both make a living and advance socially and economically. It is better to describe economic systems and disaggregate the roles and actions of individuals within them. |

### 3.3.2.1 Community Complexity and Conservation

Recognizing the imprecision of the term *community* entails consideration of how boundaries are constructed around areas and peoples and what such framings obscure in terms of mobility and diversity. The boundaries among and within communities are far from self-evident. Where these boundaries are drawn in conservation matter, because groups that fall within the boundaries are often considered to constitute the units of action, and benefit. Depending on how communities are viewed, some people using the land and natural resources of an area such as pastoralists, fishers, or recent migrants may be left out of the frame. They may not be present during meetings or surveys or left out because they lack formal rights. In practice, communities are composed of subgroups of people with various statuses, needs, and desires. For instance, remote dwellers belonging to an ethnic group targeted for conservation may be bypassed, even though they interact most closely with biodiversity. The assumption is that leaders in the more accessible settlements can and do represent all members of the group. Outside developers and local residents often have different ideas about what constitutes a community, and these divergences have implications for conservation projects. Conservationists, for example, defined communities in Mafia Island Marine Park in Tanzania spatially, while local residents defined their communities by social networks that stretched far beyond the village itself (Walley 2004; see also Chernela & Zanotti 2014).

The early 1980s saw the rise of various "community"-oriented conservation efforts embodied in CBNRM, CBC, and ICDPs. CBNRM and CBC were seen as antidotes to top-down, state-imposed conservation and development approaches. Anthropologists delved into this arena to document and analyze diverse approaches and their impacts.

CBNRM has had mixed outcomes. For example, in Namibia after independence, the state—with significant support of NGOs and safari and trophy-hunting businesses—brought social benefits and rights to previously disenfranchised populations through the creation of conservation conservancies. However, in neighboring Botswana, where CBNRM was also initiated, struggles over control of land, resources, and management approaches often led to negative outcomes for indigenous populations (Hitchcock et al. 2011).

Attempts to understand community heterogeneity have led to research on CBNRM that exposes the differential impacts of conservation initiatives on local residents. The Mafia Island Marine Park in Tanzania, for example, brought a school, health clinic, and tourist camp to the island, all projects that had both local support and regional opposition. These development projects increased inequality despite their philanthropic intent because of failures to see communities as comprised of many diverse groups with different needs (Walley 2004). For this reason, disaggregating conservation impacts on particular groups within communities (such as fisherfolk, shell collectors, and tourists) is important for creating policy solutions (Mascia et al. 2010). In other cases, conservation is seen to bring benefits to marginalized populations, such as in the Namibia conservancies. Under the broad narrative of success, however, the reality of conservancies is more complex as "conservation efforts have been parachuted into an existing socio-political field of institutional complexities ... that has empowered a range of new actors, generating, in turn, a series of interfaces and contestations with the pre-existing so-called 'traditional' modes and forms of organizing the use of and access to natural resources" (Hebinck et al. 2020).

Given this discussion, the terms in the left-hand column of Table 3.3 may obfuscate more than they describe. Table 3.3 provides practical suggestions for more accurate terms.

### 3.3.2.2  Institutions Underpinning Land and Resource Use

This section focuses on two types of institutions that anthropologists have studied extensively: those governing land tenure and property rights; and institutions involved in kinship, marriage, and inheritance. While the first type is clearly critical for conservation, the second is less intuitively linked to conservation; however, anthropologists have drawn important connections between how people organize families and how land and natural resources are managed.

Land tenure and property rights are structured by legal and cultural institutions that define the multiple ways that people own, access, and control land and natural resources. Customary property rights can include forms of leasing, borrowing, access, and even use rights to specific resources such as trees, tree products, livestock, or fishing grounds. Anthropologists have documented how tenure and rights systems are highly variable and adaptive, originating from the need for flexibility across seasons and biomes, and to facilitate exchange among different groups such as pastoralists, farmers, fishers, and hunters. Formal legal, or statutory, tenure systems coexist with customary rights in many parts of the world. Failure to understand customary rights and conflict between customary and statutory rights holders can lead to lack of clarity on who can use a resource, and who has the right to restrict use of that resource.

Tracing the history and evolution of tenure and rights systems is essential to understanding linked conservation and human rights dilemmas and developing right-based approaches to conservation (Box 3.6). The anthropology of enclosure documents how

---

**Box 3.6   Applications: rights and asset-based approaches**

Rights-based approaches can incorporate the notion of cultural heritage as a right, which consists of not only rights to cultural knowledge but also rights to natural resources and environmental services as part of a community's assets. When viewed from this perspective, communities that are monetarily poor begin to realize the wealth of their collective rights in land and natural resources, as well as culture. While the commitment of local populations to landscape scale conservation is critical to its success, ethnobotanist Janis Alcorn and colleagues (2006) argue that it cannot be sustained unless powerful stakeholders such as large conservation NGOs, donors, governments, and the private sector take a rights-based approach that respects and builds on heritage.

  Alcorn and colleagues describe how rights and assets approaches to conservation have been employed among communities in the Amazonian frontier. They differentiate between the two terms: an "assets-based approach assists the poor to build physical capital, financial assets, community organizations and institutions and the ability to influence policies. It also acknowledges the great value that social assets play in providing resilience to the poor. A rights-based approach, however, differs in that it involves moving beyond providing venues for participation by the poor to giving over leadership and decision-making roles to the poor" (Alcorn et al. 2006, p. 274).

  Both of these approaches have been successfully used in conservation projects. For example, asset-mapping tools have jumpstarted dialogue about conservation in Pando, Bolivia, where they used to leverage local energies and promote public deliberation at the landscape level among a variety of constituencies. The initiative continued after the initial assessment and resulted in the declaration of a grassroots PA.

---

formal and privatized tenure arrangements, including gazetted PAs, have intruded into collectively managed commons. Restriction or enforced movement of human populations in cases of enclosure and displacement are shaped by unequal power dynamics, and have implications for identity and self-determination. For instance, efforts to determine a group's "proper" location and enforce boundaries (for example, to control incursion into a conservation area) can create conflict about the group's identity, which may be complex and cross those boundaries. As well, loss of the collectively managed commons can lead to loss of group identity (Cunningham & Heyman 2004).

  In many rural areas, local populations live and farm on land that is not formally titled or deeded, a result of colonial and postcolonial regimes that decreed all land and resources property of the state, while customary tenure systems are incompletely recognized in the law. This duality creates legal space for alienation of valuable land and resources from rural people for oil, gas, mining, and forestry concessions, industrial agriculture, as well as for PAs and even ecotourism (e.g. Hughes 2001). Such alienation has implications for rural poverty, outmigration, and local resistance to conservation efforts. Further, tenure insecurity and the decline of customary work systems stemming from mobility and migration can trigger resource degradation because there are fewer personal and social incentives to manage a **common-pool resource**. CBNRM and other efforts to decentralize state management of land and natural resources like forests and fisheries perpetuated from the colonial era often do not sufficiently shift the balance of power and counter degradation trends.

To address historical and new inequities related to dispossession or lack of recognition, forms of conservation such as Indigenous and Community Conserved Areas are expanding, drawing on earlier efforts such as the creation of Ancestral Domain Claims in the Philippines. Where there is no formal documentation, ethnography and analysis of historical and genealogical records may be the best or only tools to achieve recognition of ancestral territories and to register indigenous and community-conserved areas (ICCAs 2019).

Institutions of kinship, marriage, and inheritance are also central to how individuals and groups manage environments since rules associated with them may determine who has rights to farmland or use of common-property resources such as forests or pastures. These rules have important implications for both livelihoods and for the resources involved. The differential rights of men and women, as well as those of minorities and remote dwellers, can generate inequality in decision-making, labor burden, and benefits. Working with women's groups on conservation within customary systems, for example, may be rewarding due to their energy and knowledge. However, an approach incorporating gender equality and women's empowerment is necessary if they lack the power and rights to manage lands and resources due to their gender and marital status (e.g. Schroeder 1999).

Other consequential connections between kinship and conservation are best studied using ethnographic methods such as participant observation due to the sensitive nature of the topics. The requirement to pay bridewealth in order to marry within a customary system, for example, may push young men into resource exploitation in cases where their ability to earn money is restricted and they will not inherit land until the elders pass away, as Russell and Tchamou (2001) described in their study of forest and farm management in southern Cameroon. The link between the supernatural and natural resources such as trees, watersheds, and groves serve as profound markers of kinship. Ancestors are linked to certain terrains or natural resources and when these connections are broken, by forced migration, environmental degradation, or development, the solidarity of the group can be threatened. Smith's (2005) ethnographic study of a Kenyan community in the Taita Hills documents the community's attempt to revitalize witchcraft practices in order to reinvigorate group solidarity that had been eroded by modernity and development. However, state and church assistance were eventually sought by community members to remove a controversial "witch doctor" linked to groves of trees that housed ancestors, a move that ultimately reaffirmed modern reliance on the same state power that community members had found problematic.

### 3.3.2.3 Conservation Organizations as Objects of Study

What are the subtle dynamics of power within and among conservation institutions and their "target" populations? Research into the actions and impacts of conservation organizations (e.g. Larsen & Brockington 2017) and the interactions between local and international conservation organizations (e.g. Mahanty & Russell 2002) reveals power imbalances and miscommunications. By examining the daily practices of interested actors, anthropologists probe into displays and contestations of power.

Organizational ethnographies—extended monographs that treat an organization as the target of research—represent one form of research into conservation dynamics. Ethnographic analysis undertaken at a field station of the World Wildlife Fund in Japan, for example, indicates how conservationists there attempt to reconcile tensions between local and global environmentalisms in their quotidian practices (Claus 2020). Staff members engage in acts of translation, promoting the same work using very different terminology in domestic and transnational venues. They also engage in acts of circumvention, for example creating new organizations to undertake projects that are categorized as "conservation" domestically, but would not be considered so in mainstream global environmentalism, while still finding ways to fund that work. These challenges permeate the work of transnational organizations that seek to respond to domestic and international audiences who have very different ideas about what conservation should look like and who should be responsible for enacting it.

In addition to formal organizational studies, other research assesses the interactions of various networks in international arenas. The relative power that various stakeholders have at events where political power is differentially wielded—such as the United Nations Climate Change conferences and the World Parks Congresses—is a growing area of inquiry given the rising prominence of such global events (e.g. Brosius 2004). These organizational ethnographies assess how environmental knowledge is produced, by whom, and what the consequences of asymmetric power relationships are for conservation outcomes. Kiik (2019) argues, however, that anthropologists have not done enough to study conservationists as subjects in their complexity and lays out a roadmap for the ethnography of "Conservationland" by attending to the perspectives and diversity of conservation professionals and institutions, their transnational social worlds, naturalist worldviews, and emotional lives.

Even as scholars convincingly argue that little attention has been paid to conservationists, transnational organizations have been studied much more than small-scale, local conservation institutions. This is surprising considering that local environmental groups often carry out the work on the ground in internationally funded conservation. Local environmental defenders worldwide are at the forefront of conservation and environmental justice while often being attacked and murdered. Are there characteristics that make these local conservationists effective? How do they create and sustain constituencies? In what ways do they form alliances to fight powerful interests and threats?

### 3.3.3 Power, Political Economy, and Conservation

Anthropological insights help clarify the relations of power and economic forces that shape conservation practice and outcomes. The anthropological study of political economy has origins in the works of Karl Marx, Max Weber, and many political economy theorists who followed them, most recently within the cross-cutting field of political ecology (see also Chapter 5, Human Geography). Many anthropologists working in conservation and NRM have also been significantly influenced by agrarian and peasant studies and by the study of common-property NRM. The crosscutting field of political ecology, as well as the economic and political subdisciplines within anthropology (Table 3.1), comprise diverse theoretical viewpoints but share in common a focus on power and political economy.

Anthropologists have demonstrated how power is grasped and maintained not only by physical and political means but also through control of **discourse**—the vocabulary and symbols describing, and circumscribing, an issue or concern. Forsyth and Walker (2008) describe how the framing of environmental problems in Thailand circumscribes and assigns responsibility for soil degradation and soil erosion in ways that are less reflective of actual causal relationships and more reflective of who is able to control and shape the ways that problems are understood and responded to (see Section 5.3.2.2 in Chapter 5 for more on environmental discourses). Power relations also shape economies: markets and other economic arenas deeply impact the distribution and very nature of power.

Conservation too is a societal institution that allocates access to resources and benefits, and thus it is enmeshed in political economies at all levels. For example, the push for CBNRM allowed for some increases in conservation decision-making by communities, while ecoregional approaches (Box 3.7), in contrast, privileged international NGOs as the main actors who defined and "planned" landscapes across multiple social boundaries (Brosius & Russell 2003).

---

**Box 3.7   Methods: ecoregions, regional analysis, and multisited ethnography**

Recognizing that species and ecological functions do not respect societal boundaries and require connectivity, conservation efforts in the 1990s moved beyond a focus on PAs and ICDPs to embrace ecoregional, landscape, and seascape approaches. This evolution was also a backlash against ICDPs, seen to be ineffective at reducing threats to ecosystems and biodiversity at the needed scale.

Anthropological critiques of ecoregional approaches noted a lack of integration with the body of knowledge and practice in regional analysis used by anthropologists, economists, geographers, and landscape historians. Contrary to stereotypes that anthropologists' work is constrained to villages or remote locations, some have long studied larger regions and wider networks in research on market and food supply chains, regional NRM systems, and the global political economy. Regional analyses and multisited ethnography are useful tools for addressing recent conservation trends.

One critique of ecoregional approaches is that cultural, administrative, and market boundaries that impact ecosystems may not be well captured; differing nomenclature and units of analysis underlie this disconnect. While landscape conservation initiatives focus on identifying, connecting, and "managing" ecological units, anthropologists typically first identify socially defined units, such as ethnic group territories, villages, markets, neighborhoods, or associations, and consider how these are or can be connected spatially, economically, or politically.

Effective conservation requires coordination of this information and these perspectives. Conservation action across large-scale ecological units must necessarily coordinate across multiple management institutions and "stakeholders." What makes sense ecologically thus requires deep knowledge of institutions and linkages among them, acquired through sound research methods. Such research may uncover the fact that many of the pressures and actors impacting biodiversity are located outside of a given ecoregion. Substantial progress has been made in interdisciplinary collaboration for the study of common-property NRM, as well as in the science of coupled human–natural systems, now being supported by the US National Science Foundation. This

| **Box 3.7** | **(Continued)** |
| --- | --- |

progress was fostered because the interconnectedness of social and environmental dimensions was clearly identified from the outset of the research program.

Operationally, anthropologists have postulated that employing large-scale ecological units as a basis for planning may disadvantage local- and even national-level institutions to the detriment of conservation outcomes, unless planning involves participation across those levels (Brosius & Russell 2003). Effective conservation necessitates respect for local property, assets, rights, and values. Hence, while conservation scientists continue to study and map connections and trends at appropriate ecological scales, many anthropologists have concluded through their research that, to be effective, conservation action needs to be local (e.g. Brosius et al. 2005).

Adapting the anthropological method of multisited ethnography also has the potential to bolster efforts to address "drivers" of biodiversity loss (again these drivers are often *outside* a conservation area). Take the example of wildlife trafficking: Collaboration of the American Museum of Natural History and colleagues in Vietnam covers multiple sites and actors (NSF 2015) to tease out ways in which actors involved in the wildlife trade are connected, revealing incentives that entice new actors and illuminating how "illegal" trade is linked to structures within the "legal" economy. In her studies of the "underground economy" of Zaire (now Democratic Republic of Congo) anthropologist Janet MacGaffey (1991) revealed how social networks facilitated financing, labor recruitment, and access to valuable resources such as ivory and timber that are traded locally and internationally. Their research shows how ethnographic methods are essential for understanding the workings of clandestine operations.

Anthropologists also work to bridge gaps between ecoregions and human territories through maps and remote sensing. Pioneering work in aerial photography (Vogt 1974) spread across the discipline to understand land use and land changes in Africa (e.g. Fairhead & Leach 1996). Other initiatives involve community mapping of indigenous peoples' territories over large areas, often harmonizing local maps with satellite imagery and official government maps to achieve policy and human-rights objectives.

An extensive body of work that fully integrates ethnography and GIS using remote sensing comes from Emilio Moran, Eduardo Brondizio, and their colleagues and students at the Center for Analysis of Social–Ecological Landscapes. They set out to integrate analysis of satellite imagery with detailed ethnographic and historical studies, in part to understand patterns of intensification, deforestation, and forest succession in the Amazon. Combining ethnographic attentiveness and increasingly precise environmental monitoring tools holds great promise for developing fine-tuned multiscalar and transdisciplinary approaches (see Chapter 5 for more on remote sensing and GIS).

### 3.3.3.1 Colonialism and Neocolonialism

The direct appropriation of land and natural resources for the benefit of the colonial powers and those settling in metropoles was an essential element of colonization. One example is the many PAs in Africa (Figure 3.5) originally set up by colonial administrators and militarized to keep out former landowners and resource users (Box 3.8).

**Figure 3.5**    Park guards on patrol in W National Park, Benin. Photo by Daniel C. Miller.

---

**Box 3.8    Debate: conservation refugees in Central Africa**

Conservation refugees exist on every continent but Antarctica, and by most accounts live far more difficult lives than they once did, banished from lands they thrived on for hundreds, even thousands of years.

(Dowie 2005, p. 161)

War, development, and mining are often blamed for forcing people to flee from their lands. Yet, as Dowie argues above, environmental conservation has also been pinpointed as the cause of the eviction of millions of people. Conservation refugees often originate with PA establishment, enlargement, or the stepped up enforcement of conservation policies. Anthropologists have documented the effects of the expansion of PAs and subsequent attempts to resettle local residents. In Central Africa, the large average size of PAs and the predominantly rural populations make this issue particularly problematic (West et al. 2006).

Conservation refugees suffer physical, economic, and social displacements when they are evicted from their lands. These resettlements occur because it is believed that people within parks are a threat to biodiversity and, further, that resettlement will reduce pressure on habitats and wildlife. In Cameroon's Korup National Park, the resettlement of a village rested on the erroneous assumption that the local people were "poaching" animals in the park (Tiani & Diaw 2006). Once relocated, the village's traditional governance systems rapidly eroded. The impact on reducing hunting pressure was questionable since, in addition to the absence of the indigenous communities, the weak protection offered by park authorities translated into fear that others would move into the vacated space and claim hunting territory. In another case, the removal of villagers from the Mkomazi Game Reserve in Tanzania resulted in the collapse of livestock markets due to loss of pasturelands that the population depended on for their livelihood (Brockington 2001). These examples illustrate that physical displacement is

**Box 3.8** **(Continued)**

often intertwined with economic and social displacements (Table 3.4), and the environmental rewards in many instances are ambiguous.

**Table 3.4** Potential direct and indirect social costs and benefits of protected area establishment (Mascia and Claus 2009 / John Wiley & Sons).

**Governance**

Decreased/increased resource control

Property lost*/gained

User rights lost/gained

Conflict resolution mechanisms weakened/strengthened

**Economic well-being**

Employment lost*/gained

Income lost*/gained

Assets lost*/gained

Consumption reduced/increased

**Health**

Health diminished*/enhanced

Food availability reduced*/increased

Nutrition status diminished/enhanced

Psychological well-being diminished/enhanced

Health services reduced/increased

**Education**

Public services lost*/gained

Human capital lost*/gained

Education opportunities lost/gained

**Social capital**

Social networks degraded*/increased

Social status lost*/increased

Partnerships and alliances lost/increased

Trust lost/gained

**Culture**

Cultural space lost*/gained

Local knowledge lost/gained

Sense of place diminished/enhanced

Norms and values undermined/reinforced

*Highlighted by Cernea's (2000) framework of physical displacement risks

*(Continued)*

---

**Box 3.8 (Continued)**

Conservation displacements are often blamed on the international environmental NGOs, although NGOs argue that they lack the authority to evict people from their lands and point instead to national governments. This issue has led to heated debates between social scientists and conservation organizations. In the case of Central Africa, the Wildlife Conservation Society asserted that "not a single individual has been removed from the protected areas created in central Africa over the past decade" (Curran et al. 2009, p. 30). A social scientist working with different data—collected with fine-grained social science research methods and deploying an expanded definition of resettlement (Table 3.4)—argued that in fact thousands of people had been displaced (Schmidt-Soltau 2009).

In areas where human rights are not respected, there are structural and systemic inequalities between conservation planners and local residents, whereby the latter lack governmental representation or even basic land security. These inequalities appear to enable displacement. Some argue that NGOs are responsible for conservation refugees because of their preference for creating "fortress-style" national parks as well as their ahistorical representation of PA territories, as Brockington and Igoe describe,

> Conservation NGOs ... have significant influence in representing protected areas as peopleless spaces, conserved free from the influence and despoiling activities of people. This is important because part of the disempowerment of dispossession and eviction is the obliteration of former residents from their landscapes, from their homes and past. Celebrating and proclaiming former homelands as wilderness denies people's place in these landscapes. It thereby reduces the political space available to them as they attempt to reclaim lost lands.
>
> (Brockington & Igoe 2006, p. 445)

Further, in Central Africa, the combination of weak states and colonial legacies may make it particularly hard to plan adequately for displacement. Practicing anthropologists build on social scientists' attempts to insert safeguards and compensation for local communities into large-scale multilateral development that involves resettlement and associated impacts (e.g. Cernea 2000). It is important to note that costs and risks, as well as benefits, are unequally borne, even within one group, necessitating customized social safeguards developed using ethnographic methods.

---

Colonial legacies remain in international relations, where powerful nations and interests have moved from overt territorial control toward control over global commodities, financial systems, as well as ideas and values (e.g. the growing influence of neoliberal political and economic thinking since the 1980s). McAnany and Yoffee (2010 p. 14) observe that "the inequities of colonialism are an ongoing process played out internally in terms of access to education and political voice, as well as internationally in the arenas of resource distribution and political clout."

Within conservation, colonial legacies manifest in models such as private game ranches and PAs that cater to (largely white, Western) safari tourists and trophy hunters. More broadly, anthropologists have argued that such legacies are discernible in the prominence of transnational NGOs in conservation and the dominant approach of attacking "poaching" and wildlife trafficking through the militarization of conservation (Larsen & Brockington 2017). The term "poaching" has a problematic history. Colonial elites tended to label indigenous hunting as poaching, in contrast to supposedly more "gentlemanly" trophy hunting by white colonizers, thereby criminalizing indigenous practices. There are concerns that contemporary conservation discourse surrounding poaching ignores the complex drivers of hunting and is used to justify the increasing use of force and weapons to protect wildlife.

Meager resources flow to communities and local institutions in areas of significant conservation value compared to the resources provided to international NGOs, parastatals, and the private sector, leading to poor conservation results (e.g. the case of Fiji documented by Siwatibau & Lees 2007). Local project "participants" are often tasked to generate their own resource flows or work voluntarily. In remote and minority areas, people lack political **agency** to change this situation. Also, much of what people do for a living in remote communities might be criminalized or tightly controlled by conservation projects. On the other hand, while conservation is a marginal concern in most countries, especially compared to extractive industries and agribusiness, it can represent an important source of support to communities facing threats to land and resources depending on approaches employed.

Colonial legacies are not just about Western domination. Indonesia provides a case of internal colonialism. Under President Suharto, the policy of *transmigrasi* (transmigration) flooded indigenous territories with migrants from densely populated areas, inciting conflict and fragmenting the social and natural landscape. Today, despite decades of investment in social and community forestry, social scientist Nining Liswanti (quoted in Shahub 2018) finds that "many forest-dependent people—natives and migrants both—do not know what social forestry is. They don't know the legal basis, rights and obligations, processes, or whether or not there is tenure security for the forests their communities manage." And when they are able to chart their own paths, community members may have differing priorities as they are incentivized by commodity production and upward mobility that they see others achieving.

The challenge for anthropologists—and conservationists—working amidst colonial legacies is finding ways to expose oppressive relationships while building more equitable, just, and sustainable forms of conservation. The next section details a few of these efforts undertaken by anthropologists.

### 3.3.3.2 Anthropologists and International Conservation

Anthropologists have worked in, studied, and often clashed with conservation efforts. In Melanesia, anthropologists sought to collaborate with conservationists to counter what they observed as devastating effects of logging, mining, and oil and gas extraction on clans and their territories (e.g. the creation of the Committee of Concerned Pacific Scholars in 1993). Differences soon emerged, however, as anthropologists became concerned that externally driven conservation initiatives were operating in an imperialist fashion, as West (2006) describes in *Conservation Is Our Government Now*. In subsequent work, West and her long-time collaborator John Aini created conservation projects that focus on conservation with communities in Papua New Guinea in a way that builds on local concerns and needs as well as support a new generation of conservation scientists.

Anthropologists working in Africa had similar reactions to what they felt were imperialistic approaches to conservation. As property rights in Africa are not typically vested in rural communities, dispossession became a major concern. For instance, in southern Africa, large areas of rangeland once used by pastoralists were seized for the purpose of wildlife conservation, based on the premise that pastoralist practices were harmful to the land and wildlife. A number of anthropologists and their collaborators studying African pastoralists and (former) hunter-gather peoples have documented how these peoples contribute to, and clash with, conservation efforts (e.g. Little 1996; Fay 2013). This alienation of land from local people for PAs, among other uses, led anthropologists to research the phenomenon of "conservation refugees" (Box 3.8). Scholars working in West and Central Africa also disputed widely circulating narratives of how environmental degradation was perpetrated and accelerated by local actors and documented customs that support conservation (Fairhead & Leach 1996; Peterson 2003).

In South and Central America, many anthropologists have focused on indigenous peoples, supporting them to obtain rights to land, and to have a voice in conservation. Underpinning this approach is research on landscape history. The archaeology and ethnography of the Amazon reveal how once densely populated areas were depopulated due to colonialism, disease, and other factors, hitherto becoming the "wildernesses" beloved of conservationists (e.g. de Souza et al. 2018). This depopulation rendered the people and their terrains invisible. There is evidence that recent rural depopulation has had negative environmental impacts such as increasing fire incidences due to unchecked natural fires (Uriarte et al. 2012).

Many anthropologists do more than carry out research. Helping secure the rights of the Kayapó peoples to their lands was the life's work of anthropologist and ethnobotanist Darrell Posey. By 2008, the Kayapó were able to control, legally and physically, a continuous block of the Amazonian forest totaling 28.4 million acres (11.5 million hectares), at that time the planet's largest block of tropical forest protected by a single indigenous group (Posey 2008). In his work, Posey has remarked that conservation paradigms needed to change, arguing that ecosystems that environmentalists want to "save" depend on their management by indigenous and local peoples. This achievement is under constant threat, however, by dams, pollution, and extractive industry. Anthropologists Janet Chernela and Laura Zanotti (2014) find additional challenges for international collaboration with the Kayapó due to lack of outside organizations' understanding of people's allegiances and exchange and marriage networks, which are often more important for identity than a group's physical territory. This finding illustrates the importance of questioning notions of "community" and even the very term (Table 3.3).

Chapin and colleagues' (2005) community mapping initiatives, where they assisted many indigenous groups in South and Central America to obtain legal title to their territories, boosted indigenous conservation efforts in the process. Anthropologists from the region continue to fight for indigenous rights in conservation through groups such as Survival International and Cultural Survival, and within government institutions, such as the Fundação Nacional do Índio in Brazil.

In Asia, anthropological research in Borneo exposed the risks to indigenous people that paradoxically arose from both logging that put targets on the backs of local activists and efforts to "save the rainforest" that limited their options for economic and societal development (Brosius 1999). Other scholars have documented the richness of indigenous NRM systems in the face of rapid and large-scale environmental and social transformation (e.g. Tsing 1993; Padoch & Peluso 2003; Lowe 2006). In the Philippines, pioneering anthropologist, ecologist, and pastor Delbert Rice (2018) over decades helped secure the rights of upland indigenous peoples to their lands and forests, and constantly reminded lowlanders of the ecosystem services that their upstream indigenous neighbors sustain.

On the subcontinent, a rich literature concerning indigenous practices focuses on the link between conservation and the sacred, contrasting with Western conservation paradigms. Echoing Posey, Apffel-Marglin (1998, p. 233) concludes from the study of sacred groves in India, "the wilderness is a world in which humanity is involved. Excluding people from natural areas in order to preserve these lands is a recent idea imported from the West; it has not worked well in India and the rest of tropical Asia ... More biodiversity preserves and more nature and wildlife parks cannot be an answer either in India or anywhere else."

In sum, anthropologists provide critical information and tools to improve conservation and bring long-term commitment to effect needed policy changes and strengthen local capacity. These contributions, however, often challenge dominant conservation paradigms and as such may go unrecognized or diminished by mainstream conservation.

### 3.3.3.3 The Role of Markets and Commodities in Neoliberal Conservation

Shaped by political power and policies, markets differentially impact populations. Untangling these impacts has been an important feature of research within both environmental and economic anthropology. Conservation approaches today increasingly emphasize the creation or extension of markets and the commoditization of the environment to support the costs of conservation and to create incentives to "change behavior" away from what are seen to be destructive practices. A key tenet of such neoliberal thought and action is the portrayal of markets as neutral or beneficial to all parties involved. Anthropological research empirically assesses such claims by looking at the reverberations of changing market structures on various populations.

Anthropologists have documented how even "benign" industries such as ecotourism can have negative impacts on local populations (and on ecosystems). For example, ecotourism in Zimbabwe provided the justification for white South Africans and Europeans to grab land for trophy hunting from small-scale subsistence hunters (Hughes 2001). In Tanzania, the extension of Tarangire National Park promised to bring ecotourism benefits to nearby Maasai communities, but benefits were sparse and unevenly distributed, which reinforced economic inequalities and created conflicts (Igoe 2004). Ironically, ecotourism in this region is premised on its portrayal to Western tourists as pristine nature, which dismisses the generations of pastoralists who shared and shaped the land along with wildlife. In fact, the systematically low revenues of ecotourism ventures disadvantage Maasai when compared with pastoralism and other livelihood options that preceded these conservation "solutions" (Homewood et al. 2009).

Anthropologists also study markets that develop and expand around global commodities and the way these commodities impact land use and local economies, such as cocoa, coffee, and palm oil. Market analysis, in some cases of an entire value chain (e.g. sugar, tea) sheds light on land-use change as well as how this change leads to new social formations (Mintz 1986; Besky 2014). The insight for conservation from these studies is that land use is not the only thing that transforms in the wake of new markets; the way in which people organize, manage, and even think will change. During the "cocoa crisis" in southern Cameroon, for instance, drastic declines in farmgate prices for cocoa brought about profound shifts in social organization and gender relations such as the increased economic importance of women's food crops and youth rebellion against the high cost of bridewealth, as well as increased deforestation as forest was cleared for food crops (Russell & Tchamou 2001).

Research into the marketing of endangered species has followed that of other clandestine markets (e.g. Scheper-Hughes 2004). Some anthropologists focus on understanding demand and local markets in nations such as China and Vietnam while others look at the role of local communities in wildlife trafficking in source areas (e.g. Cooney et al. 2017). An important body of work assesses the *cultural values* that drive markets for ivory and other products of endangered species (Drury 2009).

Markets for environmental services are deployed to finance conservation by linking "consumers" of services, such as water provisioning and carbon sequestration, with stewards of that resource. REDD+, the international initiative to financially incentivize reducing carbon emissions through avoided deforestation and forest degradation, has sparked considerable concern and debate within anthropology and allied disciplines. While much of the global policy dialogue has been about measuring forest carbon sequestration accurately and assuring the flow of funds, anthropological inquiry focuses on the degree to which investments have just and equitable benefit-sharing systems. Anthropologists cautioned that REDD+, like other payments for ecosystem services approaches, where countries prepare to receive large sums of money for protecting forests or other "services," will exacerbate or drive power imbalances that will prevent these funds from reaching those with the greatest forest dependence (Sullivan 2009).

### 3.3.4 Managing Environments over Time and Space

Anthropology studies the use and management of natural resources over time and within specific forms of social organization. Given the immense diversity of ecosystems and human settlements, this is an exceptionally rich area of research. While early anthropologists looked at how environmental variables such as climate and topography shaped or even determined human evolution and societal variability, today's anthropologists home in on the cultural systems that arise to manage environments. This research describes the vast array of NRM systems, builds theories about resource sustainability, and investigates causes and consequences of societal collapse. Probing the relationships that people have with their surroundings brings the past to bear on present problems.

#### 3.3.4.1 Systems and Technologies

NRM comprises ways that societies organize to manage lands, as well as aquatic and terrestrial natural resources, to produce goods for subsistence and market and to conserve species and habitats. In addition to looking at conservation areas and systems, anthropologists study NRM systems of farming, foraging, hunting, and fishing, along with extractive industries such as mining and logging, including how these systems intersect with local and external conservation actions. NRM systems are studied within the context of culture, identity, and power, and anthropological subjects are often at the social—and spatial— margins of the state: nomadic pastoralists intersecting with global cattle markets and competing for terrain with PAs; farming communities adopting cocoa or coffee cultivation amidst commodity booms and busts; or indigenous people coping with conversion of their lands to oil palm plantations. Table 3.5 depicts the diversity of anthropological inquiry into NRM.

This section addresses the anthropological study of four NRM sectors and themes critical to conservation: (i) agriculture; (ii) PAs; (iii) technology change; and (iv) the impact of climate change.

**Table 3.5** Natural resource management across major theoretical frameworks.

| Major theoretical tradition | Corollaries | Approach to natural resource management (NRM) |
|---|---|---|
| Natural history | Historical particularism (Boas) | Catalogs material culture, describing practices |
| | Holistic ethnography Multisite ethnography | Describes NRM system components and how they evolved and interrelate |
| | Causal explanation | Uncovers causes of behavior through fine-grained ethnographic fieldwork |
| Comparative analysis of behavior and traits | Sociobiology | Models behavior based on natural selection |
| | Structuralism | Analyzes myths and symbols contrasting nature and culture |
| | Formal ethnobiology analysis | Creates typologies and models of taxonomies to reveal structure |
| | Formal linguistic analysis | Analyzes terms for species and ecosystems, showing cultural history and borrowings |
| Comparative analysis of cultures and systems | Ethnology | Compares practices and management systems across one or multiple dimensions |
| | Marxist and neomarxist | Reveals material effects of power dimensions within NRM practices and systems |
| | Political ecology | Reveals both hidden and overt power elements and their impacts; gender, agrarian change dynamics |
| Archaeology | Culture history | Describes ancient cultures' use of land and natural resources |
| | Comparative | Analyzes patterns of resource use in relation to social and economic factors as well as "outcomes" of different strategies |

### 3.3.4.1.1 *Agriculture*

Agriculture (with animal husbandry) is considered among the top threats to biodiversity and is the dominant human use of land and natural resources. Anthropological research focuses on describing bodies of agricultural knowledge as well as situating agriculture in wider political and economic systems. For instance, swidden cultivation—commonly called "slash-and-burn" agriculture—has historically intrigued anthropologists as an intricate and adaptive practice (H. Conklin 1957) and most recently attracted attention because of its demonization as a major cause of deforestation within REDD+. Yet it is not monolithic: as small-scale cultivators rotate their fields, they can contribute to the retention and, in some cases, enhancement of agrobiodiversity (Nazarea 1998). Similarly, practices such as raising cattle and growing oil palm are major threats when they become **industrialized**, but they are integral parts of smallholder farming systems (Netting 1993). Anthropologists track the evolution of these systems and their intersections with markets, industry, and land appropriation.

### 3.3.4.1.2 *Protected Areas*

PAs are the most prominent NRM systems deployed for conservation. There is a wide range of PAs, from globally recognized World Heritage Sites to locally managed marine areas, local parks, and sacred groves that may have no formal protected status. Informally or formally, people live within most PAs and depend on their resources. It has been argued that many PAs were created in areas where conservation may have not been most needed but where political resistance was weak as the lands, and peoples, are marginal (Joppa & Pfaff 2009). Some anthropologists and allied scholars find that when PAs follow colonialist and now neoliberal modes of acquiring and managing land and resources, they lack accountability and local legitimacy (displacement from PAs is discussed in Box 3.8). Some worry that PAs may protect valuable resources not only for the sake of conservation, but for corporate and elite use and benefit, for example fencing off areas with valuable mineral reserves from local use (Brockington et al. 2008). Others, such as ethnobotanist Gary Martin (2004), concentrate on strengthening new forms of conservation, such as Indigenous and Community Conserved Areas, which seek to empower disenfranchised populations as well as to conserve (see also Stevens 2014).

### 3.3.4.1.3 *Technological Change*

Tracking technology change is one lens anthropologists employ to study how NRM systems evolve and the implications for ecosystems and species. Clearly some technologies are unique to specific ecosystems. For example, although both deployed for fishing, harpoons used in the Arctic seas differ from hooks or seines used in streams. Yet technologies cross boundaries and in the process dramatically change NRM practices and social structures associated with them, creating complex and often rapid impacts on people and their environments.

In other cases, attempts to "improve" local NRM systems with new technologies can have unintended consequences. For example, anthropologists have studied how colonial powers in Africa introduced new technologies, crops, and forms of agricultural intensification that transformed not only landscapes but also social relations. Many farmers switched from food crops to cash crops, leading to environmental degradation and hunger.

Today as agricultural scientists seek to reduce the footprint of farming as well as greenhouse gas emissions, it is important to explore how "sustainable intensification" technologies such as agroforestry (Figure 3.6) or improved fallows may fail to account for the labor burden and lack of profitability of such practices for poor farmers, or for potential perverse impacts. For instance, in Guinea, intensive onion farming was hypothesized to reduce deforestation in the context of a landscape conservation project, as revenue from the onions would substitute for forest farming revenue and thus forest farming would decrease. But intensification might also increase women's labor burden and possibly accelerate deforestation if it became highly profitable. Research using ethnographic methods can provide invaluable insights into such approaches through observing changes in labor and market patterns and interviewing a range of actors.

The loss of traditional technologies when societies become dependent on imported or manufactured goods has been linked to loss of knowledge about natural resources themselves, and this loss may increase vulnerability (Blaikie 1994). But returning to what may previously have been sustainable NRM systems is often not possible. Research on the restoration of historically robust irrigation systems in Tanzania found that local support and long-term funding were not enough to restore these systems (Sheridan 2009). Rural populations may find it difficult if not impossible to recuperate NRM systems, yet they may be unable to move toward "modern" technologies where there is underinvestment, land insecurity and conflict, youth disaffection, and stagnating gender roles.

### 3.3.4.1.4   Climate Change
Climate change and the policy and technology responses to it are increasingly impacting NRM systems, including shifts in land use, management regimes, cropping seasons, pest

**Figure 3.6**   A nutmeg nursery for an agroforestry project in Aceh Seletan, Indonesia. Photo by Diane Russell, 2017.

and disease load, and water supply. Mitigation efforts may involve local labor-intensive actions such as intensification of agriculture, as discussed above, or promoting tree planting. Strategies for adaptation and "resilience" may be based on traditional NRM technologies but fail to incorporate rapid cultural and social change. Strategies to alter land use toward conservation or climate change mitigation and adaptation must thus take into account not just the outward manifestations of resource degradation but also the social, economic, and political structures that underpin degradation, many of which operate at different scales or in different geographical locations. Environmental degradation can be driven by economic depredation: for instance, low farmgate prices and insecure land tenure can lead to expansion of the agricultural frontier.

Archaeologists Susan Crate and Mark Nuttal (2009, p. 11) argue that we need to think even more broadly: "On a temporal scale, the effects of climate change are the indirect costs of imperialism and colonization—the 'non-point' fall-out for peoples who have been largely ignored. ... Climate change is environmental colonialism at its fullest development—its ultimate scale—with far-reaching social and cultural implications." The world's wealthy countries and individuals consume vastly more resources than the poor. Thus, they are disproportionately responsible for $CO_2$ emissions and climate change (see also Chapter 8 for dependency theory and the "development of underdevelopment"). In short, the poorest are suffering for the lifestyles of the wealthiest. Anthropological research allows us to adopt a broad and deep approach to understanding these power relations and to tackle a range of climate change and NRM issues, from large-scale planning decision frameworks and land-use policies to incorporating detailed knowledge of local NRM systems.

### 3.3.4.2 Land Histories

Anthropology's holistic perspective is evident in historical ecology, a field of inquiry that incorporates long-term archaeological data with ethnographic methods to build understanding of past and present landscapes. This analytical perspective recognizes that humans do not merely adapt to their environments; they also change them. That is, human cultures and physical environments mutually influence each other in a complex feedback loop. Land histories can help explain present-day land use and enforce land-use or property-rights claims. Historical data shed light on how current landscapes have been anthropogenically modified and are similarly important for understanding long-term ecological processes such as soil formation. They provide a practical perspective on restoration ecology and realistic assessment of conservation potential, and pinpoint potential alternative land uses and management systems such as sustainable small-scale farming systems.

Research into land histories has encouraged critical reflection regarding long-held assumptions, such as the myth of the pristine America. Landscapes once viewed as "wild" have increasingly been recognized as anthropogenically formed as the presence of human-mediated disturbances such as fire, pathogens, and viruses provide evidence for past human settlement in such landscapes. Wetland sediment cores and animal middens house pollen which provide paleoecological and archaeological evidence regarding species ranges and extinctions, resource management practices, ecosystem extent, and the paths of plant and animal species introductions. This information regarding the ebb and flow of civilizations in particular landscapes sheds light on the impact of political changes on the land (e.g. Richards 1986) and the role of culture and perception in environmental interactions (e.g. Crumley & Marquardt 1987). Such material evidence, when supplemented with ethnographic and historical data, is useful not only for resource management. It has been used to redress the exclusion of indigenous peoples from their lands by addressing the differing legal, moral, and cultural frameworks for restitution of these losses.

### 3.3.4.3 Resource Management and "Collapse"

Over decades, anthropologists and archaeologists have investigated human population growth and the rise of the state in relation to the management, and mismanagement, of natural resources. This research looks at both extremes of human–environmental outcomes: "sustainability" and the so-called collapse of civilizations.

Changing social demographics are one factor shaping outcomes. Calculating the balance between population growth rates and natural resource use is exceedingly difficult, since there is no simple relationship between these variables. Population growth, including in-migration, tied to technological or resource use intensification can magnify environmental impacts. But population growth itself does not automatically lead to overuse of resources. In some cases, population growth can spur agricultural innovation, which can reduce pressure on natural systems. In other cases, population growth is coupled with increased consumption or land grabbing that forces those with limited financial capital to scramble for more marginal lands. The struggle for these marginal lands can play out between conservationists and the people whose farming or pastoral systems are considered a threat to biodiversity. Counterintuitively, *depopulation* can also lead to environmental degradation because humans can play a key role in regulating natural systems such as fire regimes and clearing for new growth (Figure 3.7).

Political change, or lack thereof, factors into environmental change, as technologies of resource use and extraction are embedded in political and economic systems. For instance, archaeologists have postulated that natural resources in some societies were exhausted due to overconsumption by political elites as they consolidated power, built empires, or waged war (Railey & Reycraft 2008). The failure of elites to adopt technological innovation also can

**Figure 3.7** The foothills of the Blue and John Crow Mountains in Jamaica are degraded despite, or perhaps because of, decreasing population densities. Photo by Diane Russell, 2017.

lead to socioenvironmental change. Norse Greenland is an example of a society that failed to adapt during the Little Ice Age. While nearby Inuit were tapping into marine resources, Norse leadership maintained their cultural preferences for farming and hunting, declining to pursue Inuit marine technologies (Dugmore et al. 2007). The Norse made a choice *not* to change, a choice that ultimately led to the deterioration of their Greenland settlements.

Does environmental degradation drive the decline of civilizations? A thought-provoking collection by archaeologists analyzes several cases of purported collapse, including well-known ones such as Easter Island (Rapa Nui) and the Mayan civilization. The authors conclude, "Although change is inevitable, and living through some kinds of change is difficult, painful or even catastrophic, 'Collapse'—in the sense of the *end* of a social order and its people—is a rare occurrence" (McAnany & Yoffee 2010, p. 11, emphasis in original). Even the Norse whose settlements disappeared may have migrated elsewhere (Berglund 2010). This new archeological research on social collapse indicates that overexploitation by elite classes in pursuit of continued dominance and wealth exacerbated or even created these crises. These findings reinforce the need for a holistic approach to conservation informed by deep understanding of how power relations shape environmental outcomes.

## 3.4 Future Directions

Anthropologists can attest that globalization, large-scale environmental change, and the machinations of elites to attain and retain valuable resources are not new phenomena, and as such many insights from earlier research remain highly relevant. But having been tasked with laying out some emerging directions for the field of conservation anthropology, we anticipate that the field will increasingly pursue research into changing climates and rapid loss of biodiversity, increased technological connectivity, and the challenges and opportunities of evidence-based approaches. As a multifaceted historical discipline grounded in the study of diversity, anthropology can help other sciences grasp the nature of heterogeneous, fractal, nonequilibrium "systems" that emerge with globalization, migration, and social media. Shifts in ideas and actions happen rapidly and asymmetrically as information and influence speed around the globe. Can the anthropological focus on the plasticity and creativity of social formations help conservation science to understand these new realities?

### 3.4.1 Responding to the Climate and Biodiversity Crises

Anthropologists like Donna Haraway (2016) question the wisdom of renaming the current geological epoch as the Anthropocene. Even as this name is perceived as engendering recognition of how influential humanity is on even the Earth's geological systems, the term is devastatingly oblivious to the importance of nonhuman actors and environments. By taking the Anthropocene not as a given but rather in querying who is using this term and to what ends, anthropologists question what its implications are for conservation and the way we talk about nature. Biological diversity is now found more in intensively managed landscapes than in "the wild" (Ellis & Ramankutty 2008) and it's not appropriate to assume that conservation can be something that happens away from urban and peri-urban areas. To that end, anthropological research will increasingly probe the boundaries of conservation by studying "rewilding" efforts, industrial environments, and techno-natures. To do so, we predict that anthropologists will engage in new and renewed efforts to contribute to socioecological modeling.

Climate change profoundly affects all people, but especially those in vulnerable areas and those directly dependent on natural resources. Though anthropologists have historically documented the effects of changing climates in human evolution and within small-scale societies, new directions in climate research are appearing as ethnographers increasingly probe into climate change-related disasters. Such research assesses the workings of international policy bodies, adaptations associated with shifting global food systems, and global-scale responses to disaster events. Anthropological methods are highly suited to both understanding and communicating the effects and challenges of climate change to diverse audiences (e.g. Furman & Bartels 2017).

The crisis of biodiversity loss only recently received media attention on par with climate change, with research showing dramatic declines, for example, in pollinators and insects in general (e.g. Tollefson 2019). This crisis reaches beyond PAs and indeed beyond conservation into sustaining food systems and life itself. Anthropological insights and tools help make connections among changing land uses, power relations, market dynamics, and how technologies and solutions are conceived of and packaged. Who would benefit or lose out for example in E.O. Wilson's (2016) "half-earth" strategy that seeks to set aside and create vast "wild" areas designed to mitigate this existential threat?

### 3.4.2 Bringing Anthropology into the Conservation Evidence Discourse

Much of the attention in the conservation evidence field has focused on "rigorous" methodologies such as impact evaluations and systematic reviews in journals such as *Environmental Evidence* and *Conservation Evidence*. Anthropologists have contributed a tremendous body of knowledge to and about conservation, yet within the conservation evidence sphere this knowledge can be perceived as "anecdotal" if it is recognized at all. Anthropologists rightly problematize depiction of evidence as "neutral," but the underlying aims of this initiative—to promote accountability and transparency in how conservation is practiced and evaluated—can significantly address common anthropological critiques of conservation practice. Questions that have emerged include: How do conservationists accurately and honestly represent multiple perspectives about conservation outcomes? How can conservationists gauge the implications of deploying different evidence bases for outcomes, not only for social actors but also for species and ecosystems? More broadly, how can conservationists and scholars respect the need for evidence while critiquing specific bodies of evidence? We anticipate that anthropologists, skilled in examining a problem from multiple perspectives, will have a lot to offer.

### 3.4.3 Studying the Impact of New Technologies and Social Media

As discussed in this chapter, technology is never neutral. It is developed and wielded in a social context, in ways that can benefit or disadvantage certain groups. New technologies can shift discourses about conservation and even generate large-scale change. Anthropologist Shafqat Hussain (2019) studied how conservationists are using camera traps to track snow leopard populations in northern Pakistan. Using observations of herders and other key informants, he found that snow leopards are more common around towns than in "wilderness" areas, yet the camera traps are not located near towns. This placement skews conservation strategy toward a traditional approach of shoring up PAs and supports the claim that the leopards are in drastic decline (hence the need for more conservation funding).

Other examples of how technology is unexpectedly deployed comes from drones and social media platforms. Drones may be efficient in the fight against "poachers," but they might also threaten the human rights and even the lives of those "caught" in PAs. Another area of study is how social media platforms such as WhatsApp affect intergenerational relations, as young people may become more grounded in their social networks than in place-based affinities. Emerging research suggests a reconfiguration of the idea of "remoteness," a perennial theme differentiating peoples and places in both anthropology and conservation (Keleman Saxena 2016). This kind of deep reflection on technologies can help to tease out unintended consequences, ensure that they do not harm people, and explore ways they could actually work to promote socially just conservation.

## For Further Reading

1 Anthropology and Environment Society Blog: https://aesengagement.wordpress.com.
2 *Ethnobiology* (Anderson, Pearsall, Hunn, Turner, eds. 2011, Wiley-Blackwell, Malden, MA).
3 *Environmental Anthropology: A Historical Reader*, vol. 10 (Dove & Carpenter 2008, Wiley-Blackwell, Malden, MA).
4 *The Environment in Anthropology: A Reader in Ecology, Culture and Sustainable Living* (Haenn & Wilk, eds. 2005, New York University Press).
5 *Groundwork for Community-based Conservation: Strategies for Social Research* (Russell & Harshbarger 2003, AltaMira Press, Walnut Creek, CA).
6 *The Nature of Whiteness: Race, Animals, and Nation in Zimbabwe* (Suzuki 2017, University of Washington Press).
7 *Friction: An Ethnography of Global Connection* (Tsing 2005, Princeton University Press).

## References

AAA (American Anthropological Association) (1998) AAA statement on race. https://www.americananthro.org/ConnectWithAAA/Content.aspx?ItemNumber=2583 (accessed May 27, 2020).

AAA (American Anthropological Association) (2012) Principles of professional responsibility. http://ethics.americananthro.org/category/statement (accessed October 1, 2020).

Agrawal, A. (1995) Dismantling the divide between indigenous and scientific knowledge. *Development and Change* 26: 413–439.

Akimichi, T. & Ruddle, K. (1984) *The Historical Development of Territorial Rights and Fishery Regulation in Okinawan Inshore Waters.* Senri Ethnological Studies, no. 17. Suita, Osaka, Japan: National Museum of Ethnology.

Alcorn, J.B., Carlo, C., Rojas, J. et al. (2006) Heritage, poverty and landscape-scale biodiversity conservation: an alternative perspective from the Amazonian frontier. *Policy Matters* 14: 272–285.

Apffel-Marglin, F. (1998) Secularism, unicity and diversity: the case of Haracandi's grove. *Contributions to Indian Sociology* 32: 217–235.

Asad, T. ed. (1973) *Anthropology & the Colonial Encounter*. London: Ithaca Press.

Balée, W.L. (1999) *Footprints of the Forest: Ka'apor Ethnobotany – The Historical Ecology of Plant Utilization by an Amazonian People*. New York: Columbia University Press.

Bateson, G. (1972) *Steps to an Ecology of Mind*. Chicago, IL: University of Chicago Press.

Berglund, J. (2010). Did the medieval Norse society in Greenland really fail? In: *Questioning Collapse: Human Resilience, Ecological Vulnerability, and the Aftermath of Empire* (ed. P.A. McAnany & N. Yoffee), 45–70. Cambridge, UK: Cambridge University Press.

Berkes, F., Colding, J. & Folke, C. (2000) Rediscovery of traditional ecological knowledge as adaptive management. *Ecological Applications* 10: 1251–1262.

Berkes, F., Hughes, T.P., Steneck, R.S. et al. (2006) Globalization, roving bandits, and marine resources. *Science* 311: 1557–1558.

Berlin, B. (1992) *Ethnobiological Classification: Principles of Categorization of Plants and Animals in Traditional Societies*. Princeton, NJ: Princeton University Press.

Besky, S. (2014) *The Darjeeling Distinction: Labour and Justice on Fair-Trade Tea Plantations in India*. Berkeley, CA: University of California Press.

Biersack, A. (1999) Introduction: from the "new ecology" to the new ecologies. *American Anthropologist* 101: 5–18.

Blaikie, P. (1994) *At Risk: Natural Hazards, People's Vulnerability and Disasters*. New York: Routledge.

Brockington, D. (2001) Women's income and the livelihood strategies of dispossessed pastoralists near the Mkomazi Game Reserve, Tanzania. *Human Ecology* 29: 307–338.

Brockington, D., Duffy, R. & Igoe, J. (2008) *Nature Unbound. Conservation, Capitalism and the Future of Protected Areas*. London: Earthscan.

Brockington, D. & Igoe, J. (2006) Eviction for conservation: a global overview. *Conservation and Society* 4: 424–470.

Brosius, J.P. (1999) Green dots, pink hearts: displacing politics from the Malaysian rain forest. *American Anthropologist* 101 (1): 36–57.

Brosius, J.P. (2004) Indigenous peoples and protected areas at the World Parks Congress. *Conservation Biology* 18: 609–612.

Brosius, J.P. & Russell, D. (2003) Conservation from above: an anthropological perspective on transboundary protected areas and ecoregional planning. *Journal of Sustainable Forestry* 17: 39–66.

Brosius, J.P., Tsing, A.L. & Zerner, C. eds. (2005) *Communities and Conservation: Histories and Politics of Community-Based Natural Resource Management*. Lanham, MD: AltaMira Press.

CBD (Convention on Biological Diversity) (1992) Article 8(j) – traditional knowledge, innovations and practices. https://www.cbd.int/traditional (accessed June 2, 2020).

Cernea, M. (2000) Risks, safeguards and reconstruction: a model for population displacement and resettlement. *Economic and Political Weekly* 35 (41): 3659–3678.

Chapin, M., Lamb, Z. & Threlkeld, B. (2005) Mapping indigenous lands. *Annual Review of Anthropology* 34: 619–638.

Chernela, J. & Zanotti, L. (2014) Limits to knowledge: indigenous peoples, NGOs, and the moral economy in the Eastern Amazon of Brazil. *Conservation and Society* 12 (3): 306–317.

Cinner, J.E. & Aswani, S. (2007) Integrating customary management into marine conservation. *Biological Conservation* 140: 201–216.

Claus, C.A. (2020) *Drawing the Sea Near: Satoumi and Coral Reef Conservation in Okinawa.* Minneapolis, MN: University of Minnesota Press.

Conklin, B.A. & Graham, L.R. (1995) The shifting middle ground: Amazonian Indians and eco-politics. *American Anthropologist* 97: 695–710.

Conklin, H.C. (1957) *Hanunóo Agriculture: A Report on an Integral System of Shifting Cultivation in the Philippines*, vol. 2. Rome: Food and Agriculture Organization of the United Nations.

Cooney, R., Roe, D., Dublin, H. et al. (2017) From poachers to protectors: engaging local communities in solutions to illegal wildlife trade. *Conservation Letters* 10: 367–374.

Crate, S.A. & Nuttall, M. eds. (2009) *Anthropology and Climate Change: From Encounters to Actions.* Walnut Creek, CA: Left Coast Press.

Crumley, C.L. & Marquardt, W.H. (1987). Regional dynamics in Burgundy. In: *Regional Dynamics: Burgundian Landscapes in Historical Perspective* (ed. C. Crumley), 609–623. San Diego, CA: Academic Press.

Cultural Survival. (2000) Protecting indigenous intellectual property rights: tools that work. *Cultural Survival Quarterly Magazine*, September. https://www.culturalsurvival.org/publications/cultural-survival-quarterly/protecting-indigenous-intellectual-property-rights-tools (accessed June 13, 2020).

Cunningham, H. & Heyman, J. (2004) Introduction: mobilities and enclosures at borders. *Identities: Global Studies in Culture and Power* 11 (3): 289–302.

Curran, B., Sunderland, T., Maisels, F. et al. (2009) Are Central Africa's protected areas displacing hundreds of thousands of rural poor? *Conservation & Society* 7: 30–45.

de Souza, J.G., Schaan, D.P., Robinson, M. et al. (2018) Pre-Columbian earth-builders settled along the entire southern rim of the Amazon. *Nature Communications* 9: art. 1125. https://doi.org/10.1038/s41467-018-03510-7.

Dove, M.R. (2006) Indigenous people and environmental politics. *Annual Review of Anthropology* 35: 191–208.

Dowie, M. (2005) Conservation refugees: when protecting nature means kicking people out. *Orion Magazine Online.* https://orionmagazine.org/article/conservation-refugees (accessed October 1, 2020).

Drury, R. (2009) Reducing urban demand for wild animals in Vietnam: examining the potential of wildlife farming as a conservation tool. *Conservation Letters* 2: 263–270.

Dugmore, A., Keller, C. & McGovern, T.H. (2007) Norse Greenland settlement: reflections on climate change, trade, and the contrasting fates of human settlements in the North Atlantic islands. *Arctic Anthropology* 44: 12–36.

Ellen, R. (2007). Local and scientific understanding of forest diversity on Seram, Eastern Indonesia. In: *Local Science versus Global Science: Approaches to Indigenous Knowledge in International Development* (ed. P. Sillitoe), 41–74. New York: Berghahn Books.

Ellis, E.C. & Ramankutty, N. (2008) Putting people in the map: anthropogenic biomes of the world. *Frontiers in Ecology and the Environment* 6: 439–447.

Fairhead, J. & Leach, M. (1996) *Misreading the African Landscape: Society and Ecology in a Forest-Savanna Mosaic*. Cambridge, UK: Cambridge University Press.

Fay, D. (2013) Neoliberal conservation and the potential for lawfare: new legal entities and the political ecology of litigation at Dwesa–Cwebe, South Africa. *Geoforum* 44: 170–181.

Forsyth, T. & Walker, A. (2008) *Forest Guardians, Forest Destroyers: The Politics of Environmental Knowledge in Northern Thailand*. Seattle, WA: University of Washington Press.

Frazier, J.G. (1993) *The Golden Bough: The Roots of Religion and Folklore*. New York: Gramercy Books.

Furman, C. & Bartels, W.-L. (2017) Climate histories in black and white: contextualizing climate services through anthropology. *Practicing Anthropology* 39 (1): 17–22.

Gezon, L. (1997) Institutional structure and the effectiveness of integrated conservation and development projects: case study from Madagascar. *Human Organization* 56 (4): 462–470.

Gladwin, T. (1970) *East Is a Big Bird: Navigation and Logic on Puluwat Atoll*. Cambridge, MA: Harvard University Press.

Grove, R. (1996) Indigenous knowledge and the significance of south-west India for Portuguese and Dutch constructions of tropical nature. *Modern Asian Studies* 30 (1): 121–143.

Haraway, D. (2016) *Staying with the Trouble: Making Kin in the Chthulucene*. Durham, NC: Duke University Press.

Hardin, R. & Remis, M.J. (2006) Biological and cultural anthropology of a changing tropical forest: a fruitful collaboration across subfields. *American Anthropologist* 108: 273–285.

Hathaway, M. (2010) The emergence of indigeneity: public intellectuals and an indigenous space in Southwest China. *Cultural Anthropology* 25: 301–333.

Hebinck, P., Kiaka, R.D. & Lubilo, R. (2020). Navigating community conservancies and institutional complexities in Namibia. In: *Natural Resources, Tourism and Community Livelihoods in Southern Africa* (ed. M.T. Stone, M. Lenao & N. Moswete), 64–77. New York: Routledge.

Hitchcock, R.K., Sapignoli, M. & Babchuk, W.A. (2011) What about our rights? Settlements, subsistence and livelihood security among Central Kalahari San and Bakgalagadi. *The International Journal of Human Rights* 15 (1): 62–88. https://doi.org/10.1080/13642987.2011.529689.

Homewood, K., Kristjanson, K. & Trench, P.C. eds. (2009) *Staying Maasai?: Livelihoods, Conservation and Development in East African Rangelands*. New York: Springer.

Hughes, D.M. (2001) Rezoned for business: how eco-tourism unlocked black farmland in eastern Zimbabwe. *Journal of Agrarian Change* 1: 576–599.

Hunn, E. (1996) Columbia Plateau Indian place names: what can they teach us? *Journal of Linguistic Anthropology* 6: 3–26. https://doi.org/10.1525/jlin.1996.6.1.3.

Hussain, S. (2019) *The Snow Leopard and the Goat: Politics of Conservation in the Western Himalayas*. Seattle, WA: University of Washington Press.

Hymes, D. (1972) *Reinventing Anthropology*. New York: Pantheon Books.

ICCAs (Indigenous and Community Conserved Areas) (2019) www.iccaconsortium.org (accessed May 5, 2019).

Igoe, J. (2004) *Conservation and Globalization: A Study of National Parks and Indigenous Communities from East Africa to South Dakota*. Belmon, CA: Wadsworth.

ILO (International Labor Organization) (1989) C169 – Indigenous and tribal peoples convention, 1989 (No. 169). https://www.ilo.org/dyn/normlex/en/f?p=NORMLEXPUB:1210 0:0::NO::P12100_ILO_CODE:C169.

Joppa, L.N. & Pfaff, A. (2009) High and far: biases in the location of protected areas. *PLOS ONE* 4 (12): e8273. https://doi.org/10.1371/journal.pone.0008273.

Kalland, A. (2009) *Unveiling the Whale: Discourses on Whales and Whaling*. New York: Berghahn Books.

Keleman Saxena, A. (2016) Dealing with data in socio-environmental field research. *EnviroSociety Blog*, October 31. https://www.envirosociety.org/2016/10/digital-environments-dealing-with-data-in-socio-environmental-field-research/#more-1110.

Kiik, L. (2019) Conservationland: toward the anthropology of professionals in global nature conservation. *Critique of Anthropology* 39: 391–419.

Larsen, P.B. & Brockington, D. eds. (2017) *The Anthropology of Conservation NGOs: Rethinking the Boundaries*. Cham, Switzerland: Palgrave Macmillan.

Li, T.M. (2000) Articulating indigenous identity in Indonesia: resource politics and the tribal slot. *Comparative Studies in Society and History* 42: 149–179.

Little, P.D. (1996) Pastoralism, biodiversity, and the shaping of savanna landscapes in East Africa. *Africa* 66: 37–51.

Lowe, C. (2006) *Wild Profusion: Biodiversity Conservation in an Indonesian Archipelago*. Princeton, NJ: Princeton University Press.

MacGaffey, J. (1991) *The Real Economy of Zaire: The Contribution of Smuggling and Other Unofficial Activities to National Wealth*. Philadelphia, PA: University of Pennsylvania Press.

Maffi, L. ed. (2001) *On Biocultural Diversity: Linking Language, Knowledge, and the Environment*. Washington, DC: Smithsonian Books.

Mahanty, S. & Russell, D. (2002) High stakes: lessons from stakeholder groups in the biodiversity conservation network. *Society & Natural Resources* 15: 179–188.

Martin, G.J. (2004) *Ethnobotany: A Methods Manual*. London: Routledge.

Mascia, M.B. & Claus, C.A. (2009) A property rights approach to understanding human displacement from protected areas: the case of marine protected areas. *Conservation Biology* 23: 16–23.

Mascia, M.B., Claus, C.A. & Naidoo, R. (2010) Impacts of marine protected areas on fishing communities. *Conservation Biology* 24: 1424–1429. https://doi.org/10.1111/j.1523-1739.2010.01523.x.

McAnany, P.A. & Yoffee, N. eds. (2010) *Questioning Collapse: Human Resilience, Ecological Vulnerability, and the Aftermath of Empire*. Cambridge, UK: Cambridge University Press.

McGoodwin, J.R. (1990) *Crisis in the World's Fisheries: People, Problems and Policies*. Stanford, CA: Stanford University Press.

Mintz, S.W. (1986) *Sweetness and Power: The Place of Sugar in Modern History*. New York: Penguin Books.

Moore, H.L. (1988) *Feminism and Anthropology*. Cambridge, UK: Polity Press.

Nazarea, V.D. (1998) *Cultural Memory and Biodiversity*. Tucson, AZ: University of Arizona Press.

Nazarea, V.D. (2006) Local knowledge and memory in biodiversity conservation. *Annual Review of Anthropology* 35: 317–335.

Netting, R.M. (1993) *Smallholders, Householders: Farm Families and the Ecology of Intensive, Sustainable Agriculture*. Stanford, CA: Stanford University Press.

NSF (National Science Foundation) (2015) Saving the slow loris. https://www.nsf.gov/discoveries/disc_summ.jsp?cntn_id=135668 (accessed June 13, 2020).

Orlove, B.S. & Guillet, D.W. (1985) Theoretical and methodological considerations on the study of mountain peoples: reflections on the idea of subsistence type and the role of history in human ecology. *Mountain Research and Development* 5: 3–18.

Padoch, C. & Peluso, N.L. (2003) *Borneo in Transition: People, Forests, Conservation, and Development*. Kuala Lumpur: Oxford University Press.

Parathian, H.E., McLennan, M.R., Hill, C.M. et al. (2018) Breaking through disciplinary barriers: human–wildlife interactions and multispecies ethnography. *International Journal of Primatology* 39: 749–775.

Paulson, N., Laudati, A., Doolittle, A. et al. (2012) Indigenous peoples' participation in global conservation: looking beyond headdresses and face paint. *Environmental Values* 21 (3): 255–276.

Penna-Firme, R. & Brondizio, E. (2007) The risks of commodifying poverty: rural communities, Quilombola identity, and nature conservation in Brazil. *Revista Habitus-Revista do Instituto Goiano de Pré-História e Antropologia* 5: 355–373.

Peterson, R.B. (2003) Central African voices on the human/environment relationship. In: *This Sacred Earth: Religion, Nature, Environment*, 2nd ed. (ed. R.S. Gottlieb), 151–157. New York: Routledge.

Posey, D. (2008). Indigenous management of tropical forest ecosystems: the case of the Kayapó Indians of the Brazilian Amazon. In: *Environmental Anthropology: A Historical Reader* (ed. M.R. Dove & C. Carpendter), 89–101. Malden, MA: Wiley-Blackwell.

Railey, J.A. & Reycraft, R.M. (2008) *Global Perspectives on the Collapse of Complex Systems*. Albuquerque: Maxwell Museum of Anthropology, University of New Mexico.

Rappaport, R.A. (1968) *Pigs for the Ancestors*. New Haven, CT: Yale University Press.

Rice, D.E. (2018) *Life in the Forest: Ikalahan Folk Stories*. Quezon City, Philippines: New Day Publishers.

Richards, P. (1986) *Coping with Hunger: Hazard and Experiment in an African Rice-Farming System*. London: Allen and Unwin.

Riley, E. (2018) The maturation of ethnoprimatology: theoretical and methodological pluralism. *International Journal of Primatology* 39: 705–729.

Robbins, P. (2003) Beyond ground truth: GIS and the environmental knowledge of herders, professional foresters, and other traditional communities. *Human Ecology* 31: 233–253.

Russell, D. & Tchamou, N. (2001). Soil fertility and the generation gap: the Bënë of southern Cameroon. In: *People Managing Forests: The Links between Human Well-Being and Sustainability* (ed. C.J. Pierce Colfer & Y. Byron), 229–249. Washington, DC: Resources for the Future.

Scheper-Hughes, N. (2004) Parts unknown: undercover ethnography of the organs-trafficking underworld. *Ethnography* 5 (1): 29–73.

Schmidt-Soltau, K. (2009) Is the displacement of people from parks only 'purported', or is it real? *Conservation & Society* 7: 46–55.

Schnegg, M. (2018) Institutional multiplexity: social networks and community-based natural resource management. *Sustainability Science* 13: 1017–1030.

Schroeder, R.A. (1999) *Shady Practices: Agroforestry and Gender Politics in the Gambia.* Berkeley, CA: University of California Press.

Shahub, N. (2018) In Indonesia, social forestry gets socialized: a new guidebook helps community members and policymakers understand complex social forestry schemes. *Forests News*, May 9. https://forestsnews.cifor.org/55625/ in-indonesia-social-forestry-gets-socialized?fnl=.

Sheridan, M. (2009) The environmental and social history of African sacred groves: a Tanzanian case study. *African Studies Review* 52 (1): 73–98.

Sillitoe, P. ed. (2007) *Local Science vs. Global Science: Approaches to Indigenous Knowledge in International Development.* New York: Berghahn Books.

Siwatibau, S. & Lees, A. (2007) *Review and Analysis of Fiji's Conservation Sector.* Hollywood, FL: Austral Foundation. https://www.sprep.org/att/IRC/eCOPIES/countries/fiji/10.pdf.

Smith, J.H. (2005) Buying a better witch doctor: witch-finding, neoliberalism, and the development imagination in the Taita Hills, Kenya. *American Ethnologist* 32 (1): 141–158.

Stepp, J.R., Castaneda, H. & Cervone, S. (2005) Mountains and biocultural diversity. *Mountain Research and Development* 25: 223–228.

Stepp, J.R., Cervone, S., Castaneda, H. et al. (2004) Development of a GIS for global biocultural diversity. *Policy Matters* 13: 267–270.

Stevens, S. ed. (2014) *Indigenous Peoples, National Parks, and Protected Areas: A New Paradigm Linking Conservation, Culture, and Rights.* Tucson, AZ: University of Arizona Press.

Stokols, D., Perez Lejano, R. & Hipp, J. (2013) Enhancing the resilience of human– environment systems: a social–ecological perspective. *Ecology and Society* 18 (1): art. 7.

Sullivan, S. (2009) Green capitalism, and the cultural poverty of constructing nature as service-provider. *Radical Anthropology* 3: 18–27.

Tiani, A.-M. & Diaw, C. (2006) Does resettlement contribute to conservation? The case of Ikundu-Kundu, Korup National Park, Cameroon. *Policy Matters* 14: 113–127.

Tollefson, J. (2019) Humans are driving one million species to extinction. *Nature News*, May 6. https://www.nature.com/articles/d41586-019-01448-4.

Tsing, A.L. (1993) *In the Realm of the Diamond Queen: Marginality in an Out-of-the-Way Place.* Princeton, NJ: Princeton University Press.

Tsunemoto, T. (2001) Rights and identities of ethnic minorities in Japan: indigenous Ainu and resident Koreans. *Asia-Pacific Journal on Human Rights and the Law* 2: 119–141.

Uriarte, M., Pinedo-Vasquez, M., DeFries, R.S. et al. (2012) Depopulation of rural landscapes exacerbates fire activity in the western Amazon. *Proceedings of the National Academy of Sciences of the United States of America* 109 (52): 21546–21550. https://doi.org/10.1073/ pnas.1215567110.

Vogt, E.Z. (1974) *Aerial Photography in Anthropological Field Research.* Cambridge, MA: Harvard University Press.

Walley, C.J. (2004) *Rough Waters: Nature and Development in an East African Marine Park.* Princeton, NJ: Princeton University Press.

West, P. (2006) *Conservation Is Our Government Now.* Durham, NC: Duke University Press.

West, P., Igoe, J. & Brockington, D. (2006) Parks and peoples: the social impact of protected areas. *Annual Review of Anthropology* 35: 251–277.

Wilson, E.O. (2016) *Half-Earth: Our Planet's Fight for Life.* New York: Liveright Corporation.

# 4

# Economics and Conservation

*Stephen Polasky*

## 4.1 Defining Economics

The discipline of economics studies the allocation of scarce resources to meet desirable but often competing goals. Though economists typically focus on the production and consumption of commodities traded in markets, economics is in fact much broader. In the book *Nature: An Economic History*, Vermeij (2004, p. 13) states, "[T]he human economic system is built on the same principles of competition for locally scarce resources as are all other economic systems composed of living things and their environments." There are numerous similarities between the discipline of economics, which focuses on interactions of people in market economies, and the discipline of ecology, which focuses on interactions of species in **ecosystems** (Tilman et al. 2005).

The most basic division within economics is between microeconomics, which studies the decisions of individuals and firms and their interactions, and macroeconomics, which studies the aggregate behavior of the economy to analyze economic growth and **development**, unemployment, and inflation. Microeconomics analyzes the prices and quantities of goods and services traded in markets, where markets link the decisions made by producers of goods and services ("supply") with the decisions made by consumers of goods and services ("demand"). Though economists typically focus on market interactions, they also study the operation of governments and social interactions outside the marketplace. Microeconomics has been applied to a wide range of environmental and conservation topics, including how to estimate the value people place on the natural world, as well as how the environmental impacts of production and consumption often do not factor into decision-making, and how such omissions can be corrected. Macroeconomics is relevant to conservation because it studies economic growth and can be used to address the question of whether economic growth is sustainable on a finite planet.

Several subfields within economics specialize in topics particularly relevant for conservation, namely ecological economics, environmental economics, and natural resource economics. Ecological economics is a transdisciplinary area of inquiry incorporating approaches from many fields besides economics to analyze the interactions and coevolution of humans and nature (Costanza, Cumberland et al. 1997). The central themes

*Conservation Social Science: Understanding People, Conserving Biodiversity*, First Edition.
Edited by Daniel C. Miller, Ivan R. Scales, and Michael B. Mascia.

of ecological economics revolve around sustainability and the long-run evolution of social–economic–ecological systems. Environmental economics analyzes policies to control pollution and methods to estimate the value of environmental improvement.

Natural resource economics analyzes the use of natural resources through time. Natural resource is a broad term that includes not only mineral and energy resources, stocks of fish, and timber, but also environmental quality, biodiversity, and natural beauty, which are important for recreation, spiritual, or cultural reasons. Economists and ecologists have increasingly used "**ecosystem services**" (Costanza, d'Arge et al. 1997; Daily 1997; MEA 2005) or "nature's contributions to people" (Díaz et al. 2018) rather than "natural resources" to discuss the broad set of contributions of ecosystems and biodiversity in generating benefits for people.

### 4.1.1 Major Themes and Questions in Economics

Perhaps the most basic question in economics, stemming from the very definition of the discipline, is how to allocate scarce resources to meet multiple desirable but often competing goals. One application of economic methods to conservation is how to allocate limited budgets to maximize **biodiversity conservation** outcomes (Ando et al. 1998; Naidoo et al. 2006; Murdoch et al. 2007). Economic approaches have also been applied to find the most important areas for conservation or restoration (Nelson et al. 2009; Bateman et al. 2013; Ouyang et al. 2016), and to show that the value of conservation often exceeds the value of exploitation (Balmford et al. 2002; Nelson et al. 2009; Polasky et al. 2011).

Economics is a behavioral science, and understanding human behavior is a crucial component in trying to design effective conservation programs (Cowling 2014; Byerly et al. 2018). Economists study how people respond to both financial and non-financial incentives (Ferraro & Price 2013; Wichman et al. 2016). The rise of behavioral economics, which integrates ideas from psychology and other disciplines to yield new insights about individual behavior and the limits of classical economic theory (Camerer et al. 2004; Thaler & Sunstein 2008; Ariely 2009), has been particularly influential. Behavioral economics has made significant inroads into the analysis of conservation and environmental policy over the past decade (Carlsson & Johansson-Stenman 2012; Byerly et al. 2018; Schill et al. 2019).

Economics focuses on understanding market outcomes resulting from interactions among producers and consumers of goods and services. This topic is important for conservation because the major causes of biodiversity decline stem directly or indirectly from the production, consumption, and trade of goods and services. The top threats to biodiversity as defined by the Intergovernmental Science-Policy Platform on Biodiversity and Ecosystem Services are (i) the loss of habitat from land conversion to human-dominated uses, (ii) overharvesting of species, (iii) climate change, (iv) pollution, and (v) introduction and spread of invasive species (IPBES 2019). These threats are the direct result of economic activity (loss of habitat from land conversion, harvesting species) or are unintended byproducts of economic activity (climate, pollution, invasive species).

A major question in environmental economics is how market economies often fail to provide adequate incentives for environmental protection or conservation. Market failure related to environmental protection is caused by externalities and public goods. Externalities occur when the actions of one economic actor directly affects the welfare of others without paying a price, such as when production of goods causes pollution that harms others.

Unless there is some type of corrective policy to internalize the externality, people will generate too many negative externalities and too few positive externalities. Public goods are "non-rival" (one person's enjoyment of the good does not diminish the ability of others to enjoy the good) and "non-excludable" (if the good is available for one it is available for all) (see Chapter 6 for additional discussion of public goods).

Examples of public goods include a stable global climate, local air and water quality, and the conservation of species. Public goods are typically underprovided because there is an incentive to free ride: why pay the cost to provide a public good when you can freely enjoy the good provided by others? Policies to internalize externalities or provide incentives for public goods include payments for ecosystem services (PES), cap-and-trade systems, and environmental taxes. These policies can be used to provide incentives to conserve biodiversity (Engel et al. 2008; Salzman et al. 2018). For example, cap-and-trade policies can be used to limit the negative impact of land development on biodiversity (Panayotou 1994).

Economists have analyzed the effectiveness of various conservation policies, such as the US Endangered Species Act (Brown & Shogren 1998; Innes et al. 1998; Polasky & Doremus 1998; Langpap et al. 2018; Epanchin-Niell & Boyd 2020) and the Convention on Trade in Endangered Species (Barbier et al. 1990; van Kooten & Bulte 2000). They have developed statistical approaches to show the effects on outcomes of interest (Angrist & Pischke 2009, 2010) and applied these methods to assess the impacts of policy interventions (Ferraro & Pattanyak 2006; Ferraro & Hanauer 2014a; Ferraro et al. 2019). Economists also have analyzed the impacts of protected areas on conservation outcomes (Andam et al. 2008; Ferraro & Hanauer 2014b) and the impacts of PES (Jack et al. 2008; Jayachandran et al. 2017; Jack & Jayachandran 2019), among other policies.

Another major question in environmental economics is how to assess the benefits of environmental improvement and biodiversity conservation. To address these questions, economists have developed methods of "non-market valuation." Goods and services traded in markets have market prices that signal the relative value of these goods and services. Many things that have value, such as clean air, clean water, and biodiversity, are not marketed goods and services and thus their value is not directly observable. Economists use nonmarket valuation to assess how people value aspects of the environment, such as nearby open space (Lutzenhiser & Netusil 2001; Irwin 2002), clean air (V. Smith & Huang 1995), clean water (Carson & Mitchell 1993), and diverse species (Loomis & White 1996; Richardson & Loomis 2009).

Economics also addresses issues related to long-run growth and development. Rapid economic growth, especially since the 1950s, has reduced poverty rates and increased living standards for much of the world's population (World Bank 2018), but it also has transformed the global environment in ways that threaten future human prospects (Vitousek et al. 1997; Steffens et al. 2007; IPCC 2014, 2018; IPBES 2019; Díaz et al. 2019; see also Chapters 5 and 8 for more on the idea of development and its environmental impacts). Human economic activity fueled by growth has driven the increasing rate of loss of biodiversity (Wilcove et al. 1998; MEA 2005; IPBES 2019).

Though ecological economics covers a very broad range of topics and ideas, from the laws of thermodynamics and energy flows to questions of social equity and justice, a central focus is sustainability and the long-run evolution of social–economic–ecological systems. Ecological economists focus on the crucial role that nature plays in supporting the economy and human well-being, and the imminent threats posed by biodiversity loss, climate change, and global environmental change, along with issues of power and inequality (Costanza 1991; Daly 1999; Raworth 2017; Spash 2020).

Natural resource economists also address issues of sustainability and long-run prospects. Natural resource economics make a distinction between renewable resources—such as timber and fish—that can regenerate (as long as they are not harvested to extinction), and nonrenewable resources—such as oil, coal, and minerals—that have a fixed stock (at least at time scales short of geologic time). For renewable resources, sustainable use requires limiting harvest to no greater than the regeneration rate. For nonrenewable resources such as oil or natural gas, where consumption reduces remaining reserves, there is no level of sustainable use. Sustainability then requires finding a renewable substitute such as wind and solar energies to replace nonrenewable fossil-fuel energy.

Despite powerful reasons for conservation researchers and practitioners to engage with economists, there is often skepticism toward economics in the conservation community, with some going so far as to view the discipline of economics as one of the main sources of conservation problems. For example, Hall et al. (2000) claim that the discipline of economics is antithetical to conservation because it ignores biological or physical reality in thinking that there can be infinite economic growth on a finite planet. Czech (2000, p. 2) states that "[i]n the history of academia, neoclassical economics has produced the policies most problematic to wildlife conservation."

In reality, economists who work in ecological economics, environmental economics, and natural resource economics take seriously the notion of limited resources and environmental degradation (e.g. Arrow et al. 1995, 2004; Heal 1998; Polasky et al. 2019). These economists also address issues of how to use policy to reform market systems to reduce environmental impacts of economic activity and improve conservation outcomes (Hanley et al. 1997; Polasky et al. 2005; Fisher et al. 2014; Tietenberg & Lewis 2018). Though the vast expansion of economic activity over the past several centuries has been the main cause of the decline of biodiversity, the discipline of economics can potentially help to reconcile economic activity with conservation, as well expressed by ecologist Roughgarden (2001, p. 87):

> Economists can deal conceptually with limits to growth perfectly well. But they may not be convinced that the limits are where ecologists say they are. Economists have long known how to fold into the price of an item all the costs of its production. A company that pollutes the environment can sell a product at an artificially low price because the public pays the cleanup. But the cost of the cleanup, called the social cost, should be fed back to the company .... Dealing with ecology does pose some new challenges for economics, but it is polite to know which these are. It is rude to assume that economists haven't considered the environment at all. In fact, they are often on our side, so let's keep them there.

## 4.2 A Brief History of Economics

Modern economics began with the publication of *An Inquiry into the Nature and Causes of the Wealth of Nations* by Adam Smith in 1776. Smith explained how markets coordinate the independent decisions of numerous producers and consumers and how producers and consumers pursuing their own self-interests yields beneficial outcomes for all. A market economy is an economic system in which economic decisions and the pricing of goods and services are guided by the interactions of individual citizens and businesses. This idea of the "invisible hand" of the market guiding actions without any form of top-down management became a central argument of economists who believe in free and unfettered markets with minimal government oversight.

Classical economists following Smith analyzed many of the central issues in economics including economic growth and development, the creation of wealth and alleviation of poverty, trade patterns, inequality, and population growth. An important classical economist from a conservation perspective was Thomas Malthus (1766–1834). According to him, an unchecked population grows geometrically (i.e. doubling every generation and resulting in exponential growth) while agricultural production grows arithmetically (i.e. linear growth) so that eventually demand for food outstrips supply. Malthus thought that population pressures would drive wages down to subsistence levels with starvation and disease limiting population growth. In recent times, Malthus's ideas have tended to have a greater influence on environmentalists than on economists. **Neo-Malthusians** emphasize the environmental threats from the increasing demands of a growing population on rising pollution and diminishing per capita resources, while economists tend to emphasize technological advances and substitution possibilities that will overcome these threats (see Chapters 3, 5, and 8 for more discussions of the relationships between population growth and resource use).

Other important classical economists include David Ricardo, Jeremy Bentham, John Stuart Mill, Karl Marx, and William Stanley Jevons. Ricardo developed the idea of comparative advantage. A country has a comparative advantage when it can produce a particular good at a lower opportunity cost than its trading partners, where opportunity costs measures what the country would have to give up in other goods in order to produce more of this particular good. Comparative advantage is important in the theory of trade, with countries specializing in goods for which they have a comparative advantage (see Chapter 8 for a discussion of World-Systems Theory and a contrasting view of global trade).

Bentham, and later Mill, developed the idea of utilitarianism, which ranks different outcomes in terms of which generate the greatest amount of happiness (utility). Economists use the idea of "utility function" to describe how individuals rank the outcome-based welfare of the individual under each outcome. Mill also wrote about the value of the natural world to humans and expressed concerns about the threats posed by economic growth.

Marx gave rise to a large body of thought about the structure and historical evolution of the capitalist system. Marx analyzed the conflict between owners of capital and workers, and argued that the capitalist system sows the seeds of its own destruction through declining rates of profits and increasing concentrations of wealth (see Chapter 8 for more on Marx's ideas). Like Malthus, Marx has had greater influence outside economics, in other social sciences and among political leaders of socialist parties, than he has within modern economics.

Jevons published *The Coal Question* in 1865, in which he noted that coal was essential to the economy but was a finite resource that would eventually run out, highlighting the problem of dependence on nonrenewable resources. Jevons also noted that increased efficiency in the use of resources could result in increased resource use because greater efficiency would lower cost and increase demand, now known as the "Jevons paradox."

Neoclassical economics grew out of classical economics and became dominant in the late nineteenth century. Alfred Marshall published his *Principles of Economics* in 1890, which brought the concept of marginal analysis to the forefront of economic theory. According to this concept, a business seeking to maximize profits will produce to the point where marginal revenue (the additional revenue gained by selling another unit) equals marginal cost (the additional cost of producing another unit). An individual seeking to maximize utility will buy a product to the point where the marginal benefit from an additional unit of consumption equals the price they have to pay to purchase an additional unit. In neoclassical economics, benefit (utility) determines value. In equilibrium, market price equates the marginal benefit for consumers (demand) with the marginal cost of production for companies (supply) for each product in the economy. Throughout the twentieth century, neoclassical economists developed a rich set of theories, empirical methods, and data, and applied them to an increasingly wide range of topics, including environment and natural resources, healthcare, education, family structure, poverty, inequality, sustainable development, and virtually every area of modern life.

The spread of economic thought, however, has not been without controversy and dissent. For example, there are vigorous debates among economists over the proper role of government in regulating the economy. Followers of John Maynard Keynes, a British economist prominent during the Great Depression, emphasize the need for government stimulus to increase demand during economic downturns, and advocate for a more active government role in regulating the economy. However, followers of the Chicago School, which originated at the University of Chicago in the 1930s, emphasize the self-correcting nature of markets and think society is better off letting markets operate largely free of government interference. In addition, others outside economics grew critical of the expansion of economic thinking into realms where they felt it should not go. Political philosopher Michael Sandel (2012) argues that the spread of market logic into such areas as organ transplants or pollution control changes the way people think about such issues, crowding out moral or ethical concerns with largely detrimental consequences.

The rise of modern environmental economics and natural resource economics as subfields within economics dates to the 1950s and 1960s. Notable contributions to this rise occurred earlier in the twentieth century, including the work of Arthur Pigou who laid out the logic of externalities and the "Pigouvian tax" to correct for externalities. Such a tax would be imposed on market activities that generate costs (e.g. pollution) not included in the market price, so as to correct undesirable or inefficient market outcomes (i.e. market failures). Ramsey (1928) laid out the theory of optimal savings and investment that determines the discount rate to apply to future benefits and costs (which is discussed later in this chapter), while Hotelling (1931) solved the problem of how the market would allocate a nonrenewable resource over time.

It was not until mid-century, however, that a body of thought began to coalesce that would form the bases of modern environmental and natural resource economics. Concerns over the adequacy of natural resources led to the formation of Resources for the Future (RFF), which gathered a nucleus of influential economists. Early influential studies revisited the question of the adequacy of natural resource (Barnett & Morse 1963), and formulated a compelling argument about the rising value of conservation due to the increasing scarcity of nature (Krutilla 1967). RFF economists also pioneered nonmarket valuation methods and incorporated the values of environmental improvement into bene-fit-cost analysis. The growth of the environmental movement in the 1960s following the publication of *Silent Spring* by Rachel Carson (1962) along with concerns about air and water pollution, and the 1973–74 oil crisis spurred rapid growth in environmental and resource economics.

These same events spurred thinking that would ultimately result in the formation of eco-logical economics. In 1966, Boulding published his ideas on the economics of "spaceship earth" (Boulding 1966) which implied clear limits on natural resources and the ability of the environment to absorb pollution. Ehrlich (1968) wrote about the threat posed by exploding human population growth, Georgescu-Roegen (1971) wrote about embedding the economy within limits imposed by biophysical context, and Meadows et al. (1972), for the Club of Rome, wrote about *The Limits to Growth*. Drawing from these works, Daly (1977) examined the dangers of unrestrained economic growth and argued for the need to shift to a steady-state economy. In the wake of these influential works, ecological eco-nomics coalesced into its own field in the late 1980s and early 1990s with the formation of the International Society for Ecological Economics in 1988, and the start of the journal *Ecological Economics* in 1989.

## 4.3 Concepts, Approaches and Areas on Inquiry in Economics

### 4.3.1 Analysis of Market and Nonmarket Values

#### 4.3.1.1 Concepts of Value

*Value* is a central term in economics with a specific meaning. However, value has many meanings. Sociologists, environmental philosophers, conservation biologists, and econo-mists often use different notions of value (see Chapter 3 for a discussion of non-economic values). In economics, the value of an object to an individual refers to the benefits the person derives from that object. For example, eating a meal, watching a movie, or viewing a beautiful bird in its natural habitat can generate benefits. Value is not restricted to objects with current use. Economists use the term *option value* to describe something that might have value in the future, such as the possibility that research may show that a certain natural compound is useful in fighting disease. Economists also recognize the benefit a person may get from just knowing that something exists (existence value). Many people value ecosystem preservation even in remote places that they are unlikely ever to visit, such as the north slope of Alaska or the Amazon rain forest.

The notion of value as the benefit that a person derives from the use, possible future use, or existence of something differs from some notions of value in environmental ethics and conservation biology. When valuing nature (Figure 4.1), some environmental philosophers and conservation biologists argue that nature has *intrinsic value*: value is inherent in nature itself regardless of whether or not humans benefit (Norton 1987; Rolston 1988). In eco-nomics, nature has *instrumental value*: it is valuable because it contributes to human well-being. Instrumental versus intrinsic value are fundamentally different approaches to

**Figure 4.1** What is the value of an elephant: ivory, ecotourism, or intrinsic? Photo by Stephen Polasky.

ethics and the proper relationship between humans and the natural world, which can make it difficult for those who hold one view to communicate with those holding a different view. Arguments over the proper framework for value, like arguments over religion, cannot be settled by gathering more information or better facts.

The two approaches to value—intrinsic and instrumental—underlie a major debate in conservation. According to the instrumental view, nature is valuable because it provides ecosystem services, defined as the goods and services provided by ecosystems that contribute to human well-being (Daily 1997). *Ecosystem services* were a central theme in the Millennium Ecosystem Assessment, a major international effort to assess the global status and trends of biodiversity (MEA 2005). The International Science-Policy Platform on Biodiversity and Ecosystem Services (IPBES 2019) used the related phrase *nature's contributions to people* (Díaz et al. 2018).

Attempts to quantify nature's contributions to people has greatly increased the integration of economics and the natural sciences (Box 4.1). Economists have long studied how the supply of natural resources (Figure 4.2) and improvement in environmental quality (reduction of pollution) are valuable to people. Over the past several decades, ecologists and economists have worked more closely to study the broad range of links between ecosystem processes and human well-being (Daily 1997; MEA 2005; NRC 2005; TEEB 2010; Díaz et al. 2018, 2019; IPBES 2019). For example, researchers have shown how changes in land use affect habitat for pollinators, and in turn how this affects the supply of pollination to agricultural crops (Ricketts et al. 2004; Garibaldi et al. 2016).

Some conservationists are critical of the ecosystem services approach, arguing instead that conservation should rest on ethical or moral arguments. According to this view, nature's intrinsic value provides a stronger rationale for conservation than do appeals to what nature can do for people (Ehrenfeld 1988; McCauley 2006; Redford & Adams 2009). For example, in his critique of the ecosystem services approach, McCauley (2006) argues

**Figure 4.2**  Do we care about the forest or the trees? Photo by Trevor Littlewood (cc-by-sa/2.0), https://www.geograph.org.uk/photo/4900679.

---

**Box 4.1  Applications: Land use, biodiversity, and ecosystem services**

Several analyses have applied economic frameworks to consider the question of land use and land management to produce marketed commodities and non-marketed ecosystem services, and to conserve biodiversity. Balmford et al. (2002) compared the value of ecosystem services from keeping the natural ecosystem relatively intact (for example by using reduced-impact logging, small-scale farming, and sustainable fishing) to intensive production methods that transformed the ecosystem (such as timber plantations and clear-cut harvesting, large-scale modern intensive agriculture, and shrimp farming). In all five cases they analyzed, the management approach that kept the natural ecosystem relatively intact produced higher total value of goods and services.

Polasky et al. (2008) analyzed the impacts of alternative land-use plans on biodiversity conservation and the production of marketed commodities using data for the Willamette Basin in Oregon (Figure 4.3). By considering where to locate biological reserves, farms, timber operations, and housing, they found a set of efficient land-use plans that maximized the number of terrestrial vertebrate species conserved on the landscape for any given level of economic return, tracing out an "efficiency frontier." They found that some land-use plans that simultaneously conserved more species and generated higher economic returns than the current land use. Though there are tradeoffs between higher economic returns and biodiversity conservation, starting from the current inefficient outcome it is possible to increase both. Nelson et al. (2009) expanded the analysis to include the value of a set of ecosystem services. Paying for ecosystem services resulted in land-use choices that increased both biodiversity conservation and the total value of marketed goods and ecosystem services produced by the landscape.

Over the past decade, analyses of biodiversity conservation and multiple ecosystem services have expanded greatly. Ecosystem service analysis integrating economics and natural sciences building from detailed landscape-level analysis has broadened to national-level analyses (Bateman et al. 2013; Lawler et al. 2014; Ouyang et al. 2016) and to the entire globe (Chaplin-Kramer et al. 2019).

---

*(Continued)*

**Box 4.1 (Continued)**

**Figure 4.3** View of the Oregon Coast range featuring Marys Peak, Willamette Basin, Oregon. Photo by J.A. Burlock (cc-by-sa/4.0) https://commons.wikimedia.org/w/index.php?curid=110933654.

that (i) nature is not always benevolent, so what is good for people and what is good for biodiversity conservation do not always align; and (ii) human values and particularly market prices are volatile, so they are an unreliable foundation for long-term conservation.

In contrast, those who favor an ecosystem services approach argue that conservationists have used ethical and moral arguments for a long time without much success. They contend that people will conserve only what they think is valuable and improves their well-being. There are also different sets of moral arguments about the importance of providing people with the means to escape poverty and make a decent living.

Though the instrumental and intrinsic value approaches have different foundations, they may come to the same conclusion about conservation policy in practice (Polasky et al. 2012). For example, some people may wish to preserve a pristine watershed because of its intrinsic value while others may wish to preserve it because it provides them with clean drinking water. What really matters for conservation is what people choose to do, not their stated intentions or motivations (Polasky & Segerson 2009). By focusing on decisions rather than the motivations, common ground can emerge in conservation policy. In addition, instrumental and intrinsic value arguments need not be mutually exclusive. Conserving nature can be valuable on moral and ethical grounds as well as generating tangible benefits for people (Reyers et al. 2012).

### 4.3.1.2 Demand Analysis: Using Market Prices to Assess Values

Economists are interested in the practical question of measuring the value of goods, and comparing the relative value of different goods. The economic approach to measuring value is based on observations of how much people are willing to pay for a good or service at a given price when it is available for purchase in markets. This basic approach can be extended to valuing aspects of nature that are not available for sale.

The starting point of economic analysis of value are the benefits that an individual derives from use or existence of an object. The standard approach in economics assumes that individuals know what they like (have well-defined preferences) and will always choose an alternative they like best (behavior is rational). An individual demand curve summarizes information about how much an individual would choose to purchase of a good for sale at various prices (Figure 4.4). The demand curve is typically downward sloping because the amount of a good the individual would normally choose to buy declines as the price increases. For example, if the price of a loaf of bread increases, an individual may not be able to afford to buy as much bread or may decide to buy something else to eat instead. The demand curve shows how much an individual is willing to pay for a good. A rational individual who buys a good for a given price must value the item at least as much as the price, or otherwise he or she would not have bought it. Spending more money on bread will leave less money to spend on fruits and vegetables or entertainment, so if a rational individual buys bread, the person must value it more than what the person gives up in fruits or vegetables or entertainment.

A person may be willing to pay a lot more for a good than what it actually costs to buy. For example, a hungry person may be willing to pay a lot for a loaf of bread, far more than its price. The area under the demand curve measures the total benefit associated with consumption of the good. The difference between the demand function (willingness to pay) and the price (what the individual actually pays) is called "consumer surplus." Consumer surplus measures the increase in value to the individual, measured in monetary terms, from consuming a good at a given price versus not consuming the good at all (see Figure 4.4).

We can go from an individual demand curve to a market demand curve by adding up the total across all individuals of how much each individual would like to buy at each price. Publicly available data exists for market prices and quantities of goods for a large number of marketed goods. Economists use this data to estimate demand curves and consumer surplus empirically.

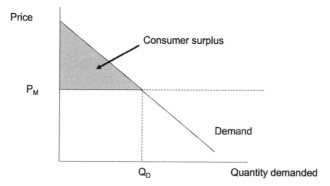

**Figure 4.4**   The demand curve. The horizontal axis measures the quantity demanded of the good. The vertical axis measures the price of the good. The demand curve is downward sloping, showing that an increase in quantity will be demanded as price falls. If the market price is $P_M$, the quantity demanded will be $Q_D$. Consumer surplus, the difference between the demand curve and the price, is shown as the shaded triangle.

### 4.3.1.3 Nonmarket Valuation

If everything of conservation value were traded in markets, conservationists could use demand curves to generate estimates of consumer surplus generated by conservation, all measured in a common monetary metric of value. However, most of what is of value in conservation is not traded in markets and therefore does not have an observable demand function. One cannot simply go to the store and purchase clean air or the existence of a species. Over the past half century, economists have developed methods of "nonmarket valuation" that extend market valuation to things other than marketed goods, including aspects of environmental quality, ecosystem services, and biodiversity (see Freeman et al. 2014 for a comprehensive treatment of nonmarket valuation theory and methods).

There are three broad types of nonmarket valuation methods: (i) revealed preference, (ii) stated preference, and (iii) cost-based. Revealed and stated preferences generate willingness-to-pay estimates from actual behavior or responses to survey questions. Cost-based methods use estimates of the costs of replacing ecosystem processes. Since these methods focus on costs rather than benefits, some economists view these methods with suspicion. However, they can generate valid estimates of value in certain circumstances, as described in the following paragraphs.

#### 4.3.1.3.1 *Revealed Preference Methods*

Revealed preference methods use observed behavior to infer values about non-marketed objects of interest. For example, what is the value of clean air or access to open space? Though these are not things purchased directly in markets, they may influence the price of market goods, such as housing. By observing the price of housing or other market goods along with how prices vary with environmental quality, revealed preferences methods can uncover the values associated with various aspects of environmental quality.

The hedonic property price method uses data on property values and the attributes of the property to describe the correlation between various attributes and property values. For example, how much more would a house be worth if it were located in a neighborhood with high environmental quality rather than a neighborhood with low environmental quality? How much people are willing to pay for a house depends on characteristics of the house itself (size, number of rooms, age, quality of construction, etc.), neighborhood characteristics (quality of local schools, neighborhood crime rate, tax rates, distance to the urban center, etc.), and environmental characteristics (distance to parks and open space, air quality, etc.). Multiple regression analysis using data on house prices along with characteristics of the house, neighborhood, and environment, allows an analyst to isolate how changes in each characteristic change property value while holding all other characteristics constant (see Box 4.2).

---

**Box 4.2   Applications: Using property prices to value nature**

In a study in Portland, Oregon, Mahan et al. (2000) used the hedonic property price method to see how proximity to wetlands, lakes, rivers, and parks (Figure 4.5) influenced property values. The study used data on the sale prices of over 14,000 houses sold between June 1992 and May 1994. The study also used extensive data on housing characteristics, neighborhood characteristics, and environmental characteristics

**Box 4.2   (Continued)**

**Figure 4.5**   Oak Bottoms Wildlife Refuge, Portland, Oregon. Photo: Generalrelative / Wikimedia Commons / CC BY-SA 4.0.

(distance to nearest wetland, size and type of nearest wetland, distances to nearest stream, river, lake, and park). They found that people in urban areas do indeed value nature. Reducing the distance to the nearest wetland by 1,000 feet, evaluated at the average house value and an initial distance of one mile, increased house value over $400. Reducing the distance to the nearest stream by 1,000 feet increased house value by over $250, while reducing the distance to the nearest lake by 1,000 feet increased house value by over $1600. Similarly, Lutzenhiser and Netusil (2001), in another study in Portland, found that being within 1,500 feet of a protected natural area increased house values by over $10,000. Proximity to urban parks and golf courses also positively influenced housing values.

The travel cost method uses data on how often people visit recreation sites and the time and expense involved. Travel cost is the sum of the monetary cost of travel time (evaluated using a fraction of the wage rate) plus entrance fees and other expenses. Recreation sites close to where a person lives take less time to visit so tend to have a lower "price." A willingness-to-pay function for visits, analogous to a demand curve for a marketed good, can be generated by using data on the number of trips as a function of the travel cost. Economists have used the travel cost method to estimate the value of access to recreational sites (Champ et al. 2009; Freeman et al. 2014), and how a change in environmental quality affects the value of visits (Egan et al. 2009).

Revealed preference methods use observed behavior for decisions with real consequences and often provide convincing evidence about the value of nature. However, these methods do not apply to all values of nature. For example, when people value the existence of species but do not live near or plan to visit where the species occurs, then neither the hedonic property price nor the travel cost methods are applicable. If people do not have full information about choices when they make their decisions, then their choices may not accurately reflect their values (Caplin & Dean 2015). For example, people visiting a recreational site may be unaware of the presence of toxic chemicals. Had they known, they may not have chosen to visit the site. A revealed preference study based on choices made with poor information may give biased results on the value people place on nature.

### 4.3.1.3.2 Stated Preference Methods

Stated preference methods, also called choice experiments or contingent valuation, pose questions to people about which alternative they prefer from a set of choices. By varying the attributes of alternatives and seeing how this affects the choices people make, the results from a stated preference study can be used to estimate the willingness to pay for various environmental attributes. Stated preference methods can also be used to assess how different environmental attributes rank vis-à-vis other environmental attributes, for example, whether water quality is more or less important a concern than species conservation, or within species conservation whether conserving salmon or spotted owls is of greater concern. Carson (2011) contains a bibliography and summary of major issues in contingent valuation studies for nonmarket valuation.

Stated preference studies have been used to value a wide range of environmental goods including the value of conserving biodiversity. Richardson and Loomis (2009) compiled contingent valuation studies of the value of protecting endangered species in the US. Annual household willingness to pay ranges from a low of $8.32 to avoid extinction of the striped shiner to over $300 for a 50% increase in the population of Western Washington and Puget Sound saltwater fish. For the most part, these studies show higher values for more charismatic species (e.g. bald eagles versus striped shiners) and higher values for greater potential losses. However, estimates are not always consistent. The value for the same or similar species can differ widely across studies, e.g. $98 per household in Washington state per year for a 600% increase in salmon (Loomis 1996) versus $308 per household per year in Washington state for a 50% increase for all migratory fish (Layton et al. 2001).

Use of the contingent valuation method has been controversial (Portney 1994; Kling et al. 2012). Critics claim that stated preference methods do not generate reliable estimates of value (Diamond & Hausman 1994; Hausman 2012; McFadden & Train 2017). Stated preference methods may suffer from "hypothetical bias" in which people respond differently to a question posed about a hypothetical situation that has no real consequences compared to an actual choice with real consequences (Diamond & Hausman 1994). Responses to contingent valuation surveys can be insensitive to large changes in the scale of the benefits provided (Kahneman & Knetsch 1992) but are sensitive to how questions are framed. As noted earlier, though, psychologists and behavioral economists have also found that framing affects all decisions, not just hypothetical choices. Proponents of stated preference claim that carefully done studies that follow best practices generate reliable estimates of value (Hanemann 1994; Carson 2012). In addition, stated preference methods may be the only methods available to estimate non-use values. Debates about the reliability of stated preference methods in general and contingent valuation in particular are likely to continue.

### 4.3.1.3.3 Cost-based Methods

Some nonmarket estimates of value use the cost incurred when the environment is degraded. Avoided cost and replacement cost are two types of cost-based methods commonly used. Avoided cost equals the amount of money that does not need to be paid because of what is provided by nature. For example, the flood mitigation value supplied by a wetland that stores water during periods of high water could be assessed by looking at the differences in flood damages with versus without the wetland (Watson et al. 2016). Avoided cost has been used to value coastal marshes, mangroves, and other ecosystems that provide protection to coastal communities from storms and tsunamis (Das & Vincent 2009).

Replacement cost estimates the cost of providing a good by an alternative means. For example, clean drinking water can be supplied through ecosystem processes that filter nutrients and pollutants. Urban or agricultural development may diminish filtration services and require human-engineered water purification. In the case of New York, the replacement cost of building and operating a water treatment facility to provide the same service provided by intact watersheds in the Catskill Mountains that provide water to the city was six to eight billion dollars (Chichilnisky & Heal 1998; NRC 2000). Of course, protecting the watersheds also provides other benefits beyond water quality improvements. The avoided cost estimate therefore provides a lower bound on the value of watershed protection.

Cost-based methods measure the cost of replacing what nature provides by some alternative means or the damages associated with not providing the service, but do not directly measure value of what nature provides. These methods can generate valid estimates of the value, however, as long as certain conditions are met. For replacement cost to be a valid measure of value, the human-engineered alternative must provide an equivalent quality and quantity of the service that is the lowest cost alternative, and people would have to be willing to pay the cost of this alternative method to provide the service (Shabman & Batie 1978). In the case of avoided damages, proper care needs to be taken to account for other ways to avoid damages, such as moving structures out of the floodplain.

### 4.3.1.4   Critiques of Economic Valuation

Criticisms of economic valuation applied to nature range from practical concerns about whether such methods generate reliable estimates of value to more deep-seated concerns about the rationality of consumer choice, and ethical and moral concerns. Some critics find the whole exercise of valuing nature wrong on a basic level (see Box 4.3). Others claim that using tools designed to measure tradeoffs among products in their role as consumers is not suited to measure environmental values in their role as citizens (Sagoff 1988). Still others think that environmental protection and biodiversity conservation is a moral or ethical issue that should not be subject to economic valuation (Ehrenfeld 1988; Rolston 1988; Sandel 2012). A related critique is that economic valuation changes the framing from an ethical choice into a choice about which commodities to buy, which fundamentally changes how people value the choice (Sandel 2012). Moral or ethical consideration may be especially important for issues of conserving biodiversity (for example, whether other species have a right to exist) and sustainability issues where the well-being of future generations, who are not around to defend their interests, is at stake.

Critics of the revealed and stated preference methods question whether choices made by people, real or hypothetical, really say anything meaningful about value. In economics, the assumption of rational choice means that observed choices made by an individual reveal what the individual prefers so that choice shows what they value most. If, however, individuals are not fully rational in this sense, then observed choices may not reveal what is most valuable.

**Box 4.3    Debates: The value of the Earth**

Costanza, d'Arge, and colleagues (1997) took on the ambitious task of trying to determine the global value of natural capital and ecosystem services (Figure 4.6). Their
paper has been widely cited but has also drawn sharp criticism from economists. The
paper used existing estimates from prior literature to estimate the value of 17 ecosystem services for 16 types of ecosystems. For each ecosystem service in each ecosystem type, they multiplied an estimate of the productivity per hectare by an estimate
of the value of the service per hectare. They summed the value of each service per
hectare over all the services to get value per hectare for each ecosystem type. They
then multiplied this number times the number of hectares of that type of ecosystem
to estimate the global value for the ecosystem type. Finally, summing the values of
all ecosystem types generated an estimate of $33 trillion per year as the value of the
Earth's ecosystem services. An updated version of the paper raised the estimated value
to $125 trillion per year even with the loss of habitat that reduced ecosystem services
in the intervening 17 years (Costanza et al. 2014).

   Economic critics were quick to point out what they viewed as fatal flaws in
this approach, leading them to conclude that estimates of global value had little
scientific merit. For example, Bockstael et al. (2000) argued that taking an estimate
of the value of a small-scale change at a particular location cannot be used to
estimate the value of large-scale changes for an entire ecosystem type. For one
thing, the value of the last remaining fragment of a rare ecosystem type would
likely be of great value even if the per hectare value had not been that large when
that ecosystem type was common. The value of ecosystem services also depends on
ecological and economic context. For example, the value of storm protection from
coastal marshes is likely to be far higher in areas that protect cities and towns than
in uninhabited stretches of coastline.

**Figure 4.6**   What is the Earth worth? Photo: Renan / Adobe Stock.

| Box 4.3 (Continued) |
| --- |
| The $33 trillion value (or $125 trillion later) is obviously quite large. In fact, the value of the entire world income as measured by gross domestic product (GDP) for all countries at that time was only $18 trillion. One critic, however, thought this estimate was not nearly large enough. Because humans could not exist without the life-support system provided by the biosphere, Toman (1998) called the $33 trillion estimate of the value of the Earth "a serious underestimate of infinity." |

Psychologists, behavioral economists, and other social scientists, are skeptical of the assumption of rational choice that underlies the link economists make from choice to value. Experiments show cases where the choices of individuals violate the assumptions of rational choice (Kahneman et al. 1982; Ariely 2009). Tversky and Kahneman (1981) show that framing the same choice between the same two alternatives as a gain in one case and as a loss in the other case leads to a different choice about which alternative people prefer. In another experiment, people were offered a choice between the chance of winning a modest amount of money with a high probability and the chance of winning a large amount of money with a low probability. The majority of people choose the chance of winning a modest amount of money with a high probability. However, when asked which of the two alternatives was of greater value, people said the chance of winning a large amount of money with a low probability (Lichtenstein & Slovic 1971). Psychologists have also argued that people do not have fixed preferences as assumed by economists, but in fact construct preferences on the spot, and therefore are subject to shifts with context (Slovic 1994; see also Chapter 7, Psychology). If preferences and values depend on context, then the method used to study values can have an important influence on reported values.

### 4.3.2 Markets, Market Failure, and Policy

#### 4.3.2.1 Supply

The basic building blocks of economic analysis are supply and demand. The theory of demand has already been discussed. This section develops the theory of supply—how much of a good will be produced at a given price. The starting point for understanding supply is the behavior of producers. Economists typically assume that producers are motivated by profit. Profit is the difference between revenue earned from selling the good and cost of producing the good. For example, a farmer growing corn for sale on the market gains revenue from sale of the corn, but must pay for seed and fertilizer, tractors and other machinery, wages for labor, and land rent or property taxes. While farmers might use their own labor to grow crops and also own the land on which they grow their crops, there is still an "opportunity cost." Instead of farming themselves, farmers could earn wages by working elsewhere and rent their lands to other farmers.

For markets with a large number of small producers, economists typically assume that the market is "perfectly competitive." In a perfectly competitive market, individual producers take the market price as given and do not think they can affect the market price by what they do. In the case of the market for corn, each farmer is small relative to the size of the overall market. Changing the amount produced by an individual farmer will have a negligible effect on the overall market price.

For a "price-taking" producer, the additional revenue from selling one more unit of the good, called the marginal revenue, is simply the price of the good. Producing more output takes more input, which increases the cost of production. The marginal cost is the additional cost of producing another unit of output and tends to increase as the amount produced becomes larger. In trying to produce more output, the farmer will face declining returns from application of additional units of fertilizer or additional amounts of labor and machinery inputs on an acre of land. To maximize profit, the farmer will choose the level of output where marginal revenue (price) equals marginal cost (Figure 4.7). If marginal cost is less then marginal revenue, a producer can increase profit by selling an additional unit. If marginal cost is greater than marginal revenue, the producer would do better by cutting back and avoiding production where revenue does not cover costs. The amount of profit (also called "producer surplus") is shown as the difference between revenues and costs and is equal to the shaded area in Figure 4.7 minus any fixed costs of operation that are not captured in the marginal cost curve.

The amount of output a farmer will choose to supply is a function of the market price. Higher prices (as represented by price $P_M$ in Figure 4.7) will mean that marginal revenue will equal marginal cost at a higher level of output. The supply function shows how much output the producer will choose to produce at any given price. The supply function of the individual producer is simply the marginal cost curve for the producer because the producer finds the quantity at which marginal cost equals marginal revenue (price) in order to maximize profits. The market supply function is the aggregation of all individual producers' supply functions (analogous to the market demand function being the aggregation of the demand function of all individual consumers).

### 4.3.2.2 Market Equilibrium

In hunter-gatherer societies or small isolated agricultural villages, there is not much need for separate considerations of production (supply) and consumption (demand). The same people who gather or grow the food also eat the food. In **industrialized** societies, however, most of us do not grow all our own food or make everything we use. We depend on others to provide us with the things we need. In turn, each of us helps to supply others with what

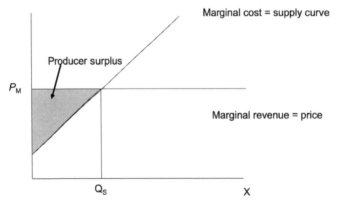

**Figure 4.7** The supply curve for a firm. The horizontal axis measures the quantity supplied of the good. The vertical axis measures the price of the good. The profit-maximizing choice of output occurs where marginal cost equals marginal revenue (price), and the supply curve is equal to the marginal cost curve. Producer surplus, the difference between revenue and cost, is shown as the shaded area.

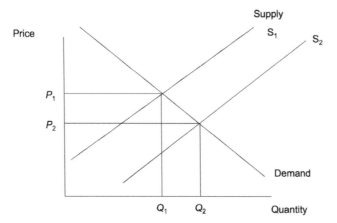

**Figure 4.8** Market equilibrium. Market equilibrium occurs at the price for which quantity demanded equals quantity supplied. Given the initial supply curve, $S_1$, equilibrium market price is $P_1$ and equilibrium quantity is $Q_1$. When supply increases from $S_1$ to $S_2$, a lower equilibrium price will be established ($P_2$) at a higher level of quantity ($Q_2$).

they need. But how does all this function? How do producers and consumers know what each other wants and coordinate so that supply equals demand?

The answer to this question is through the operation of markets. Markets bring together producers who supply goods with consumers who demand goods. In market equilibrium, price will be set where quantity demanded equals quantity supplied (Figure 4.8). Market price signals to producers how valuable the good is to consumers. Similarly, market price signals to consumers what it costs to produce the good. When supply and demand are not equal there will be pressure for the market to adjust. For example, good weather that yields large harvests for farmers shifts the supply curve to the right (from $S_1$ to $S_2$ in Figure 4.8). At the initial price, $P_1$, there is excess supply. Farmers anxious to sell their output lower their price. Declining prices work to both increase demand and decrease supply until a new equilibrium is reached at price $P_2$ where quantity demanded and quantity supplied both equal $Q_2$ (Figure 4.8).

Following Adam Smith, economists are fond of saying that the market is like an "invisible hand" guiding the decisions of countless producers and consumers and bringing their decisions into balance so producers make what consumers want and consumers want to buy what producers make. Markets are an efficient way to provide information across a vast decentralized web of producers and consumers.

### 4.3.2.3 Markets and Efficiency

Some economists are ardent admirers of the ability of markets to balance supply and demand without any form of coercion, oversight, or government control. Under ideal conditions, when consumers and producers follow their own self-interest, they also satisfy the needs of everyone in society in an efficient manner. Or as Adam Smith (1937, p. 14) explained it: "It is not from the benevolence of the butcher, the brewer, or the baker that we expect our dinner, but from their regard to their own interest. We address ourselves not to their humanity but to their self-love, and never talk to them of our own necessities, but of their advantages."

Modern economic theory shows that under certain conditions market outcomes achieve an efficient outcome as Adam Smith envisioned. Economists use the notion of "Pareto efficiency," named after economist Vilfredo Pareto, to analyze whether an outcome is efficient. An outcome is Pareto efficient if everyone involved is better off. A market trade in which someone is better off and someone else is worse off would be Pareto inefficient. When producers and consumers are price takers, are fully informed, and do not affect others except through market outcomes and prices, market equilibrium will be Pareto efficient. This remarkable result provides the philosophical underpinning for the support of markets by economists.

Society often cares about other objectives beyond the efficient allocation of resources, including sustainability, justice, fairness, and equality. Even when markets are efficient, market outcomes may yield an unequal distribution of income (Piketty 2014; World Inequality Lab 2018) and do not provide any guarantee of being sustainable (Costanza 1991; Arrow et al. 2004, 2012). While many on the right of the political spectrum are strong advocates of unregulated markets, many on the left distrust markets and think markets must act within political and social constraints in order to ensure just, equitable, and sustainable outcomes.

#### 4.3.2.4 Market Failure: Public Goods and Externalities

There are many reasons for thinking that markets will fall short of delivering efficient outcomes in real-world settings. Producers may be large relative to the size of the market and can manipulate prices by changing how much they supply. Producers and consumers may be ill-informed about market conditions and fail to achieve gains from trade. Most directly relevant for conservation issues are cases of externalities and public goods where prices fail to reflect the full costs or benefits of production or consumption. All of these real-world complications are examples of market failure, where the unregulated market will fail to achieve an efficient result. With market failure, political or social institutions are needed in addition to markets (or instead of markets) to achieve an efficient outcome.

Markets fail to provide adequate incentives for the conservation of biodiversity or the protection of environmental quality due to the existence of externalities and public goods. Examples of externalities include cases where a business generates air or water pollution that affects people downwind or downstream or where a landowner cuts down a forest or drains a wetland resulting in reduced ecosystem services such as flood mitigation, water purification, or habitat provision. With negative externalities in production, producers do not pay the full costs of production (Figure 4.9). Producers face marginal private costs (MPCs), which do not include externalities, while the marginal social costs of production (MSCs) include all costs, both private and external. If firms face the full costs of production they produce at $Q^*$, where price equals MSC. When firms ignore external costs, market equilibrium occurs at $Q_m$, where price equals MPC. By ignoring the external cost, producers produce too much and there is a net loss in value because costs exceed benefits (shown by the shaded area in Figure 4.9).

Public goods are closely related to externalities. If one person undertakes actions to generate a public good, that person generates a positive externality for everyone else who benefits from the public good. A useful way to look at potential problems caused by public goods and externalities is to consider the degree to which goods are non-rival and non-excludable (see Table 4.1). At one end of the spectrum are pure private goods (rival and excludable) for which markets work well. At the other extreme are pure public goods

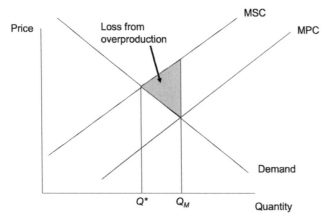

**Figure 4.9** Negative externality. Socially optimal production ($Q^*$) occurs where demand equals marginal social cost (MSC). Market equilibrium without a policy intervention ($Q_M$) occurs where demand equals marginal private cost (MPC). Including the external costs generates MSC.

**Table 4.1** Four types of goods.

| | | Subtractability of use | |
| --- | --- | --- | --- |
| | | High (rival in consumption) | Low (non-rival in consumption) |
| Difficulty of excluding potential beneficiaries | High (non-excludable) | *Common-pool resources*: groundwater basins, lakes, irrigation systems, fisheries, forests, grazing lands, etc. | *Public goods*: peace and security of a community, national defense, knowledge, fire protection, weather forecasts, etc. |
| | Low (excludable) | *Private goods*: food, clothing, automobiles, etc. | *Club goods*: theaters, private clubs, private parks, daycare centers, etc. |

*Source:* Adapted from Ostrom 2010, Figure 1.

(non-rival and non-excludable), which are typically underprovided or not provided at all by markets. For example, in the case of air quality, an individual or firm bears the full cost of reducing emissions but only gets a small share of the benefits of cleaner air because others who live in the area also benefit. **Common-pool resources**, such as marine fisheries or open grazing land, are rival in consumption but non-excludable. Common-pool resources tend to suffer from the "tragedy of the commons" (see Chapter 6, Political Science, for a critique of the idea of a tragedy of the commons). Each resource user gets the private benefit of use but ignores the costs imposed on others from degradation of the resource. For common-pool resources and pure public goods the inability to exclude potential beneficiaries leads to too little conservation. On the other hand, club goods, like a private park or concert, are excludable but non-rival. Though there is an incentive to produce these goods, typically their provision will not be efficient. If there is no cost of crowding, the marginal cost of an additional user is zero but producers will set positive prices in order to maximize profit.

#### 4.3.2.5    Addressing Market Failure through Government Policy or Social Action

From an economic perspective, externalities and public goods lie at the heart of understanding why biodiversity conservation and environmental protection are so challenging. Biodiversity conservation is most like a pure public good, and the unregulated market will systematically underprovide conservation. Addressing the problems of externalities and public goods requires some type of government policy or other social intervention to correct for market failure. Such interventions include direct regulation (prohibitions and sanctions), policies to create incentive-based mechanisms to internalize externalities (payments for ecosystem services, environmental taxes, cap-and-trade systems), self-regulation by communities, and voluntary programs.

##### 4.3.2.5.1    *Government Regulation*

Governments can enact laws regulating activities that threaten biodiversity conservation, public health, or welfare, with sanctions for violations. The Convention on International Trade in Endangered Species (CITES) bans international trade in endangered species, making importing or exporting endangered species or parts of endangered species a criminal offense. However, making actions illegal is not the same thing as preventing actions. CITES has not stopped trade in endangered species but merely driven it into the (illegal) black market (Barbier et al. 1990; van Kooten & Bulte 2000). Consumers who use endangered species in traditional medicines, from rhino horn to bear bile, are willing to pay high prices for these products. Even heavy sanctions have failed to depress demand for animal parts used in traditional medicine. CITES has been somewhat more effective in other cases. Elephants are hunted for their ivory, which is made into items meant to be displayed. The demand for ivory became much less popular once trade in ivory was banned (Who wants to advertise that they are a criminal?). However, even in the case of elephants, rampant poaching exists and elephant populations are still declining (Chase et al. 2016). In general, whether laws will be effective in practice depends on the willingness and ability of governments to enforce the law and on the incentives of firms and individuals to obey the law.

##### 4.3.2.5.2    *Incentive-based Mechanisms*

Disenchantment with overly rigid and often inefficient regulatory approaches led to a push toward using incentive-based mechanisms in conservation and environmental protection. Incentive-based mechanisms work by changing relative prices, or by introducing prices on previously unpriced goods. One way to internalize negative externalities is to establish a tax on the externality-generating activity (Hanley et al. 1997). If the government sets a tax equal to marginal external costs, the firms generating pollution will effectively pay the full costs of production (MSC rather than MPC; Figure 4.9). Firms facing this tax will have an incentive to produce the efficient level of output (Q*) because price will equal marginal cost at this point. Economists favor externality taxes because they provide incentives to reduce externalities by making producers pay for external costs ("polluters pay principle"). Such taxes also raise revenue, allowing the government to reduce other taxes or pay for other programs. Many European countries have high gasoline and diesel taxes in part to internalize pollution externalities. Despite these advantages, pollution taxes are not common in most countries because of strong political opposition by businesses and by consumers who face higher prices for goods.

Subsidies that pay for activities that generate positive externalities (or provide public goods) work in a similar manner as taxes on negative externalities. Paying to reduce pollution by an amount equal to the avoided marginal damages generates exactly the same incentives as a tax on marginal damages. The only difference between these two policies is who pays: under a subsidy, the government pays firms not to pollute; under a tax, firms pay the government when they pollute. Subsidy policies are easier to enact than taxes because they give away money to affected groups rather than require tax payments from these groups. Subsidy policies, however, need to be designed with great care to not introduce perverse incentives. An ill-designed subsidy policy can encourage businesses to start polluting with the aim of having the government pay them to stop.

PES is a form of subsidy policy that pays landowners or others to increase the supply of ecosystem services. PES programs bring together beneficiaries of ecosystem services with those who can supply the service in a voluntary transaction (Engel et al. 2008). Examples of PES programs include a program in Costa Rica that pays landowners to provide carbon sequestration, water purification, habitat, and scenic amenities (Pagiola 2008), and the Conservation Reserve Program in the USA that pays farmers to retire land from crop production to plant perennial vegetation that reduces erosion, increases carbon sequestration, and provides habitat (Kirwan et al. 2005). Another example is Reducing Emissions from Deforestation and Forest Degradation (REDD+), which is based on the idea of paying developing countries to maintain forests to sequester carbon rather than release it to the atmosphere through deforestation. Maintaining forests also provides benefits for biodiversity conservation. Despite the merits of providing incentives to prevent deforestation, REDD+ has many critics. Some of the important design issues with REDD+ are (i) ensuring fair treatment of indigenous people living in forested areas; (ii) ensuring additionality: payments go only to forests that would have been lost without payments; and (iii) preventing leakage: displacement of deforestation from areas covered by the program to other areas (Angelsen 2008).

Another incentive-based policy approach is cap-and-trade where the government establishes a limit on how much of a certain activity can take place (cap) and issues permits for allowable activity, which can be traded. Cap-and-trade approaches have been used in the USA to reduce sulfur dioxide air emissions (Ellerman et al. 2000) and in Europe to reduce carbon dioxide ($CO_2$) emissions (Convery et al. 2008). Similar programs have been used to limit land development or habitat loss (Panayotou 1994; Carroll et al. 2008) and in fisheries management, called individually transferable quotas (ITQs) or catch shares, to limit harvests (Costello et al. 2008). A difficulty with tradable development rights is establishing equivalent trades. Unlike carbon emissions that have an identical effect no matter where emissions take place, the location and spatial pattern of habitat is vitally important in biodiversity conservation. In principle, trading ratios that account for differential benefits can be designed, though having enough information and administrative capacity to implement them in practice can be problematic.

Both price-based mechanisms (taxes and subsidies) and quantity-based mechanisms (cap-and-trade) can internalize externalities and generate an efficient outcome. The main differences between these approaches arise in the details of implementation, and with uncertainty about the benefits or costs of reducing externalities. In the context of pollution reduction, a tax on emissions fixes the price faced by firms but the quantity of emissions is uncertain. With a cap-and-trade system, the quantity of emissions is fixed but the price is uncertain. Weitzman (1974) compared the performances of price-based and quantity-based systems under uncertainty and showed that the preference for priced-based or quantity-based mechanisms depends on the relative slopes of marginal benefit and marginal costs curves. For conservation, if there are thresholds for habitat or population size below which extinction or irreversible consequences could occur, it is best to have a quantity-based mechanism designed not to cross such thresholds.

### 4.3.2.5.3 *Social Actions: Self-regulation by Communities*

Many communities have successfully self-regulated their use of common-property resources to achieve sustainable outcomes without central government oversight (Ostrom 1990). This approach can be especially valuable in places where governments have limited capacity to enforce policies or are corrupt. Many communities dependent on communal grazing lands, forests, water resources, or fishing grounds have developed rules of use that can sustain the resource and avoid a tragedy of the commons. Just like formal government regulation, self-regulation systems have rules that specify what is allowed and what is prohibited and have sanctions for violators of the rules (Ostrom 1990). In one famous example of self-regulation, lobster fishers in Maine have been known to punish violators by destroying their equipment (Acheson 1988).

Another example of self-governance is the use of negotiation and mutually agreeable bargains to overcome cases of externalities. Coase (1960) explained that as long as there are clear property rights and negotiating costs are low, individuals and firms can bargain with each other to produce an efficient outcome. For example, a person negatively affected by pollution could bargain with the business responsible for the pollution and agree to pay them to reduce pollution. Alternatively, if the business did not have the right to cause harm, the business could instead offer payment to neighbors to allow pollution to occur. Through this sort of bargaining, Coase argued that markets would arrive at solutions acceptable to all parties involved without the need for expensive and inefficient regulation or taxation.

Self-regulation works best in cases where a small number of parties are in frequent contact with one another. Many cases of externalities and public goods, however, involve large numbers of people who are not in frequent contact (e.g. global climate change, biodiversity conservation, air pollution in major cities). These latter types of public goods and externalities typically require some sort of action by governments to address the problem.

### 4.3.2.5.4 *Voluntary Approaches*

Frustrated with the slow pace of government action, some efforts to address social or environmental issues have gone directly to businesses or consumers. Some groups have established certification programs that inform consumers about which producers meet environmental and social standards. When consumers care about the environment and are willing to pay a premium for certified products, businesses have an incentive to produce certified products. Products produced in an environmentally sound manner can gain market share quickly if wholesalers or retailers (e.g. Walmart, Home Depot) agree to purchase and sell only certified products. Influencing wholesalers or retailers that control access to many consumers is far simpler than attempting to woo individual consumers.

Certification programs exist for timber products (Forest Stewardship Council), seafood (Marine Stewardship Council and Seafood Watch), electronics (Green Electronics Council), and several agricultural products (BonSucro for sugar, Round Table on Responsible Soy, Roundtable on Sustainable Palm Oil). However, certification schemes only work if they result in environmentally improved performance, if wholesalers or retailers are willing to push certified products, and if consumers are willing to pay more for certified products. There is limited evidence to date that such programs are effective (Tayleur et al. 2017; Lambin et al. 2018).

Just as there are many types of market and institutional failures leading to a lack of conservation for biodiversity and environmental protection, there are many potential solutions. While many social and policy solutions outlined earlier work well in certain circumstances, no one solution is likely to be superior in all circumstances (Ostrom 2007). In fact, crafting workable solutions for conservation and environmental protection in complex real-world situations requires multiple and interacting forms of governance to fully address problems of market and institutional failure (Ostrom 2007, 2010).

### 4.3.3    Economic Growth and Sustainability

Conservationists typically take a very long-term perspective consistent with sustaining biodiversity over evolutionary time scales. The social sciences, and economics in particular, often focus on current conditions or the very near-term future. Looking out several years is a "long-run" analysis in much of economics, and studies that go beyond 50 years into the future are rare. Therefore, it can be challenging to connect economic and social science time scales with conservation time scales. To date, the study of economic growth has largely ignored environmental change as being important to consider, but economists are increasingly recognizing this factor (Polasky et al. 2019), particularly in the economics of climate change (Nordhaus 1994; Nordhaus & Boyer 2000; Stern 2007, 2008).

#### 4.3.3.1    Economic Growth Theory
Although economists usually provide near-term views, analyzing the evolution of the economy through time is a part of macroeconomics called "growth theory." In reality, growth theory is a misleading name because it can be used to study growing, shrinking, or steady-state economies. It is, however, an accurate name in another sense because the global economy has grown rapidly over the past 200 years, as conventionally measured by increases in GDP. The size of the economy as measured in constant dollar GDP terms was 11 times larger in 2018 than in 1950, and 87 times larger than in 1820 (Maddison 2003; World Bank 2019). Therefore, most economists use growth theory to study growing (not steady-state or shrinking) economies. The underlying assumption of most macroeconomists is that economic growth will continue into the future, though many conservations question this assumption.

Growth theory focuses on the question of how much current wealth should be saved and invested for future use versus how much to use now. Saving and investing leads to greater future wealth in the form of higher levels of capital assets. Capital assets can take many forms including manufactured capital (buildings and machinery), human capital (education, skills, and experience), **social capital** (institutions and trust among members of society), and natural capital (biodiversity, environmental quality, and natural resources). Greater amounts of capital increase the capacity of the economy to provide goods and services in the future. Savings and investment, however, come at the cost of lower current consumption. Ramsey (1928) analyzed the question of how much a nation should save and invest in order to maximize benefits through time. He showed that the optimal amount of savings and investment equates the loss in marginal benefit from lower current consumption due to more savings with the marginal benefit from increased future consumption due to more investment.

Ramsey's solution for optimal saving and investment also provides justification for the discount rates an individual should apply in comparing future values with current values. In economics, there are two reasons why an individual might discount future values. Individuals who are impatient and prefer current to future benefits have a "positive rate of pure time preference." In other words, they value the present much more than the future. The other reason for discounting arises when an individual expects to have growing income through time. The marginal utility of extra consumption tends to be lower for a person who is already consuming a lot. Having enough income to buy food and shelter matters a great deal more than extra income to buy a nicer house or more expensive food. If a person will be wealthier in the future, additional future income will generate lower marginal benefits compared to additional present income, which is the same thing as saying that future benefits should be discounted.

In evaluating conservation policy where benefits may continue to accrue far into the future, the choice of the discount rate really matters. Economists, however, do not agree on what discount rate should apply. In debating climate change policy, Nordhaus (2007) argued for using the market rate of discount, which at the time was around 5.5%, while Stern (2007) argued for a far lower rate of 1.4%. These different discount rates lead to dramatic differences in how the future is valued in comparison to the present. With a discount rate of 5.5%, the present value of one million dollars in 100 years will be $4,087. However, if the future is discounted at a lower rate of 1.4%, the same sum will be worth $246,597 (more than sixty times greater). Other economists have argued that the proper rate might be even lower when one factors in uncertainty (Weitzman 1998) or the loss of biodiversity and ecosystem services not accounted for in market values (Sterner & Persson 2008).

Discounting is also important in conservation policy. Clark (1973) noted that the grim logic of maximizing the present value of returns with a high discount rate could lead to harvesting species to extinction. If the discount rate is high relative to the biological growth rate of a species, maximizing the present value of returns will result in harvesting that species to extinction. This result occurs because the rate of return on a species is lower than the rate of return on alternative investments, so it is more profitable to harvest today rather than conserve the species for future harvest. Of course, such logic sounds heartless and runs directly counter to the conservation ethic. The "efficient extinction" result occurs when species are viewed only as a source of income rather than as a living species and harvesters are sufficiently impatient. Extinction is generally not the efficient outcome when a species has existence value and is not just a source of income, or when harvesters have low enough discount rates so allowing the species to regenerate is profitable.

In reality, the threat of extinction from overharvesting is more likely to arise from market failure, for example a tragedy of the commons, than from high discount rates. The overharvesting problem is most severe for open-access resources. If harvesting is profitable even down to low population levels, open-access harvesting can drive the population to extinction. Even conservation-minded harvesters will find themselves trapped because someone else will harvest the resource even if they decide not to harvest. The solution for this type of problem is to fix the institutions and limit open access. There is evidence that fisheries with effective management have been successful in sustaining fish populations while those areas without effective management, such as the high seas that do not fall under the control of any national government, have seen continued declines in fish populations (Costello et al. 2008, 2016).

Because discounting has such a strong impact on decisions affecting the future, it has come under significant scrutiny by economists, conservationists, and others. Some critics of discounting say that decisions with potentially large impacts on the well-being of future generations have a moral dimension beyond the strictly economic determination of the rate of return on investments (Page 1977; Spash 1993). Others, however, reason that people are making savings and investment choices and that the market discount rate reflects their view of the proper tradeoff between current and future benefits (Nordhaus 2007). This debate has important implications for conservation and environmental policies, but at present no clear answers.

### 4.3.3.2 Sustainability

Given current trajectories, there is no guarantee that the economy will continue to grow, or that it will evolve in a manner consistent with maintaining environmental quality, ecosystems, biodiversity, or human well-being. Humans are now the dominant force shaping Earth systems from local land use and regional pollution, to global climate change and biodiversity loss (Vitousek et al. 1997; Steffens et al. 2007; IPCC 2014, 2018; Díaz et al. 2019; IPBES 2019). The expansion of human activity has led to a loss of natural habitat and greatly increased extinction rates, which could cause extinction of one in eight species by 2100 if current trends continue (IPBES 2019). Deforestation and burning of fossil fuels have increased atmospheric concentrations of greenhouse gases that could lead to climate change with large negative impacts on both human society and biodiversity (IPCC 2014, 2018). Global reviews of trends in ecosystem services have found that the majority of ecosystem services are declining (MEA 2005; IPBES 2019). Another survey concluded that humanity has pushed beyond planetary boundaries for sustainable management in a number of key dimensions including biodiversity loss, climate change, and nitrogen cycles (Rockström et al. 2009). Fundamental changes in the Earth systems threaten current and future human well-being (Díaz et al. 2019; IPBES 2019). The realization that humanity depends on nature and that human actions are rapidly changing nature has led to urgent calls for transforming society and our economic system (Díaz et al. 2019; IPBES 2019).

The study of long-run trends and future prospects for the economy and the environment has been a hotly debated topic going back to the time of Malthus and before. Recent debates over future prospects have pitted more optimistic economists on one side versus more pessimistic conservationists on the other (Box 4.4). Economists tend to have greater faith in

---

**Box 4.4   Debates: The bet**

The debate between the optimistic vision of the future held by many economists and the pessimistic vision of the future held by many conservationists is illustrated by a famous bet between economist Julian Simon and ecologist Paul Ehrlich (Tierney 1990; Sabin 2013). In 1980, Ehrlich and Simon agreed to bet on whether the average price of five metals (chrome, copper, nickel, tin, and tungsten) would increase or decrease over the following decade. Ehrlich argued that prices should rise because population pressure and ever increasing appetites for more goods would cause increased demand for a limited supply of these resources. Of even greater concern in Ehrlich's view was the continuing burden that economic growth was placing on the environment. In his view, unless society made fundamental changes, the future of humanity and of biodiversity looked grim. Simon argued that prices of the metals should fall because innovation would work to reduce the demand for these metals even as population and the economy grew. Simon (1981) went so far as to say that natural resources were essentially infinite because of the ingenuity of people (the "ultimate resource") to make more productive use of natural resources so economic growth could continue forever.

   The 1980s turned out to be a period when most resource prices fell, including the prices of the five metals included in the bet, so Simon won the bet. Prices fell because of the discovery of new reserves and innovations that led to substitution of alternatives for these metals, such as fiber optic lines replacing copper wires (Tierney 1990).

   What, if anything, do we learn from the bet about long-term prospects? Most natural resource prices fluctuate through time in something of a boom-and-bust cycle. Had the bet taken place several decades earlier or later, Ehrlich could have won the bet. The 1980s bet illustrated that innovations can reduce demand or increase the economically producible supply of resources so prices of resources can fall at least for a period of time. For example, hydraulic fracturing (fracking) has allowed a major increase in natural gas and oil production in the USA and resulted in lower energy prices. Critics, however, say that increased production from fracking has come at a high environmental cost and that, potentially, the most important resource constraints involve environmental quality and climate change. Increasing population pressure with a finite resource base will cause increased scarcity and trigger price increases unless it is offset by sufficiently fast technological progress. Past price trends and rapid progress in much of the twentieth century is no guarantee that such trends will continue in this century or beyond.

---

markets to signal relative scarcity, and in new technology to find solutions to overcome scarcity. Price signals supply the incentive for innovation and for substitution away from scarce goods. Sufficient innovation and substitution can lead to continued expansion of the economy even with essential nonrenewable resources (Solow 1974; Stiglitz 1974).

   In contrast, conservationists stress the limits to growth imposed by finite resources and the threats to long-run sustainability from environmental degradation and loss of biodiversity (Meadows et al. 1972, 2004; Rockström et al. 2009). Conservationists tend to have less faith in market outcomes and often think that market economies are the central force driving us toward future environmental catastrophes (Czech 2000; Hall et al. 2000; Klein

2015). Important components of natural capital fail to have prices or are systematically undervalued by market economies, as previously discussed. Further, complex systems like ecosystems have the potential to cross thresholds that dramatically change system behavior and can lead to rapid and sudden collapse (regime shifts) that are often difficult to predict and difficult to reverse (Scheffer et al. 2001, 2012). Given the complexity of system dynamics and the potential for unexpected regime shifts, prices may well fail to provide accurate signals of future conditions or to direct current actions in ways to avoid detrimental shifts.

On the other hand, optimists, including some economists, point out that living standards have improved markedly for most people around the globe even while some measures of natural capital have declined. Improvements in agriculture have meant more food production per capita despite a rising population and limited land base. Improvements in healthcare and access to clean water have improved life expectancy despite declines in water quality in some countries. Overall, measures of human well-being have shown dramatic improvement over recent decades. For example, the human development index (HDI), which combines measures of human health (life expectancy), education (adult literacy rate), and income (gross national income per capita), improved in all but three countries between 1970 and 2011, with rapid increases in some countries such as China (UNDP 2011). Some economists go further to make the case that economic growth itself may lead to improving environmental quality or other components of natural capital. For example, empirical studies that relate local air and water qualities to per capita income found that air and water qualities decline with income at low to moderate levels of per capita income, but then improve with income from moderate to high per capita income (Selden & Song 1994; Grossman & Kreuger 1995). This inverted U-shaped relationship between economic growth and environmental degradation is referred to as an Environmental Kuznets Curve, a concept that derives from the work of Kuznets (1955), who proposed an inverted U-shaped relationship between economic development and income inequality (as opposed to environmental quality). However, whether the Environmental Kuznets Curve holds for all environmental measures, and what it means for economic growth and sustainability, remains the subject of considerable debate (Dinda 2004; Sarkodie & Strezov 2019).

Sustainability addresses overarching questions about whether living standards, biodiversity, and environmental quality can be maintained over the long run. Various definitions of sustainability exist. "Strong sustainability" requires maintaining important dimensions of natural capital, including biodiversity and ecosystem processes, which many ecological economists view as being irreplaceable. Their loss or degradation would diminish future prospects and would mean that society is not on a sustainability path. "Weak sustainability" is defined in terms of maintaining human well-being rather than maintaining any specific form of capital. Mainstream economists tend to favor the notion of weak sustainability with its focus on the overall capital requirements needed to sustain human well-being. If different forms of capital can provide the same level of well-being, substituting more of one form of capital for declines in another satisfies weak sustainability. For example, under weak sustainability it is acceptable to use up nonrenewable resources as long as sufficient investment is made in technology that results in substitutes that are as useful and inexpensive (e.g. solar energy as a substitute for oil). If, however, biodiversity or ecological processes are essential for sustaining human well-being, these components of natural capital must be maintained to achieve either strong or weak sustainability.

Several researchers have attempted to measure whether society is meeting the requirements of weak sustainability. Economists have shown that non-declining "inclusive wealth," which measures the total value of all capital assets, gives future generations the capability of being at least as well off as the current generation (Arrow et al. 2004, 2012). An increase in inclusive wealth means that society has a larger base of productive assets with which to improve future standards of living and well-being. Measuring how inclusive wealth changes through time requires (i) measuring changes in manufactured, human, social, and natural capital and (ii) assessing the price (value) of a unit of each type of capital. In principle, the price of a capital asset is equal to the present value of the flow of benefits generated by the capital asset. For example, the value of land is determined by the present value of the rents that it is expected to generate over time. A major difficulty in trying to measure inclusive wealth is the lack of prices for many forms of capital, including most forms of natural capital. Getting prices for natural capital requires the use of nonmarket valuation to estimate these prices (see section 4.3). Ambitious attempts to measure inclusive wealth empirically have found positive trends in most countries despite declines in many measures of natural capital (World Bank 2011, 2018; Arrow et al. 2012). However, empirical attempts to measure inclusive wealth are still just beginning, and current measures of inclusive wealth do not include considerations of biodiversity or ecosystem services other than carbon sequestration (Polasky et al. 2015).

## 4.4 Future Directions

Economics provides a very powerful set of tools applicable to conservation issues. These issues include how people value nature, how best to allocate limited conservation resources, how markets operate, how to use policy to correct market failure to provide better conservation outcomes, and how economic growth and environmental change may affect sustainability and long-term conservation goals. Conservationists who are serious about making a difference should study enough economics to gain a working understanding of economics and to apply its lessons to conservation. This task is not always easy. Economists often write in ways that seem designed to befuddle outsiders, with the use of specialized jargon and a large dose of specialized mathematical methods. Members of both disciplines need to work together to find common ground and move in a positive direction. In addition, much of economics seems, at least on the surface, to be alien to conservation thinking, such as the analysis of economic growth versus sustainability. But economics offers valuable insights into activities that threaten biodiversity as well as how to think about human behavior and incentives to motivate better conservation outcomes. Economic analysis is particularly useful for understanding production, consumption, and trade and is useful in understanding how to minimize the conflicts between these activities and conservation. Economics also offers a variety of policy approaches, including payments for ecosystem services, taxes on pollution, and cap-and-trade policies to limit pollution or habitat conversion. These approaches can offer practical solutions to many conservation problems. In fact, it is fair to say that successful conservation in an increasingly crowded, human-dominated world will require reconciling and integrating economics and conservation science.

Looking forward, the essential question facing humanity is whether it is possible to achieve sustainable development, in which all people have access to a decent standard of living and ecosystems and biodiversity, on which all life depends, are conserved (Guerry et al. 2015). Currently, global society is not on a path to meet these goals (IPBES 2019). Some have suggested that meeting this challenge will require a new type of economic thinking that accounts for both the value of meeting human needs but doing so in ways that respect planetary boundaries and moves away from current notions of economic growth revolving around increases in GDP (Raworth 2017). However, many tools within economics can help with the sustainable development challenge. Success will require that these tools be employed in different ways and that a greater, more consistent, and creative focus be applied to conservation questions. Tallis et al. (2018, p. 6) showed that "the biophysical limits of a finite planet by themselves may not constrain more sustainable development. Rather, it is the complex interactions between social, economic, political, and biophysical systems that make sustainable development such a daunting challenge." Meeting sustainable development challenges will require harmonizing the economy with the environment, which will require integrating conservation thinking into economics, and economic thinking into conservation.

## For Further Reading

1  Are we consuming too much? (Arrow et al. 2004, *Journal of Economic Perspectives* 18: 147–172).
2  *Nature's Services: Societal Dependence on Natural Ecosystems* (Daily, ed. 1997, Island Press).
3  *A Field Guide to Economics for Conservationists* (Fisher, Naidoo & Ricketts 2014, Roberts and Company).
4  *The Measurement of Environmental and Resource Values: Theory and Methods*, 3rd ed. (Freeman, Herriges & Kling 2014, Resources for the Future Press).
5  *Valuing Ecosystem Services: Toward Better Environmental Decision-Making* (National Research Council (NRC) 2005, National Academies Press).
6  *Governing the Commons: The Evolution of Institutions for Collective Action* (Ostrom 1990, Cambridge University Press).
7  The role of economics in analyzing the environment and sustainable development (Polasky et al. 2019, *Proceedings of the National Academy of Sciences* 116: 5233–5238).
8  *Doughnut Economics: Seven Ways to Think Like a 21st Century Economist* (Raworth 2017, Chelsea Green Publishers).
9  Towards a more dynamic understanding of human behaviour in the Anthropocene (Schill et al. 2019, *Nature Sustainability* 2: 1075–1082).
10  *Environmental and Natural Resource Economics*, 11th ed. (Tietenberg & Lewis 2018, Prentice Hall).

# References

Acheson, J.H. (1988) *The Lobster Gangs of Maine*. Hanover, NH: University Press of New England.

Andam, K.S., Ferraro, P.J., Pfaff, A. et al. (2008) Measuring the effectiveness of protected area networks in reducing deforestation. *Proceedings of the National Academy of Sciences of the United States of America* 105: 16089–16094.

Ando, A., Camm, J., Polasky, S. et al. (1998) Species distributions, land values, and efficient conservation. *Science* 279: 2126–2128.

Angelsen, A. (2008) *Moving Ahead with REDD: Issues, Options and Implications*. Bogor, Indonesia: CIFOR.

Angrist, J.D. & Pischke, J.-S. (2009) *Mostly Harmless Econometrics: An Empiricists Companion*. Princeton, NJ: Princeton University Press.

Angrist, J.D. & Pischke, J.-S. (2010) The credibility revolution in empirical economics: how better research design is taking the con out of econometrics. *Journal of Economic Perspectives* 24 (2): 3–30.

Ariely, D. (2009) *Predictably Irrational Revised and Expanded Edition: The Hidden Forces that Shape Our Decisions*. New York: Harper Perennial.

Arrow, K.J., Bolin, B., Costanza, R. et al. (1995) Economic growth, carrying capacity, and the environment. *Science* 268: 520–521.

Arrow, K.J., Dasgupta, P., Goulder, L.H. et al. (2004) Are we consuming too much? *Journal of Economic Perspectives* 18 (3): 147–172.

Arrow, K.J., Dasgupta, P., Goulder, L.H. et al. (2012) Sustainability and the measurement of wealth. *Environment and Development Economics* 17: 317–353.

Balmford, A., Bruner, A., Cooper, P. et al. (2002) Economic reasons for conserving wild nature. *Science* 297: 950–953.

Barbier, E.B., Burgess, J.C., Swanson, T.M. et al. (1990) *Elephants, Economics and Ivory*. London: Earthscan.

Barnett, H. & Morse, C. (1963) *Scarcity and Growth: The Economics of Natural Resource Availability*. Baltimore, MD: Johns Hopkins University Press.

Bateman, I.J., Harwood, A.R., Mace, G.M. et al. (2013) Bringing ecosystem services into economic decision-making: land use in the United Kingdom. *Science* 341: 45–50.

Bockstael, N.E., Freeman, A.M., Kopp, R.J. et al. (2000) On measuring economic values of nature. *Environmental Science and Technology* 34 (8): 1384–1389.

Boulding, K.E. (1966) The economics of the coming Spaceship Earth. In: *Environmental Quality in a Growing Economy: Essays from the Sixth RFF Forum* (ed. H. Jarrett), 3–14. Baltimore, MD: Johns Hopkins University Press.

Brown, G.M., Jr. & Shogren, J.F. (1998) Economics of the Endangered Species Act. *Journal of Economic Perspectives* 12 (3): 3–20.

Byerly, H., Balmford, A., Ferraro, P.J. et al. (2018) Nudging pro-environmental behavior: evidence and opportunities. *Frontiers in Ecology and the Environment* 16 (3): 159–168.

Camerer, C.F., Loewenstein, G. & Rabin, M. eds. (2004) *Advances in Behavioral Economics*. Princeton, NJ: Princeton University Press.

Caplin, A. & Dean, M. (2015) Revealed preference, rational inattention, and costly information acquisition. *American Economic Review* 105 (7): 2183–2203.

Carlsson, F. & Johansson-Stenman, O. (2012) Behavioral economics and environmental policy. *Annual Review of Resource Economics* 4 (1): 75–99.

Carroll, N., Bayon, R. & Fox, J. (2008) *Conservation and Biodiversity Banking: A Guide to Setting Up and Running Biodiversity Credit Trading Systems*. London: Earthscan.

Carson, R.T. (1962) *Silent Spring*. Cambridge, MA: The Riverside Press.

Carson, R.T. (2011) *Contingent Valuation: A Comprehensive Bibliography and History*. Northampton, MA: Edward Elgar.

Carson, R.T. (2012) Contingent valuation: a practical alternative when prices aren't available. *Journal of Economic Perspectives* 26 (4): 27–42.

Carson, R.T. & Mitchell, R.C. (1993) The value of clean water: the public's willingness to pay for boatable, fishable, and swimmable quality water. *Water Resources Research* 29 (7): 2445–2454.

Champ, P., Boyle, K.J. & Brown, T.C. eds. (2009) *A Primer on Nonmarket Valuation*, 2nd ed. Boston, MA: Kluwer.

Chaplin-Kramer, R., Sharp, R.P., Well, C. et al. (2019) Global modeling of nature's contribution to people. *Science* 366: 255–258.

Chase, M.J., Schlossberg, S., Griffin, C.R. et al. (2016) Continent-wide survey reveals massive decline in African savannah elephants. *PeerJ* 4: e2354.

Chichilnisky, G. & Heal, G.M. (1998) Economic returns from the biosphere. *Nature* 391: 629–630.

Clark, C. (1973) Profit maximization and the extinction of animal species. *Journal of Political Economy* 81 (4): 950–961.

Coase, R. (1960) The problem of social cost. *Journal of Law and Economics* 3 (1): 1–44.

Convery, F., Ellerman, D. & de Perthuis, C. (2008) The European carbon market in action: lessons from the first trading period. *Journal for European Environmental Planning & Law* 5 (2): 215–233.

Costanza, R. ed. (1991) *Ecological Economics: The Science and Management of Sustainability*. New York: Columbia University Press.

Costanza, R., Cumberland, J., Daly, H. et al. (1997) *An Introduction to Ecological Economics*. Boca Raton, FL: St Lucie Press.

Costanza, R., d'Arge, R., de Groot, R. et al. (1997) The value of the world's ecosystem services and natural capital. *Nature* 387: 253–260.

Constanza, R., de Groot, R., Sutton, P. et al. (2014) Changes in the global value of ecosystem services. *Global Environmental Change* 26: 152–158.

Costello, C., Gaines, S. & Lynham, J. (2008) Can catch shares prevent fisheries collapse? *Science* 321: 1678–1681.

Costello, C., Ovando, D., Clavell, T. et al. (2016) Global fishery prospects under contrasting management regimes. *Proceedings of the National Academy of Sciences of the United States of America* 113 (18): 5125–5129.

Cowling, R.M. (2014) Let's get serious about human behavior and conservation. *Conservation Letters* 7: 147–148.

Czech, B. (2000) The importance of ecological economics to wildlife conservation: an introduction. *Wildlife Society Bulletin* 28 (1): 2–3.

Daily, G.C. ed. (1997) *Nature's Services: Societal Dependence on Natural Ecosystems*. Washington, DC: Island Press.

Daly, H.E. (1977) *Steady-State Economics.* San Francisco, CA: W.H. Freeman.

Daly, H.E. (1999) *Ecological Economics and the Ecology of Economics: Essays in Criticism.* Northampton, MA: Edward Elgar.

Das, S. & Vincent, J.R. (2009) Mangroves protected villages and reduced death toll during Indian super cyclone. *Proceedings of the National Academy of Sciences of the United States of America* 106: 7357–7360.

Diamond, P. & Hausman, J. (1994) Contingent valuation: is some number better than no number? *Journal of Economic Perspectives* 8: 45–64.

Díaz, S., Pascual, U., Stenseke, M. et al. (2018) Assessing nature's contributions to people. *Science* 359: 270–272.

Díaz, S., Settele, J., Brondízio, E.S. et al. (2019) Pervasive human-driven decline of life on Earth points to the need for transformative change. *Science* 366: eaax3100.

Dinda, S. (2004) Environmental Kuznets curve hypothesis: a survey. *Ecological Economics* 49 (4): 431–455.

Egan, K.J., Herriges, J.A., Kling, C.L. et al. (2009) Valuing water quality as a function of water quality measures. *American Journal of Agricultural Economics* 91 (1): 106–123.

Ehrenfeld, D. (1988) Why put a value on biodiversity? In: *Biodiversity* (ed. E.O. Wilson), 212–216. Washington, DC: National Academy Press.

Ehrlich, P. (1968) *The Population Bomb.* New York: Sierra Club-Ballantine.

Ellerman, A.D., Joskow, P.L., Schmalensee, R. et al. (2000) *Markets for Clean Air: The U.S. Acid Rain Program.* New York: Cambridge University Press.

Engel, S., Pagiola, S. & Wunder, S. (2008) Designing payments for environmental services in theory and practice: an overview of the issues. *Ecological Economics* 65 (4): 663–674.

Epanchin-Niell, R. & Boyd, J. (2020) Private-sector conservation under the US Endangered Species Act: a return-on-investment perspective. *Frontiers in Ecology and the Environment* 18 (7): 409–416.

Ferraro, P.J. & Hanauer, M.M. (2014a) Advances in measuring the environmental and social impacts of environmental programs. *Annual Review of Environment and Resources* 39: 495–517.

Ferraro, P.J. & Hanauer, M.M. (2014b) Quantifying causal mechanisms to determine how protected areas affect poverty through changes in ecosystem services and infrastructure. *Proceedings of the National Academy of Sciences of the United States of America* 111: 4332–4337.

Ferraro, P.J. & Pattanyak, S.K. (2006) Money for nothing: a call for empirical investigation of biodiversity conservation investments. *PLoS Biology* 4 (4): e105.

Ferraro, P.J. & Price, M.K. (2013) Using nonpecuniary strategies to influence behavior: evidence from a large-scale field experiment. *Review of Economics and Statistics* 95 (1): 64–73.

Ferraro, P.J., Sanchirico, J.N. & Smith, M.D. (2019) Causal inference in coupled human and natural systems. *Proceedings of the National Academy of Sciences of the United States of America* 116 (12): 5311–5318.

Fisher, B., Naidoo, R. & Ricketts, T. (2014) *A Field Guide to Economics for Conservationists.* Greenwood Village, CO: Roberts and Company.

Freeman, A.M., III, Herriges, J. & Kling, C.L. (2014) *The Measurement of Environmental and Resource Values: Theory and Methods*, 3rd ed. New York: Resources for the Future Press.

Garibaldi, L.A., Carvalheiro, L.G., Vaissiére, B.E. et al. (2016) Mutually beneficial pollinator diversity and crop yield outcomes in small and large farms. *Science* 351: 388–391.

Georgescu-Roegen, N. (1971) *The Entropy Law and the Economic Process*. Cambridge, MA: Harvard University Press.

Grossman, G. & Kreuger, A. (1995) Economic growth and the environment. *Quarterly Journal of Economics* 110: 352–377.

Guerry, A., Polasky, S., Lubchenco, J. et al. (2015) Natural capital informing decisions: from promise to practice. *Proceedings of the National Academy of Sciences of the United States of America* 112: 7348–7355.

Hall, C.A.S., Jones, P.W., Donovan, T.M. et al. (2000) The implications of mainstream economics for wildlife conservation. *Wildlife Society Bulletin* 28 (1): 16–25.

Hanemann, W.M. (1994) Contingent valuation and economics. *Journal of Economic Perspectives* 8: 19–44.

Hanley, N., Shogren, J. & White, B. (1997) *Environmental Economics: In Theory and Practice*. New York: Macmillan.

Hausman, J. (2012) Contingent valuation: from dubious to hopeless. *Journal of Economic Perspectives* 26 (4): 43–56.

Heal, G. (1998) *Valuing the Future: Economic Theory and Sustainability*. New York: Columbia University Press.

Hotelling, H. (1931) The economics of exhaustible resources. *Journal of Political Economy* 39 (2): 137–175.

Innes, R., Polasky, S. & Tschirhart, J. (1998) Takings, compensation and endangered species protection on private land. *Journal of Economic Perspectives* 12 (3): 35–52.

IPBES (Intergovernmental Science-Policy Platform on Biodiversity and Ecosystem Services) (2019) *Global Assessment Report on Biodiversity and Ecosystem Services of the Intergovernmental Science-Policy Platform on Biodiversity and Ecosystem Services* (ed. E.S. Brondizio, J. Settele, S. Díaz et al.). Bonn, Germany: IPBES Secretariat.

IPCC (Intergovernmental Panel on Climate Change) (2014) *Climate Change Synthesis Report: Contribution of Working Groups I, II and III to the Fifth Assessment Report of the Intergovernmental Panel on Climate Change*. Geneva: IPCC.

IPCC (Intergovernmental Panel on Climate Change) (2018) *Global Warming of 1.5°C: An IPCC Special Report on the Impacts of Global Warming of 1.5°C Above Pre-Industrial Levels and Related Global Greenhouse Gas Emission Pathways, in the Context of Strengthening the Global Response to the Threat of Climate Change, Sustainable Development, and Efforts to Eradicate Poverty*. Geneva: World Meteorological Organization.

Irwin, E.G. (2002) The effects of open space on residential property values. *Land Economics* 78 (4): 465–480.

Jack, B.K. & Jayachandran, S. (2019) Self-selection into payments for ecosystem services programs. *Proceedings of the National Academy of Sciences of the United States of America* 116 (12): 5326–5333.

Jack, B.K., Kousky, C. & Sims, K.R.E. (2008) Designing payments for ecosystem services: lessons from previous experience with incentive-based mechanisms. *Proceedings of the National Academy of Sciences of the United States of America* 105: 9465–9470.

Jayachandran, S., de Laat, J., Lambin, E.F. et al. (2017) Cash for carbon: a randomized trial of payments for ecosystem services to reduce deforestation. *Science* 357: 267–273.

Kahneman, D. & Knetsch, J.L. (1992) Valuing public goods: the purchase of moral satisfaction. *Journal of Environmental Economics and Management* 22: 57–70.

Kahneman, D., Slovic, P. & Tversky, A. (1982) *Judgment under Uncertainty: Heuristics and Biases*. New York: Cambridge University Press.

Kirwan, B., Lubowski, R.N. & Roberts, M.J. (2005) How cost-effective are land retirement auctions? Estimating the difference between payments and willingness to accept in the Conservation Reserve Program. *American Journal of Agricultural Economics* 87: 1239–1247.

Klein, N. (2015) *This Changes Everything: Climate versus Capitalism*. New York: Simon & Schuster.

Kling, C.L., Phaneuf, D.J. & Zhao, J. (2012) From Exxon to BP: has some number become better than no number? *Journal of Economic Perspectives* 26 (4): 3–26.

Krutilla, J. (1967) Conservation reconsidered. *American Economic Review* 57 (4): 777–786.

Kuznets, S. (1955) Economic growth and income inequality. *The American Economic Review* 45 (1): 1–28.

Lambin, E.R., Gibbs, H.K., Heilmayr, R. et al. (2018) The role of supply-chain initiatives in reducing deforestation. *Nature Climate Change* 8: 109–116.

Langpap, C., Kerkvliet, J. & Shogren, J.F. (2018) The economics of the U.S. Endangered Species Act: a review of recent developments. *Review of Environmental Economics and Policy* 12 (1): 69–91.

Lawler, J.J., Lewis, D.J., Nelson, E. et al. (2014) Projected land-use change impacts on ecosystem services in the U.S. *Proceedings of the National Academy of Sciences of the United States of America* 111 (20): 7492–7497.

Layton, D., Brown, G. & Plummer, M. (2001) *Valuing multiple programs to improve fish populations*. Olympia: Washington State Department of Ecology.

Lichtenstein, S. & Slovic, P. (1971) Reversals of preference between bids and choices in gambling decisions. *Journal of Experimental Psychology* 89: 46–55.

Loomis, J.B. (1996) Measuring the economic benefits of removing dams and restoring the Elwha River: results of a contingent valuation survey. *Water Resources Research* 32 (2): 441–447.

Loomis, J.B. & White, D.S. (1996) Economic benefits of rare and endangered species: summary and meta-analysis. *Ecological Economics* 18: 197–206.

Lutzenhiser, M. & Netusil, N.R. (2001) The effect of open spaces on a home's sales price. *Contemporary Economic Policy* 19 (3): 291–298.

Maddison, A. (2003) *The World Economy: Historical Statistics*. Paris: Organisation for Economic Co-operation and Development.

Mahan, B., Polasky, S. & Adams, R. (2000) Valuing urban wetlands: a property price approach. *Land Economics* 76 (1): 100–113.

McCauley, D. (2006) Selling out on nature. *Nature* 443: 27–28.

McFadden, D. & Train, K. eds. (2017) *Contingent Valuation of Environmental Goods: A Comprehensive Critique*. Cheltamham, UK: Edward Elgar.

MEA (Millennium Ecosystem Assessment) (2005) *Ecosystems and Human Well-Being: Synthesis*. Washington, DC: Island Press.

Meadows, D., Randers, J. & Meadows, D. (2004) *Limits to Growth: The 30-Year Update*. White River Junction, VT: Chelsea Green.

Meadows, D.H., Meadows, D.L., Randers, J. et al. (1972) *The Limits to Growth*. New York: Universe Books.

Murdoch, W., Polasky, S., Wilson, K.A. et al. (2007) Maximizing return on investment in conservation. *Biological Conservation* 139: 375–388.

Naidoo, R., Balmford, A., Ferraro, P.J. et al. (2006) Integrating economic cost into conservation planning. *Trends in Ecology and Evolution* 21 (12): 681–687.

Nelson, E., Mendoza, G., Regetz, J. et al. (2009) Modeling multiple ecosystem services, biodiversity conservation, commodity production, and tradeoffs at landscape scales. *Frontiers in Ecology and the Environment* 7 (1): 4–11.

Nordhaus, W. (2007) Critical assumptions in the Stern Review on climate change. *Science* 317: 201–202.

Nordhaus, W.D. (1994) *Managing the Global Commons: The Economics of Climate Change.* Cambridge, MA: MIT Press.

Nordhaus, W.D. & Boyer, J. (2000) *Warming the World: Economic Modeling of Global Warming.* Cambridge, MA: MIT Press.

Norton, B.G. (1987) *Why Preserve Natural Variety?* Princeton, NJ: Princeton University Press.

NRC (National Research Council) (2000) *Watershed Management for Potable Water Supply: Assessing the New York City Strategy.* Washington, DC: National Academies Press.

NRC (National Research Council) (2005) *Valuing Ecosystem Services: Toward Better Environmental Decision-Making.* Washington, DC: National Academies Press.

Ostrom, E. (1990) *Governing the Commons: The Evolution of Institutions for Collective Action.* Cambridge, UK: Cambridge University Press.

Ostrom, E. (2007) A diagnostic approach for going beyond panaceas. *Proceedings of the National Academy of Sciences of the United States of America* 104: 15181–15187.

Ostrom, E. (2010) Beyond markets and states: polycentric governance of complex economic systems. *American Economic Review* 100 (3): 641–672.

Ouyang, Z., Zheng, H., Xiao, Y. et al. (2016) Improvements in ecosystem services from investments in natural capital. *Science* 352: 1455–1459.

Page, T. (1977) *Conservation and Economic Efficiency.* Baltimore, MD: Johns Hopkins University Press.

Pagiola, S. (2008) Payments for environmental services in Costa Rica. *Ecological Economics* 65 (4): 712–724.

Panayotou, T. (1994) Conservation of biodiversity and economic development: the concept of transferable development rights. *Environmental and Resource Economics* 4: 91–110.

Piketty, T. (2014) *Capital in the 21st Century.* Cambridge, MA: Harvard University Press.

Polasky, S., Bryant, B., Hawthorne, P. et al. (2015) Inclusive wealth as a metric of sustainable development. *Annual Review of Environment and Resources* 40: 445–466.

Polasky, S., Costello, C. & Solow, A. (2005) The economics of biodiversity conservation. In: *The Handbook of Environmental Economics* (ed. J. Vincent & K.-G. Mäler), 1517–1560. Amsterdam: Elsevier North Holland.

Polasky, S. & Doremus, H. (1998) When the truth hurts: endangered species policy on private land with imperfect information. *Journal of Environmental Economics and Management* 35 (1): 22–47.

Polasky, S., Johnson, K., Keeler, B. et al. (2012) Are investments to promote biodiversity conservation and ecosystem services aligned? *Oxford Review of Economic Policy* 28 (1): 139–163.

Polasky, S., Kling, C.L., Levin, S.A. et al. (2019) The role of economics in analyzing the environment and sustainable development. *Proceedings of the National Academy of Sciences of the United States of America* 116 (12): 5233–5238.

Polasky, S., Nelson, E., Camm, J. et al. (2008) Where to put things? Spatial land management to sustain biodiversity and economic returns. *Biological Conservation* 141 (6): 1505–1524.

Polasky, S., Nelson, E., Pennington, D. et al. (2011) The impact of land-use change on ecosystem services, biodiversity and returns to landowners: a case study in the State of Minnesota. *Environmental and Resource Economics* 48 (2): 219–242.

Polasky, S. & Segerson, K. (2009) Integrating ecology and economics in the study of ecosystem services: some lessons learned. *Annual Review of Resource Economics* 1: 409–434.

Portney, P.R. (1994) The contingent valuation debate: why economists should care. *Journal of Economic Perspectives* 8: 3–18.

Ramsey, F.P. (1928) A mathematical theory of savings. *Economic Journal* 38: 543–559.

Raworth, K. (2017) *Doughnut Economics: Seven Ways to Think Like a 21st Century Economist.* White River Junction, VT: Chelsea Green.

Redford, K.H. & Adams, W.M. (2009) Payment for ecosystem services and the challenge of saving nature. *Conservation Biology* 23: 785–787.

Reyers, B., Polasky, S., Tallis, H. et al. (2012) Finding common ground for biodiversity and ecosystem services. *BioScience* 62: 503–507.

Richardson, L. & Loomis, J. (2009) The total economic value of threatened, endangered and rare species: an updated meta-analysis. *Ecological Economics* 68: 1535–1548.

Ricketts, T.H., Daily, G.C., Ehrlich, P.R. et al. (2004) Economic value of tropical forest to coffee production. *Proceedings of the National Academy of Sciences of the United States of America* 101 (34): 12579–12582.

Rockström, J., Steffen, W., Noone, K. et al. (2009) A safe operating space for humanity. *Nature* 461: 472–475.

Rolston, H., III. (1988) *Environmental Ethics: Duties to and Values in the Natural World.* Philadelphia, PA: Temple University Press.

Roughgarden, J. (2001) Guide to diplomatic relations with economists. *Bulletin of the Ecological Society of America* 82: 85–88.

Sabin, P. (2013) *The Bet: Paul Ehrlich, Julian Simon, and Our Gamble over Earth's Future.* New Haven, CT: Yale University Press.

Sagoff, M. (1988) *The Economy of the Earth: Philosophy, Law, and the Environment.* New York: Cambridge University Press.

Salzman, J., Bennett, G., Carroll, N. et al. (2018) The global status and trends of payments for ecosystem services. *Nature Sustainability* 1: 136–144.

Sandel, M. (2012) *What Money Can't Buy: The Moral Limits of Markets.* New York: Farrar, Starus, and Giroux.

Sarkodie, S.A. & Strezov, V. (2019) A review on environmental Kuznets curve hypothesis using bibliometric and meta-analysis. *Science of the Total Environment* 649: 128–145.

Scheffer, M., Carpenter, S., Foley, J.A. et al. (2001) Catastrophic shifts in ecosystems. *Nature* 413: 591–596.

Scheffer, M., Carpenter, S.R., Lenton, T.M. et al. (2012) Anticipating critical transitions. *Science* 338: 344–348.

Schill, C., Anderies, J.M., Lindahl, T. et al. (2019) A more dynamic understanding of human behaviour in the Anthropocene. *Nature Sustainability* 2: 1075–1082.

Selden, T. & Song, D. (1994) Environmental quality and development: is there a Kuznets Curve for air pollution emissions? *Journal of Environmental Economics and Management* 27: 147–162.

Shabman, L. & Batie, S. (1978) The economic value of natural coastal wetlands: a critique. *Coastal Zone Management Journal* 4: 231–247.

Simon, J. (1981) *The Ultimate Resource*. Princeton, NJ: Princeton University Press.

Slovic, P. (1994) The construction of preference. *American Psychologist* 50 (5): 364–371.

Smith, A. (1937) *An Inquiry into the Nature & Causes of the Wealth of Nations*. New York: The Modern Library, Random House.

Smith, V.K. & Huang, J.-C. (1995) Can markets value air quality? A meta-analysis of hedonic property value models. *Journal of Political Economy* 103 (1): 209–227.

Solow, R.M. (1974) Intergenerational equity and exhaustible resources. *Review of Economic Studies* 41: Symposium on the Economics of Exhaustible Resources, 29–45.

Spash, C.L. (1993) Economics, ethics and long-term environmental damages. *Environmental Ethics* 15 (2): 117–132.

Spash, C.L. (2020) A tale of three paradigms: realising the revolutionary potential of ecological economics. *Ecological Economics* 169: 106518.

Steffens, W., Crutzen, P.J. & McNeill, J.R. (2007) The Anthropocene: are humans now overwhelming the great forces of nature. *AMBIO* 36 (8): 614–621.

Stern, N. (2007) *The Economics of Climate Change: The Stern Review*. Cambridge, UK: Cambridge University Press.

Stern, N. (2008) The economics of climate change. *American Economic Review* 98: 1–37.

Sterner, T. & Persson, U.M. (2008) An even sterner review: introducing relative prices into the discounting debate. *Review of Environmental Economics and Policy* 2 (1): 61–76.

Stiglitz, J. (1974) Growth with exhaustible natural resources: efficient and optimal growth paths. *Review of Economic Studies* 41: Symposium on the Economics of Exhaustible Resources, 123–137.

Tallis, H.M., Hawthorne, P.L., Polasky, S. et al. (2018) An attainable global vision for conservation and human well-being. *Frontiers in Ecology and the Environment* 16 (10): 563–570.

Tayleur, C., Balmford, A., Buchanan, G.M. et al. (2017) Global coverage of agricultural sustainability standards, and their role in conserving biodiversity. *Conservation Letters* 10 (5): 610–618.

TEEB (The Economics of Ecosystems & Biodiversity) (2010) *Mainstreaming the economics of nature: A synthesis of the approach, conclusions and recommendations of TEEB*. http://teebweb.org/publications/teeb-for/synthesis.

Thaler, R.H. & Sunstein, C.R. (2008) *Nudge: Improving Decisions about Health, Wealth and Happiness*. New York: Penguin Books.

Tierney, J. (1990) Betting the planet. *The New York Times Magazine*. December 2, 1990.

Tietenberg, T. & Lewis, L. (2018) *Environmental and Natural Resource Economics*, 11th ed. New York: Prentice Hall.

Tilman, D., Polasky, S. & Lehman, C. (2005) Diversity, productivity and temporal stability in the economies of humans and nature. *Journal of Environmental Economics and Management* 49 (3): 405–426.

Toman, M. (1998) Why not calculate the value of the world's ecosystem services and natural capital? *Ecological Economics* 25: 57–60.

Tversky, A. & Kahneman, D. (1981) The framing of decisions and the psychology of choice. *Science* 211: 453–458.

UNDP (United Nations Development Programme) (2011) *Human Development Report 2011: Sustainability and Equity: A Better Future For All.* New York: Palgrave Macmillan.

van Kooten, G.C. & Bulte, E.H. (2000) *The Economics of Nature: Managing Biological Assets.* Malden, MA: Blackwell Publishers.

Vermeij, G.J. (2004) *Nature: An Economic History.* Princeton, NJ: Princeton University Press.

Vitousek, P.M., Mooney, H.A., Lubchenco, J. et al. (1997) Human domination of earth's ecosystems. *Science* 277: 494–499.

Watson, K.B., Ricketts, T., Galford, G. et al. (2016) Economic valuation of flood mitigation services: the value of Otter Creek wetlands and floodplains to Middlebury, VT. *Ecological Economics* 130: 16–24.

Weitzman, M.L. (1974) Prices vs. quantities. *Review of Economic Studies* 41 (4): 477–491.

Weitzman, M.L. (1998) Why the far-distant future should be discounted at its lowest possible rate. *Journal of Environmental Economics and Management* 36: 201–208.

Wichman, C.J., Taylor, L.O. & von Hafen, R.H. (2016) Conservation policy: who responds to price and who responds to prescription? *Journal of Environmental Economics and Management* 79: 114–134.

Wilcove, D.S., Rothstein, D., Dubow, J. et al. (1998) Quantifying threats to imperiled species in the United States. *Bioscience* 48: 607–615.

World Bank (2011) *The Changing Wealth of Nations: Measuring Sustainable Development in the New Millennium.* Washington, DC: The World Bank.

World Bank (2018) *Poverty and Shared Prosperity 2018: Piecing Together the Poverty Puzzle.* Washington, DC: The World Bank.

World Bank (2019) *World GDP 2010 Constant Dollars.* https://data.worldbank.org/indicator/NY.GDP.MKTP.KD (accessed May 29, 2019).

World Inequality Lab (2018) *World Inequality Report 2018.* https://wir2018.wid.world.

# 5

# Human Geography and Conservation

*Ivan R. Scales and William M. Adams*

## 5.1  Defining Human Geography

*Almost everything that happens, happens somewhere. Knowing where something happens is critically important.*

(Longley et al. 2001, p. 2)

Geography's name derives from two words in ancient Greek—*geo* (Earth, land, or soil) and *grapho* (to write, draw, or describe), and ancient and modern geographers alike have understood their role as to describe the face of the Earth and, especially, to give an account of the things humanity has done to it and created on it. Whenever conservationists address human impacts in the biosphere, or conservation scientists study those impacts and their implications for nonhuman life on Earth, they draw consciously or unconsciously from ideas and ways of thinking that have long been important in geography.

One way to think about geography is that it is built around the study of space, as history is of time—and historical geography, of course, as the combination of both! Thus, *The Dictionary of Human Geography* defines geography as "[the study of] the ways in which space is involved in the operation and outcome of social and biophysical processes" (Gregory et al. 2009). This fundamentally interdisciplinary inheritance is often expressed in terms of "physical geography" (concerned with Earth surface processes) and "human geography" (concerned with social, economic, political, and cultural processes).

Physical geography includes a number of subfields. Geomorphology (the study of the shape of the Earth) is concerned with processes of weathering, erosion, and deposition and their outcomes in landforms. Other subfields include hydrology (the study of water movement), pedology (the study of soils), meteorology (the study of weather and climate), and biogeography (the study of the distribution of organisms and vegetation communities and particularly the study of past environments, for example, Quaternary paleoenvironments).

Human geography is the part of the discipline concerned with "the study of the spatial organization of human activity and peoples' relationship with their environments" (Knox & Marston 2004, p. vii). This definition captures the essential breadth of human geography, the diversity of subjects studied by human geographers, and the diversity of methods used to study them. Moreover, human geographers study society at a variety of spatial levels

*Conservation Social Science: Understanding People, Conserving Biodiversity*, First Edition.
Edited by Daniel C. Miller, Ivan R. Scales, and Michael B. Mascia.
© 2023 John Wiley & Sons Ltd. Published 2023 by John Wiley & Sons Ltd.

ranging from individuals and households to regions, nations, and the planet as a whole. Human geographers also draw on both **qualitative** and **quantitative** research methods.

Human geography is divided into many subfields (Table 5.1). Each subfield focuses on a particular aspect of human activity and has its own take on the most appropriate philosophical, **epistemological**, and **methodological** frame for its work (Cloke et al. 2005). These sub-fields of geography intersect and overlap extensively. Geographers like to argue about defini-tions and labels, but few like to be put into boxes by others, and interconnectedness is an important feature of geographical inquiry. Indeed, in response to geography's diversity, the question of what geography is or should be is debated with such intensity by geographers themselves that attempts to define the boundaries of the subject can be controversial and unsatisfactory. Some human geographers have worked within a **positivist** epistemology, as

**Table 5.1** Subfields in human geography.

| Subfield | Topics |
| --- | --- |
| Cultural and social geography | The geographical construction of identity and questions of subjectivity and belonging (e.g. relating to gender, race, class, sexuality, disability, and religion) |
| Development geography | The geography of economic, social, and cultural change in the world's poorer and less industrialized countries (e.g. international trade and financial flows, poverty, welfare and health, the impacts of capitalism and globalization on the poor, and sustainable development) |
| Economic geography | The spatial organization of economic activities (e.g. industrial location, transportation, uneven geographies of capitalism and accumulation, Fordism, globalization, core–periphery theory); the cultural economy and the culture of the firm |
| Historical geography | The geographies of places and environments in the past (e.g. the dynamics of place, space, and landscape); historiography and philosophy of historical geography; landscape, memory, and environment. |
| Feminist geography | Environment, society, and geographical space from a feminist perspective (e.g. geographic differences in gender relations, the construction of gender identities in space and place, the geography of women's lives or work, children's geographies, welfare geography, and geographies of sexuality) |
| Political geography | The relationships between space and power (e.g. territoriality, the state, nationalism, geopolitics, social movements, political economy, diaspora, elections, globalization, imperialism, governance, and peace, conflict, and security) |
| Population geography | The ways in which spatial variations in the distribution, composition, migration, and growth of populations are related to the nature of places (e.g. natality, mortality, population growth rates, demographic transitions, and migration) |
| Rural geography | Social, economic, cultural, and environmental processes in rural environments (e.g. rural settlement patterns, agriculture, and rural land use and change) |
| Urban geography | Social, economic, cultural, and environmental processes in built environments (e.g. urban planning, urban poverty, gentrification, urban segregation, and urban political ecology) |

physical geographers and other natural scientists do (see Chapter 2). This approach was particularly important in the 1960s, the heyday of geography as "spatial science" (see Section 5.1.1), and has experienced a resurgence through geographic information science (GISc) (see Section 5.3.3.3). Other human geographers have responded to ideas that influenced other social science disciplines, including Marxism (see Chapter 8), **post-structuralism** (Section 5.3.2.2), and posthumanism (Section 5.3.2.4). The resulting flowering of a theoretically intense human geography has created a rich and diverse discipline.

### 5.1.1 Major Themes and Questions in Human Geography

Each of human geography's subfields expresses the central questions of the discipline in different ways. However, at the core of the discipline lie fundamental questions about how we should understand space and place.

#### 5.1.1.1 Space

There are profound differences among human geographers in the ways they conceptualize space. A strong tradition in geography is the description and analysis of the locations of events and phenomena on the Earth's surface and near surface (Gregory et al. 2009). Therefore, at its simplest, geographers see space in terms of geometry—a three-dimensional grid in which objects and events can be located. Starting in the 1950s, such geometric notions of space were the basis of a systematic academic agenda for geography (see Section 5.2). This understanding of space has underpinned the development and practice of cartography from the days of European imperial expansion to contemporary Geographical Information Systems (GISs) (see Section 5.3.3.3). Increasingly, however, human geographers have focused on space as something **socially constructed**, in other words, perceived and imagined differently by individuals and social groups, as well as classified and divided by different political actors. According to this view, space cannot simply be thought of as something "out there"—a neutral stage on which people act. Space is bounded, defined, and made (or as many human geographers would say, "produced") through specific political and social processes. Many human geographers today focus on questions of power and knowledge that are fundamental to the way space is perceived, analyzed, partitioned, and controlled.

#### 5.1.1.2 Place

A second set of key questions center around the importance of place in shaping events. For geographers, it is a fundamental observation that *where* things happen is important; places are different from each other. Thus, despite having similar climates, southern Spain, the South African Cape, and California are different for a host of reasons (including geology, soils, past patterns of ecological change, history, economy, **culture**, and politics). Such differences are important at a range of scales, from the subcontinental down to the street corner or the forest glade. Therefore, it has been one of geography's chief concerns to explain and explore the difference between space and place.

There are simple and more complex ways to think about place. At its simplest, place can be defined as a portion of geographic space (Gregory et al. 2009). Geography has a long tradition of identifying discrete spatial units (notably the region, a staple concept in geography for at least the first half of the twentieth century) and analyzing the processes bounding them (Figure 5.1). However, there is nothing predefined about place: regions,

**Figure 5.1** Grains Gill, Borrowdale, Lake District National Park, UK. The Lake District, a small area of low mountains and lakes in northwest England, has been famous for its landscape beauty for 200 years. Romanticism (epitomized by William Wordsworth and John Ruskin) celebrated the wonder of wild nature. It became a national park in the 1950s, the second created in the UK after World War II. The idea of a "Lake District"–that it has a characteristic landscape and that this landscape is of such value that it deserves protection as a park for British citizens–demonstrates the complex ways in which space and place are culturally and socially constructed. Photo: Bill Adams

landscapes, locales, *pays* are all creations of the analyst, whether geographer or local citizen. Place, like space, is socially constructed, and everything from the cultural "sense of place" to the names on maps or the delimitation of territorial jurisdiction are both social and political.

## 5.1.2 Human Geography and Conservation

Geographical debates about space and place are fundamental to conservation. The vocabulary of conservation is full of spatial terms (e.g. protected areas, ecoregions, **biodiversity** hotspots, buffer zones, wildlife corridors). Conservationists commonly draw on spatial tools to map biodiversity, identify threats to habitats and species, and prioritize conservation action. Other conservation questions relate to the **social construction** of space and place, for example, the way a state determines whether certain areas of habitat are deemed to be conserved as "wilderness," or the power of different kinds of knowledge (e.g. scientific or indigenous) to influence conservation outcomes. Sections 5.1.2.1–5.1.2.3 provide a flavor of some of the questions important to conservation that geography could help answer.

### 5.1.2.1 Ideas of Nature and Place

1) Concern for the environment and support for conservation varies between different places—why is this? How do attitudes toward conservation differ between **industrialized** and non-industrialized countries, or in rapidly urbanizing societies, or with religious beliefs, or age and gender?

2) How do different people understand nature and how do these understandings influence the way nature is used? For example, how do artisanal fishers, or bushmeat hunters, or pastoralists think about the **ecosystems** they use? How do political and corporate leaders think about ecosystem services?

### 5.1.2.2 Nature and Society

1) What factors determine where human impacts on natural habitat are most significant, and what causes these impacts to grow, persist, or decline? For example, why does water pollution occur in some places and not others, and why is it not cleaned up? Why are some forests clear-felled instead of selectively cut?
2) Where, why, and how can economic growth contribute to sustained improvements in human well-being without environmental degradation? For example, under what conditions can societies reduce their carbon footprint or their rate of use of declining stocks of renewable resources (e.g. fish or forests)?
3) What is the relationship between wealth and environmental degradation? For example, how does national income relate to ecological footprint? Does poverty drive certain forms of environmental degradation? Is it possible for human welfare to rise while demands on living biodiversity remain stable?

### 5.1.2.3 Ideas about the Control of Nature

1) How does the way nature is classified and measured affect its conservation? For example, how might scientific and indigenous ideas about nature differ, and how far can conservation planning take account of differences? To what extent do protected area systems reflect past ideas about conservation and past values surrounding nature?
2) Where is the social impact of conservation likely to be positive and where negative, and why is this? To what extent can participatory planning of protected areas reduce negative impacts? What are the impacts of the displacement of human populations for conservation and how can they be minimized?
3) How do different political and economic systems affect biodiversity? For example, under what conditions might putting a price on biodiversity (e.g. through payments for **ecosystem services**) lead to effective conservation?

## 5.2 A Brief History of Geography

The history of geography as a discipline is complicated. Geographers have an intense concern for the shape, boundaries, and the "soul" of their discipline. The result is not a history of progressive advance or even neat and agreed paradigm shifts. Rather, the history of geography is a series of struggles among diverse epistemological, theoretical, and methodological debates.

As a field of learned inquiry, geography dates back to the nineteenth century, although cartographers (makers of maps and charts) would claim the heritage of classical Greece for their (and geography's) origins. The Royal Geographical Society was established in London in 1830 as a club for travelers and explorers, supported by wealthy individuals and made intellectually respectable by scientists (Stoddart 1986). It provided a natural home for reports of exploration and endeavor in remote regions (Figure 5.2): journeys of discovery and evangelism (Hudson 1977; Driver 2001). Geographers joined with geologists, astronomists, zoologists, and entomologists in an explosion of leisured and learned natural history in metropolitan London. Many leading field scientists were elected as fellows of the Royal Geographical Society (Charles Darwin, for example, in 1838, on his return from the voyage of the *Beagle*). Geography societies were established in New York in 1852, Edinburgh in 1884, and in Paris and Berlin in the 1920s. Cartography was a vital technology of European colonial expansion, and geography as a discipline prospered as it met imperialism's insatiable thirst for knowledge (Livingstone & Withers 1999).

As the natural sciences specialized and became more tightly organized in the second half of the nineteenth century, developing laboratory and experimental methods, geography stuck to its concern with fieldwork. In the UK, Huxley's (1877) *Physiography* set out a bold agenda for an integrated geography within the Earth and physical sciences. His lectures

**Figure 5.2** Frank Debenham, British Antarctic expedition, 1910–13. Frank Debenham (1883–1965) was a member of Captain Scott's British Antarctic Expedition (1910–13). The death of the polar party, beaten to the South Pole by Amundson's expedition, made Scott the icon of British imperial expeditionary endeavor, and also its end point. Debenham was the first director of the Scott Polar Research Institute from 1923, and in 1931 became the first Professor of Geography at the University of Cambridge. He stands by a plane table, the simple device by which the empire was mapped and made legible to colonial administrators. Photo: Herbert Ponting (National Library of New Zealand) / Wikimedia Commons / CC0 1.0.

famously started from concrete objects (e.g. a lump of coal or chalk) to develop broad accounts of the whole Earth, and indeed the solar system (Stoddart 1986). Geography still maintains natural history's emphasis on fieldwork in both teaching and research.

Geography was firmly established as a university subject in a wider range of industrialized countries by the 1930s. However, it was not until the 1950s and 1960s that serious attempts were made to define the subject beyond its self-appointed (and self-justifying) role of describing the diversity of the world. In Anglophone geography, Hartshorne's *Perspective on the Nature of Geography* (1959) set out a serious academic agenda for geography as an essentially spatial discipline, using maps to show how human and physical phenomena vary over the Earth's surface. The classic concept of *region* as "a more or less bounded area possessing some sort of unity or organizing principle(s) that distinguish it from other regions" (Gregory et al. 2009) allowed geographers to rationalize the Earth's diversity in terms of physical environments and human activities: geology, hydrology, soil, and climate set a consistent physical frame for understanding the growth of settlement, transport, economy, and society.

In the 1950s, geography underwent a "quantitative revolution" in epistemology and methods, initially in the USA, and a decade later in the UK. This reflected the desire of some in the discipline to move away from an ideographic approach (based on the description of the unique characteristics of particular regions) to a nomothetic approach (based on the search for general laws and the creation of models of physical and social processes). Physical geographers began to draw on ideas from engineering and cybernetics to think in terms of systems, measure landscapes instead of describing their evolution, and measure physical processes directly (Chorley & Kennedy 1971).

Meanwhile, human geography was reinvented as a spatial science "concerned with the formulation of the laws governing the spatial distribution of certain features on the surface of the earth" (Schaefer 1953, p. 227). The geographer's job was now to "describe spatial arrangements and associations of activities and processes in geographical space" (Nystuen 1963, p. 373), and a concerted effort was made to place quantification at the heart of geographical inquiry. Quantitative human geography drew ideas from a range of other disciplines to construct models of the spatial organization of human societies. Abstract geometric ideas about the distribution of human settlements in space, such as Von Thünen's ([1826]1966) model of agricultural land use and Christaller's (1933) Central Place Theory of settlement distribution (Figure 5.3), replaced previous descriptive approaches.

In the 1970s, quantitative geographers were quick to exploit newly developed computer hardware and software to support the statistical analysis of society. This so-called quantitative revolution did not introduce quantification to geography (there was already a long tradition of measuring phenomena and analyzing them statistically), but it did attempt to place scientific analysis at the center of geographic theory and method. Such work continues to prosper (e.g. in the analysis of disease or demography), using various forms of spatial modeling techniques and underpinned by GISc (see Section 5.3.3.3).

From the 1970s, radical human geographers began to contest what they saw as the aridity and political blindness of quantitative approaches. While statistical approaches offered powerful analytical sophistication in describing aggregate behavior, critics felt such analyses did not offer theoretical understanding or explain why people made certain decisions. Quantitative analysis *described* but could not *explain* human behavior. Critics

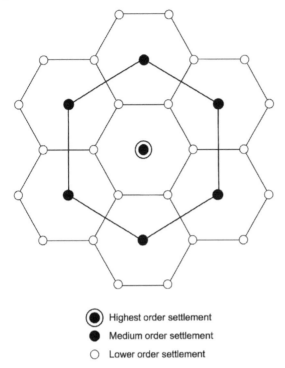

Highest order settlement
Medium order settlement
Lower order settlement

**Figure 5.3** Christaller's (1933) Central Place Theory, an example of abstract geometric models in geography. Christaller sought to explain the number, size, and distribution of settlements in a landscape (assuming a flat and homogenous surface). According to this model, the main purpose of a settlement is to provide goods and services to consumers, and this determines how settlements are distributed. Higher-order settlements (cities) provide a wider range of goods and services than lower-order settlements (villages), while lower-order settlements provide goods and services that are purchased more often.

argued that in trying to reduce social processes to abstract models, quantitative human geography risked being overly reductionist. Moreover, the focus on abstract models seemed to depoliticize human geography and lessen its relevance to real-world social and environment issues such as civil rights, poverty, environmental pollution, nuclear proliferation, and war. Many human geographers demanded new and more radical agendas that demonstrated a social relevance that quantitative methods could not provide.

The new approaches to human geography that emerged in the 1970s were accompanied by deepening debate about epistemological and **ideological** issues. Marxism promised a geography that was relevant to and engaged directly with political issues (Harvey 1973; Gregory 1978). The quantitative geography of the 1950s and 1960s had provided little insight into the processes of economic change that created inequality, and the "laws" of spatial science were simply descriptions of industrial **capitalism**. Radical geographers turned to the analysis of the geographical context of the operation of global capitalism, parallel to advances in fields such as **development** studies (Section 5.3.1.2). Through Marxian theory, human geographers became increasingly sensitive to the social (and later environmental) impacts of capitalism, from local to global scales.

In the 1970s, human geography began an in-depth engagement with the humanities (rather than the quantitative social sciences), a turn that saw the subject engaging with a wide range of philosophical traditions. These changes paralleled those in other social sciences. Humanistic approaches put people and human subjectivity at the center of geographical inquiry rather than, for example, simply analyzing the patterns that human activities create (Gregory 1994). In the 1980s, human geographers began to draw on **postmodern** social theory and post-structuralist thought (see Section 5.3.2.2). The works of European thinkers such as Jacques Derrida and Michel Foucault were read and quoted avidly by geographers as they sought further transformations of the subject, overturning both the quantitative and Marxian traditions that had dominated human geography for the previous two decades in favor of new, cultural geographies.

In the 1990s, geographers extended their understanding of space in particular through an engagement in the ideas of the French philosopher Henri Lefebvre. Contrary to the view of the quantitative geographers of the 1950s and 1960s, Lefebvre (1991) argued that space is not simply an empty, neutral pre-existing matrix in which events occur. Instead, he argued that there are different types of space. First, there is *perceived* space. Different people perceive the same space (a street, a park, a landscape) differently because they are influenced by factors such as their culture, gender, or age. Second, there is *conceived* space, where spaces are created "in theory" in the minds of experts such as cartographers, architects, and urban planners for a particular purpose: a factory designed to maximize the efficiency of workers, or a palace designed to project an image of power. Finally, Lefebvre wrote of *lived* space, in other words as the context for people's daily lives. People walk on sidewalks, drive down roads, meet people in restaurants, and shop in malls. But the way they end up using a particular space does not necessarily match what the space was conceived for. Space must therefore be understood as *produced* by the interactions between human perceptions, conceptions, and daily lives. Spaces both shape and are shaped by social relations and, as a result, rather than space being universal, every society produces its own space, be it the shopping mall of Western consumer society or the rice fields of the Ganges delta.

By the twenty-first century, human geography had been through successive waves of theoretical revolution, each seeking to transform the subject utterly. None had been completely successful, leaving a theoretically rich and internally argumentative discipline of great diversity and dynamism. Books such as the *Dictionary of Human Geography*, first published in 1981 and running to over 1000 pages by its fifth edition in 2009 (Gregory et al. 2009), and the 15-volume *International Encyclopedia of Geography* (Richardson et al. 2016), offer a chart of the complex past evolution, present dilemmas, and future prospects in the subject.

## 5.3 Concepts, Approaches, and Areas of Inquiry in Geography

As will now be clear, it is far from easy to bound geography as a discipline. David Livingstone (1992), historian of geography, suggests the subject should be understood as a tradition that evolves like a species over time. As a discipline whose history is tied to maps, routes, and the translation of distant complexities to neat generalizations, geography is diverse, energetic, and often turbulent. Geographers' long-standing interest in society and nature, and the determined transdisciplinarity of much geographical research on this issue, has much to offer those interested in the real-world challenges of conservation. Three broad areas of inquiry are particularly relevant to **conservation social science**: (i) environment and society; (ii) social nature; and (iii) cartography and power.

## 5.3.1 Environment and Society

In 1798, Malthus published *An Essay on the Principle of Population*. His basic argument was that human societies are underpinned by two "powers" (forces of nature): the power of population and the power of the land to provide food for it. According to Malthus, the "passion between the sexes" (the inability of humans to limit their reproductive behavior) meant that human populations would always outpace the ability of societies to produce enough food. Thus, famine, poverty, and war were inevitable.

Ever since it was published, Malthus' essay has divided opinions. Some see proof that Malthus was right in images of famine in sub-Saharan Africa, or in the latest reports of dwindling fish stocks in the Atlantic. Others retort that 7.9 billion people on Earth (and counting) is surely testament to humanity's talent for confounding Malthusian theories. Malthus certainly underestimated humanity's ability to transform ecosystems to suit its needs. For thousands of years, humans have modified their surrounding environments, and their environmental "footprint" has been considerable. Geographers have mapped and quantified this footprint in a variety of ways and have also shed light on the limits and constraints placed on societies by different environments.

### 5.3.1.1 Assessing Human Impacts

In his classic volume *Man and Nature*, Marsh (1864, p. 36) wrote emphatically about the destructive potential of human action, stating that "man is everywhere a disturbing agent. Wherever he plants his foot, the harmonies of nature are turned to discords." This view remains central to conservation, and considerable time and effort has been spent mapping and quantifying the extent of human impacts on the biosphere (e.g. IPBES 2019). Historical geography has offered detailed accounts of the human shaping of landscapes and ecosystems, such as Darby's (1977) work on the historical geography of England or Williams's (2003) study of historical change in global forest cover.

Geography has long played a role in understanding and mapping the scale of human transformations of the Earth's surface (e.g. Thomas 1956), a significant concern in contemporary conservation. Only relatively recently, through remote sensing, has it been possible to measure land-cover change easily and cheaply. Remote sensing can be defined as "recordings of the reflectance of visual and nonvisual wavelengths of energy from the land to an airborne camera or digital wave band sensor" (Turner 2003, p. 260). While multispectral remote sensing data can be acquired from airplanes and drones, the most commonly used remotely sensed data have been from satellites (e.g. those of the National Aeronautics and Space Administration's (NASA's) Landsat program).

Multispectral sensors record reflectance values from the Earth's surface in different bands, which correspond to different parts of the electromagnetic spectrum. Different materials (e.g. water, bare soil, grass, trees, concrete) absorb and reflect different parts of the electromagnetic spectrum, meaning that each type of land cover has its own distinct spectral signature.

Measurements in different parts of the electromagnetic spectrum allow researchers to distinguish between different types of land cover. Satellites in orbit pass over each point on the Earth's surface at regular intervals, allowing repeated images to be used for change detection.

Sensors vary depending on the type of satellite, which means land cover can be measured in many ways using different electromagnetic wavelengths and at different spatial and temporal scales. Over time, the spatial resolution of sensors (the size of objects they can detect) has improved, as well as their spectral resolution (their ability to distinguish between different wavelengths), allowing finer distinctions to be made between different objects and land-cover types.

Using different types of sensors and different analytical algorithms, remote sensing allows the measurement of various aspects of the Earth's surface, including soil type, vegetation chlorophyll, leaf and canopy properties, and temperature. Remote sensing is therefore a powerful tool for measuring changes in habitat and, given the scale of environmental problems and the difficulty of accessing some regions, researchers are increasingly turning to remote sensing. Over the last 20 years, satellite imagery has allowed researchers to measure changes in forest cover (Figure 5.4), track wildlife (some satellites have a spatial resolution of less than 50 cm, making it possible to identify large mammals such as elephants) and assess the health of coral reefs (degraded reefs are a different color and therefore have different spectral signatures). Of course, there are many problems with multispectral satellite data: cloud cover hides the Earth's surface, steep topography causes problems of shadow and shade, and many important features are hard to distinguish because of their

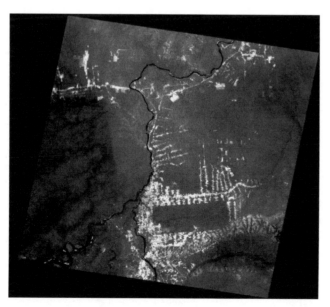

**Figure 5.4** Deforestation in Rondônia, Brazil. This Landsat image (Path 233, Row 67 with the center of the image at latitude 10°07′27″S, longitude 65°13′07″W) was taken in 1999. It shows a distinctive fishbone pattern of deforestation, the result of forest clearance following road networks as also found in many other places around the world. Landsat imagery courtesy of NASA Goddard Space Flight Center and US Geological Survey.

small size compared to the pixels that comprise the satellite image, or because ecological boundaries overlap pixels, creating mixed reflectance signals.

Although tools such as remote sensing satellites are powerful, they have their limits. Airborne reconnaissance using stereo photographs was developed for military purposes during World War II, but serious attempts to photograph Earth from the air did not start until the 1950s. In the 1960s, monochrome air photographs were used to produce maps of vegetation cover, settlements, land systems, land capability, and land use in a number of countries. Many of these images still exist as photographic negatives and prints rather than digital data, although many are buried in archives, unused by today's computer-reliant analysts. The first satellite images of Earth's orbit were produced in 1959, but satellite imagery only became available for civilian use in the 1970s, and has been widely and cheaply available only since the beginning of the twenty-first century (e.g. through Google Earth). Even now, satellite coverage of some remote, politically sensitive, and cloud-prone regions of the Earth remains poor.

One of the dangers of relying solely on remote sensing is that focusing on narrow time frames provides misleading snapshots of more complex and dynamic processes and infers long-term trends from short-term observations. In order to extend the historical depth of environmental knowledge, some geographers have drawn on other data sources. Historical geography, and the related field of environmental history, draws on a diverse range of material to gain a deeper understanding of human relationships with the environment over time. These can include oral histories, archives, written accounts, and maps. For example, in Madagascar (Kull 2000; Scales 2011) and West Africa (Leach & Fairhead 2002), such data sources have been used to challenge commonly cited estimates of forest loss and provide a better understanding of the drivers of deforestation (see also Box 5.1). Nonconventional historical data sources can shed important light on modern conservation problems by providing us with a longer view of natural resource use.

The changes in land cover revealed through remote sensing are one of the many ways in which humans have transformed the planet. Human impacts are now so substantial that some environmental scientists suggest that humans have entered a new era, the **Anthropocene**. Starting with the industrial revolution in the eighteenth century, human activity, through industrialization, urbanization, and agricultural intensification, has radically transformed the Earth (Crutzen & Stoermer 2000; Malhi 2017). Geographers have coined the term "anthromes" to describe and map **anthropogenic** biomes, defined as the ecological patterns created by the sustained interactions between humans and ecosystems (E. Ellis & Ramankutty 2008). Anthromes include human settlements (from villages to cities), cropland, grazing land, as well as seminatural areas that are inhabited but with minor land use (e.g. many forest regions). Wildlands (defined as land without human population or substantial land use) now make up less than half of the terrestrial biosphere and only 22% of global ice-free land area, with most of the Earth's land not currently in agricultural use embedded within anthromes (E. Ellis et al. 2010).

From these facts, proponents argue that conservation, which has traditionally been based on ideas of "wilderness," should focus on anthromes. Accepting the implications of the Anthropocene, and a planet whose surface is dominated by anthromes, has significant implications for the way conservation is thought about and carried out, since it challenges both assumptions about pristine nature and the traditional model of fortress conservation (see section 5.3.3.2), and calls for more attention to be paid to managed landscapes.

### 5.3.1.2 Development and Change

A major area of research in human geography concerns the ideas and practices of development (Forsyth 2005; Potter et al. 2008), an underlying driver of the Anthropocene. The word *development* was used in eighteenth-century English to convey "organic" ideas of unfolding change and growth (Crush 1995). By the nineteenth century, the word was being used to describe a process of linear progress under capitalism and Western cultural hegemony, advanced through imperialism (Cowen & Shenton 1996). The concept of development as it is now commonly understood came to prominence after World War II in a period of American political hegemony and the dismantling of European empires, following the rebuilding of a Europe left physically and economically devastated by the war (Crush 1995). Famously, US President Harry Truman called for a program of development based on the concepts of "democratic fair dealing" in his inaugural address to the US Senate in January 1949 (Escobar 1995).

So, what is meant by development? At one level, it means progress toward a goal. At first glance, this might seem obvious. However, the idea of what the goal of development should be, and the path toward it, is underpinned by powerful ideologies. Development in the twenty-first century cannot be separated from the Western idea of modernity that lies behind it (Escobar 2004). In the post-World War II world, the meaning of development soon narrowed to economic growth and modernization based on the Western model and led by the state. The purpose of development was taken to be the replication of the experiences of high-income countries across the world, classically conceived in terms of Rostow's (1960) *The Stages of Economic Growth* (Figure 5.5) from traditional society, preconditions for take-off, take-off,

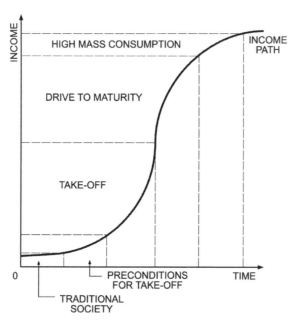

**Figure 5.5**  Rostow's (1960) Stages of Economic Growth. According to Rostow's model societies pass through a set of distinct economic stages. Rostow argued that for economies to become developed nations needed to create certain conditions for "take off." These included investing in industrialization and urbanization and becoming more integrated into the global capitalist economy.

maturity, and the age of high mass consumption. According to this model, development planning was the organized and coherent attempt to overcome constraints on economic growth through processes of industrialization, urbanization, democracy, and capitalism.

Attempts to measure development have tended to focus on quantitative economic indices, such as gross domestic product (GDP)—the financial value of all goods and services produced by an economy in a given time period—ignoring other dimensions such as social and natural capital. The conventional Western model of development has been widely challenged by a range of critics in the field of development studies. Empirical critiques have focused on the inadequacy of economic measures in capturing the multidimensional nature of poverty. The argument is that poverty is not simply about not having enough money, but also about a lack of access to healthcare, education, food, and clean water. In an effort to broaden the scope of development, more diverse indices have been developed, such as the Human Development Index (UNDP 2020). Economist Amartya Sen, in *Development As Freedom* (1999), has gone further, arguing that development is ultimately about having the freedom to choose different ways of thinking and living.

As early as the United Nations First Development Decade (1960–70), concerns about the failures of the global development project began to emerge and the gap between expectation and reality had become clear. There has since been endless debate in development studies about the nature of development and reasons it does or does not occur. Radical writers, drawing on world systems theory and Marxist theory (see Chapter 8 for more on both theories), have critiqued orthodox development for reinforcing unequal power relationships between the "developed" and "developing" worlds. They argue that as long as the global economy is dominated by Western economies, and development continues to be what is done *to* the developing world rather than *by* them, both global capitalism and development will work in favor of the rich. According to this view, development continues to fail precisely *because* it operates within a global economic system designed so that a small group of people is able to get rich at the expense of the world's poor, who are forever doomed to supply the raw materials and manual labor that fuel capitalism.

Other critics, drawing on postcolonial theory, argue that development is in fact a legacy of colonialism, replicating the same unequal power relations and underpinned by similar processes of cultural imperialism, where one group attempts to impose its own cultural values on another (Said 1979; Crush 1995). According to this view, Rostow's *Stages of Economic Growth* is not a model for solving poverty but a narrow and selective reading of European and American history, based on a belief in the cultural superiority of the West.

Since the 1980s, conservation has increasingly been involved with poverty alleviation and development. Over the last 30 years, researchers and practitioners have recognized that conservation needs to move beyond just the protection of nature through protected areas to considering the livelihoods of people living in and around them, and the impacts of poverty and development on natural resources (Adams 2020). There has also been growing awareness that conservation policy has often had considerable impact on rural livelihoods in the Global South (e.g. by restricting access to key resources). Engagement by conservation planners in poverty and development has been strongly focused on the local scale, and ranges from simply consulting communities living in proximity to protected areas to strategies of fully integrated conservation and development that hand over the management of resources to local communities (Hulme & Murphree 2001). Such approaches draw on the seductive, and sometimes problematic, idea of community (see Chapter 3).

More recently, conservation thinking about development has been heavily influenced by the **neoliberal** economic ideology of international financial institutions such as the World Bank and the International Monetary Fund. While orthodox development tended to work through state government institutions (see Chapter 6), this "new policy agenda" (Robinson 1994) increasingly emphasized the local level, **civil society**, and democracy. The neoliberal emphasis on free markets, strong property rights, and minimal state involvement has proved attractive to policy makers attempting to make conservation and development work together to protect biodiversity and reduce poverty. Perhaps conservation can finally be made to pay for itself, with nature valued in monetary terms as an ecosystem service? (See Chapter 4.) However, the influence of neoliberalism on conservation has been widely noted and critiqued by geographers as well as other social scientists (McCarthy & Prudham 2004; Igoe & Brockington 2007). Market-based incentives for conservation will be briefly explored in Section 5.3.2, and questions of neoliberal politics, the role of the state, and different forms of governance are discussed in greater depth in Chapter 6.

### 5.3.1.3 An Industrial and Urban Earth

More than half the world's population live in urban areas. This concentration of demand for natural resources, and the products and by-products of industry, has become critical to the drastic environmental transformations of the Anthropocene. Much conservation research and policy has focused on sparsely populated rural areas. Industrialized environments, urban consumption, and urban environments also need to be understood and their problems addressed by conservation planners. Important questions for conservation research and policy include (i) What impact (both positive and negative) does urbanization have on different species and ecosystems? and (ii) How can urban environments be planned to reduce biodiversity impacts and create green spaces?

Marxist analysis suggests that capitalism is incapable of dealing fully with the ecological problems it creates (Harvey 1996). One of the critical engines that has driven a significant proportion of the development/modernization process in the twentieth and early twenty-first centuries is capitalism's endless quest for profit and economic growth, which pushes firms to search for cheaper raw materials, cheaper labor, and new markets. The size and growing internationalization of corporations have restricted the capacity of national governments to regulate industries from mining to agribusiness.

The late twentieth-century transformation of industrial production in North America and Europe—involving de-skilling, automation, longer and more flexible working hours, de-unionization, and loss of job security—was accompanied by the flight of manufacturing to low-income countries in search of lower production costs, greater profits, and sustained returns for shareholders (mostly in richer countries). Relocation to countries with low wage rates, weak labor and environmental laws, and poor enforcement of those laws made sense in terms of global business strategies of maximizing returns on investment.

The profit-and-growth motive at the heart of capitalism drives much deforestation (for beef, soya, or palm oil), overfishing (on the unregulated high seas), and large-scale mineral extraction, and explains the persistence of polluting industries and sweatshops in poor countries. There is a logic to the spatial distribution of the negative environmental impacts of industrialization between high-income and low-income countries. In economic terms, it is logical to locate potentially polluting industries in countries where regulation of pollution is weak (whether because laws are not there, or are not enforced, or the capacity to measure pollution is limited), where production and clean-up costs are low, and where the poor are willing/forced to accept low levels of compensation (Low & Gleeson 1998).

The world's urban population has grown rapidly since 1950, having increased from 751 million to 4.2 billion in 2018, and is predicted to grow by 2.5 billion by 2050, with almost 90% of this growth happening in Asia and Africa (UN DESA 2019). The majority of the urban population in low-income countries lives in poverty, in makeshift informal settlements clustered around the peripheries of huge urban agglomerations, marginalized in space and beyond the reach of the formal state in terms of the provision of services such as water (Figure 5.6), waste management, and transport. Slum dwellers carry a triple health burden, debilitated by malnutrition, chronic and epidemic infectious disease, and exposed to the hazards of industrial pollution (Davis 2006).

The distribution of people across a city is the result of the workings of the urban economy and the circulation of capital (Harvey 1973). The locations of slums and industrial plants, the occurrence of pollution, and the lack of water supply reflect political, economic, and social forces. Richer and more powerful residents move to safer, cleaner, and more spacious areas, while those without a piped water supply are forced to pay high prices to water vendors for water of dubious quality (Swyngedouw 2004). These forces also work in rich countries. The environmental justice movement in the USA arose as a response to corporate and municipal decisions that located noxious facilities such as toxic waste dumps in the neighborhoods of minorities, native people, and people of color (Bullard 1990). The Appalachian open-pit coal industry also distributes human and environmental costs

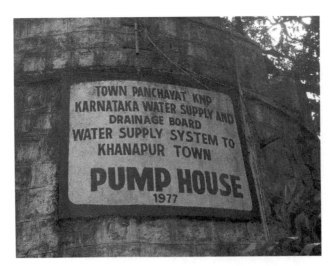

**Figure 5.6** Pump and water tank, Khanapur, Malaprabha basin, India. The supply of domestic water for drinking, cooking, and washing is a major cost for municipalities, and has in many places been subject to privatization. The availability and quality of water supply are major factors in urban poverty, and an important element in the United Nations' Sustainable Development Goals. Justice and equity are central issues in the politics of water supply. Photo: Bill Adams.

unequally, raising clear issues of injustice as well as environmental degradation (Bell 2016). For radical geographers, the **political economy** of capitalism (Section 5.3.2.1) and the distribution of human impacts on the environment are opposite sides of the same coin.

### 5.3.1.4 Beyond Malthus and beyond Determinism

Over the last 50 years, Malthus' ideas of limits to population growth have re-emerged. From Ehrlich's (1968) *The Population Bomb* to recent discussions about an optimum population and zero population growth, the idea that environmental limits and rapid population growth have catastrophic consequences (both for human populations and the planet's ecosystems) has played an important role in discussions of environmental policy.

In the early decades of the twentieth century, the idea of environmental determinism (that natural conditions, particularly climate, determine what human society could achieve) were widely accepted. Huntington's *Civilization and Climate* (1915) made popular the idea that climate shapes civilization and explained to Europeans and Americans why the temperate regions of the Earth fostered civilization and the tropics forbade it. Geographers have long abandoned such crude deterministic analyses, instead paying close attention to the role of political economy (Section 5.3.2.1), culture, and society in shaping social conditions. However, simplistic ideas about geographical limits and causes continue to maintain a hold in popular writing. The work of Jared Diamond typifies a recent wave of neo-environmental determinism, where biological models of human culture rule (Radcliffe et al. 2010). For example, in *Guns, Germs and Steel*, Diamond (1997) seeks to explain why European and Asian civilizations conquered others, arguing that the differences in technology that allowed this to occur were largely the result of environmental differences. His main thesis is that the east–west alignment of the Eurasian continent allowed the diffusion of people, ideas, goods, crops, and domesticated livestock along similar latitudes (with comparably favorable environmental conditions). Diamond argues that the north–south orientation of the African and American continents hindered such flows by creating environmental barriers, for example, the central belts of rainforest (and the zoonotic diseases they harbor) stopping the movement of domesticated animals.

In *Collapse*, Diamond (2005) analyzes why some human societies flourish, while others seemingly destroy their environments and disappear. A recurring theme is overpopulation and resource overexploitation. Although Diamond goes to considerable effort to stress that there is not necessarily a link between population growth, resource scarcity, conflict, and collapse, he shows a clear willingness to return to Malthusian theory and environmental determinism to provide an ultimate cause for societal collapse, for example, as a factor in the buildup to genocide in Rwanda in 1994 and the deforestation of Easter Island. He writes (Diamond 2005, p. 509):

> One of the main lessons to be learned from the collapses of the Maya, Anasazi, Easter Islanders, and those other past societies … is that a society's steep decline may begin only a decade or two after the society reaches its peak numbers, wealth, and power …. The reason is simple: maximum population, wealth, resource consumption, and waste production mean maximum environmental impact, approaching the limit where impact outstrips resources.

Human geographers, as well as anthropologists and archaeologists (e.g. Morris 2005), have questioned Diamond's explanations of both the successes and failures of societies, which tend to ignore or downplay the role and legacies of colonialism, imperialism, and economic exploitation in both development and environmental degradation (see also world-systems theory in Chapter 8). There is now a large body of work challenging simplistic Malthusian narratives (see Section 5.3.2.2 for more on environmental narratives). For example, in the Himalayan mountain range (Figure 5.7), policy makers have tended to assume that deforestation and soil erosion are linked to the expansion of agriculture driven by population growth and increased demand for fuelwood, timber, and grazing. In turn, this is assumed to lead to increased flooding and sedimentation far downstream in Bangladesh. However, in *The Himalayan Dilemma*, Ives and Messerli (1989) demonstrated the inadequacy of this simple scenario of a "supercrisis." Floods in Bangladesh did not increase in frequency through the twentieth century, but there were more large floods. These floods were not due to land-use change in the Himalaya but to variations in rainfall, high groundwater levels, and spring tides along river channels in Bangladesh itself (Bradnock & Saunders 2000; Hofer & Messerli 2006). What appeared to be a straightforward "conservation and degradation" story turned out to be less simple.

**Figure 5.7**   Erosion in the Himalaya. Road construction across the steep slopes of mountain regions such as the Himalaya (as here on the road from Shimla to Leh in Ladakh) is just one source of erosion. Unravelling the links between land use and environmental change is a major challenge for conservation. Photo: Ivan Scales

Geographers have drawn on the idea of risk to provide more nuanced analyses of human–environment interactions that combine the products of physical hazards with human vulnerability (Hewitt 1983; Cutter 2020). They have shown that the distribution of hazard and risk among people is the outcome of processes that determine poverty and powerlessness. Geological or meteorological events may occur independently of human action, but their effects are directly mediated through human institutions, as well as societies that are structurally unequal. This is also true of hazards in industrial and urban contexts, which include natural physical hazards, such as landslides or floods, as well as uniquely human-made hazards such as fire, a lack of drinking water and sanitation, and air and water pollution.

The question of who is at risk, and who is not, depends on their ability to evade or adapt to the hazards threatening welfare. Above all, this is affected by where they live. The poorest people tend to live on the steepest and most failure-prone slopes, the areas most liable to flood, and the most polluted urban environments. These insights are directly relevant to the growing interest in ecosystem services, the direct and indirect contributions of ecosystems to human well-being (IPBES 2019). The ability of poor people to access benefits from the supporting, regulating, provisioning, and cultural services provided by ecosystems (including areas of biodiverse habitat) is increasingly recognized as a critical issue in conservation and development. Such questions are inherently geographical (Potschin & Haines-Young 2011).

The environment constrains human opportunities and choices in many contexts, none more so than the world's drylands (Mortimore 1998). During the 1970s and 1980s, desertification dominated policy discussion about people and the environment in Africa. The received wisdom was that the agricultural practices of a growing rural African population were leading to the removal of vegetation, soil erosion, and an expansion of deserts. This simplistic, and **neo-Malthusian**, desertification narrative portrayed environmental degradation as an inevitable result of rising population density (Swift 1996). However, empirical research has since called this into question in a number of African countries, such as those in the Sahel (Mortimore 1998), Kenya (Tiffen et al. 1994) and Uganda (Carswell 2007). Local farmers, in fact, manage land with great skill, intercropping, planting crop varieties, maintaining soil fertility by integrating agriculture and livestock-keeping, and planting trees (Mortimore & Adams 1999). Dryland farmers certainly face serious challenges, but they are conserving and managing the environment as best they can (Figure 5.8).

Human geographers have shown that neo-Malthusian thinking is not a useful basis for conservation policy. Such simplified explanations risk leading to inaccurate and even dangerous conclusions about the causes of biodiversity loss and other environmental challenges. Conservation actors have overemphasized population growth at the expense of other, typically more important drivers of environmental degradation, notably consumption by wealthy populations in industrialized countries. Conservation-related research and policy must therefore pay greater attention to the range of factors that shape land-use decisions.

**Figure 5.8** Dryland agriculture in northern Nigeria. In the Sahelian zone of northern Nigeria, high rural population densities are maintained without prolonged fallow periods through the maintenance of soil fertility by managing nutrient cycles, using legume crops, and integrating agriculture and livestock keeping. Photo: Bill Adams

### 5.3.2 Social Nature

Human–environment interactions are not simply the outcome of human biological imperatives, nor purely determined by the constraints placed by environmental limits and carrying capacities. Humans have shown a remarkable ability to modify landscapes to suit both their material needs and their cultural sensibilities. Furthermore, environmental history has shown us that landscapes have been shaped by the dominant political and economic forces of the time. Nature is, therefore, both biological and social.

This section focuses on four key concepts that draw on the idea of social nature, but in significantly different ways. First, we explore the idea that environments and landscapes are shaped by human political and economic activities. While this approach takes us away from the simple biological or environmental determinism discussed in the previous section, it is still focused primarily on material processes. The second concept takes us away from material interactions to the notion of nature as socially constructed. We explore how ideas and knowledge about nature come about, and the influence that this has on both conservation policy and landscapes themselves. The third concept draws on the rich and varied field of **political ecology**, which focuses on the power relations that shape social nature. The last, and most radical, concept is the idea of hybrids and actor-networks. This conceptualization of social nature marks a departure from efforts to understand human–environment interactions that treat humans and non-humans, biotic and abiotic components of ecosystems, and culture and nature as separate and distinct entities.

### 5.3.2.1  Political Economy and Conservation

A core concern of geography has been how natural resources have been incorporated into different political and economic systems, and the impact that this has had on landscapes. In order to analyze these processes, geographers have often drawn on political economy. Broadly defined, political economy focuses on how politics affect economic processes, especially the distribution of wealth. More specifically, political economy has allowed geographers to explore the links between capitalist systems and the landscapes and ecosystems they produce.

As noted previously, geographers have critiqued Malthusian environmental narratives that focus narrowly on population growth as a driver of environmental degradation. Some geographers have drawn on political economy to explore how political and economic factors shape resource use. In northern Nigeria, for example, the development of capitalism through British colonial economic policies in the late nineteenth and early twentieth centuries played a major role in making farmers more vulnerable to famine (M. Watts 1983). The colonial government, wanting to boost the production of cash crops for export and increase tax revenues, went about transforming a regional economy historically based on subsistence agriculture and systems of reciprocal exchange. It did this primarily through the imposition of monetary taxes, which forced households into a new cash-based economy. As a result, farmers switched to groundnuts and cotton, using the income from cash crops to buy food instead of growing it. This meant that rural households were now exposed to the vagaries of global commodity markets, increasing the prevalence of hunger and famine. In times of low commodity prices, households were left unable to meet their subsistence needs. More intensive land use also put increasing pressure on soils. This shows that many processes of resource overexploitation and environmental degradation are underpinned by political and economic relationships.

Political economy matters to conservation for two reasons. First, as we have seen numerous times in this chapter, the dynamics of the global economy have major implications for the way natural resources are used. Second, **biodiversity conservation** itself has a political economy (Scales 2015). Tens of billions of dollars are raised and spent every year on conservation activities (Waldron et al. 2013). Biodiversity conservation employs people, creates new institutions, and influences the way resources are used. Furthermore, global conservation policy and global capitalism have become increasingly entwined over the last decade (Scales 2015). For example, non-government organizations concerned with conservation have formed partnerships with large corporations and have helped to design and implement market-based mechanisms such as payments for ecosystem services.

While some conservationists may argue that such schemes are simply a product of pragmatism and doing conservation in the real world, political economy reminds us that we should pay close attention to the power relations involved—who does what, who gets what (and what they do with it), and ultimately, who decides (Bernstein 2010). Market-based interventions are far from politically neutral. They have the potential to create new sources of income, for example, from selling credits for the carbon sequestered in tropical rainforests and other such payments for ecosystem services. They also are likely to have an unequal distribution of costs and benefits and lead to strong incentives for elites (local, national, or international) to take control of natural resources (Sandbrook et al. 2010). Critics of market-based conservation tools see a strong potential for "green grabbing": the appropriation of land and natural resources by elites both for environmental ends and to allow further accumulation of wealth (Fairhead et al. 2012).

### 5.3.2.2  Environmental Narratives and Discourses

Conservation biology has been labeled a mission-driven discipline (Meine & Meffe 1996). It is therefore **normative**, underpinned by a vision of the way the world should operate. One of the main ways conservation biology attempts to make this vision real is through influencing policy. In order to do so, conservation actors need to simplify complex realities so that policy makers can decide on priorities and specific courses of action. Policymakers must deal with questions such as What is driving deforestation? What are the consequences of land-cover change? and What is the best way to reduce habitat loss? In addressing such questions, and engaging in policy, conservation science contributes to the formation of broader environmental **discourses**. A discourse is the process whereby people form and share knowledge about the world around them. Discourse provides people with the language and concepts to think and communicate about a particular topic (Mills 1997). Science is powerful in shaping discourses because the "claims of science are portrayed not just as global, but universal, not just modern but eternal" (Fairhead & Leach 2003, p. 1).

Research in the social sciences increasingly shows that the production and reproduction of knowledge are central to the exercise of power (Leach & Fairhead 2002). Geographers have taken up the work of Foucault (e.g. 1980) and other post-structuralists on the relationship between discourse, knowledge, and power. According to this view, power is not just the exercise of control over others through rules or force, nor is it simply held by the state or through social structures such as class. Power can be exercised through the ability to shape, influence, and dominate the way issues are thought and talked about.

If, as is argued in constructionist views of nature (see Chapter 2), the environment is socially constructed, then power lies with those who define what the environment is and what is happening to it. For example, the act of labeling an area as "wilderness" is powerful, since it can be used to justify excluding certain people from it (see Section 5.3.3.2 for more on the power relations surrounding notions of wilderness). Those working in biodiversity conservation decide, according to their own worldviews, which forms of resource use are acceptable and which are not. These worldviews are rarely purely objective and can lead to very different policy recommendations, as shown by the heated debate in conservation over whether biodiversity is best preserved by fencing it off from humans or by using it to generate money to provide incentives to conserve biodiversity (e.g. through trophy hunting).

One way of thinking about power in conservation discourses is in terms of policy narratives (Roe 1991). A *narrative* is "a representation of a history, biography, process, etc. in which a sequence of events has been constructed into a story in accordance with a particular ideology" (*Oxford English Dictionary*). It is a story that explains a specific process or event, such as the idea that tropical deforestation is largely driven by poverty and population growth, with farmers forced to clear forest for subsistence. Narratives are powerful and seductive because they make sense. They reduce complexity and make it manageable, defining the problem and offering up possible solutions. For example, if population growth and poverty are perceived to be the major drivers of tropical deforestation, then conservation policies will tend to focus on population control and changing the land-use practices of the rural poor.

However, environmental narratives can be problematic (see Box 5.1). They can be based on unproven and culturally biased assumptions (e.g. that practices such as swidden agriculture are inherently unsustainable). As we have seen in this chapter, they often oversimplify complex human–environment interactions and have a tendency to focus narrowly on population growth and poverty as drivers of biodiversity loss. In doing so, they can lead to policies that ignore other important drivers of habitat and biodiversity loss.

---

**Box 5.1   Debates: Madagascar and the power of received wisdoms and myths in conservation science**

Madagascar is one of the most biologically diverse places on Earth. More than 80% of its species are endemic, with over 13,000 species of plants and 700 species of vertebrates (Ganzhorn et al. 2001). Its highly diverse flora and fauna are threatened by habitat loss and fragmentation, mostly due to deforestation.

Since the late 1980s a boom has occurred in the activity and spending by international conservation organizations and donors in Madagascar (Kull 2000; Corson 2016). Closely tied to the boom is a powerful environmental discourse dominated by an often repeated "fact" that Madagascar has lost up to 90% of its forest due to poverty-driven swidden (slash-and-burn) agriculture (Kull 2000; Scales 2011, 2012). Not only has the narrative of dramatic forest loss played a major role in conservation fundraising, it has also underpinned conservation policy, with both the Malagasy government and international conservation organizations focusing their efforts on protecting forests from poor farmers.

However, on closer inspection, Madagascar's deforestation narrative is problematic. The idea that Madagascar has lost 90% of its forest is derived from the "island forest" hypothesis, which emerged in the early twentieth century from the influential work of two French botanists, Henry Alfred Perrier de la Bâthie (1921) and Henri Humbert (1927). Perrier de la Bâthie and Humbert believed (based on little more than tenuous observations and inferences) that Madagascar must have once been entirely covered in forest, and that its extensive grasslands were the result of human action, primarily forest clearance for agriculture and grassland fires used to renew pasture for cattle.

---

*(Continued)*

**Box 5.1 (Continued)**

Since the 1970s, a growing body of research has challenged the island forest hypothesis. Empirical evidence shows that long before the arrival of humans, grasslands were important biomes in their own right. Paleoecological evidence from soil cores, lake sediment cores, and excavations have revealed soils characteristic of grasslands (Bourgeat & Aubert 1972), subfossils of savanna herbivores (Dewar 1984), and pollen from savanna grasses (Burney 1987, 2003). This is supported by a comparison of Madagascar's grassland flora and fauna with those from mainland Africa, which shows that they are comparably diverse and contain a significant number of endemic species (Bond et al. 2008). In other words, they have existed for sufficiently long periods to develop endemic specialists. Despite the fact that they contain numerous endemic species, and therefore merit the attention of conservationists, the island's grasslands have largely been dismissed as degraded landscapes and targeted for tree planting (Scales 2014).

As well as challenging the idea that Madagascar was entirely forested and questioning rates of forest loss, research also shows that the drivers of deforestation are more complex than simply poverty or population growth. Over the last 100 years, a range of land uses, and not simply slash-and-burn agriculture (Figure 5.9), have led to changes in forest cover. These include the cultivation of export cash crops, such as maize and sisal, on large concessions and plantations (Scales 2011). Rather than poor subsistence households, wealthy elites are often best placed to benefit from these commodity booms through control over land and the ability to hire extra labor to clear forests. In other words, political and economic factors operating at multiple spatial levels have driven forest loss in Madagascar (see Box 5.2 for more on thinking about the drivers of environmental degradation through chains of explanation).

**Figure 5.9** Burned dry forest in Menabe, western Madagascar. Standardized narratives of environmental change in Madagascar have led conservationists to assume that the dry forests of western Madagascar are suffering progressive reduction in cover due to poverty-driven agriculture. Research shows this not to be the case. Forest cover change is less continuous, and its causes are more diverse and complex. Photo: Ivan Scales

### 5.3.2.3 Political Ecology

Political ecology can be defined as "a confluence between ecologically rooted social science and the principles of political economy" (Peet & Watts 1996, p. 6), or research-based explorations of social–environmental systems that explicitly consider power (Robbins 2019). These definitions are broad, reflecting the diverse methods and theories on which political ecology draws. Political ecology is not a single body of theory nor a set group of methodologies but an approach to studying human–environment interactions with a common set of priorities and assumptions (Robbins 2019).

There are two core concerns in political ecology. First, how and why does environmental change occur? In order to answer these questions, political ecologists have drawn on a wide range of methods and data sources to measure environmental change (e.g. remote sensing, ecological surveys, historical data from archives) and understand patterns of resource use (e.g. household surveys, **ethnographic** fieldwork). Political ecologists try to understand these processes at multiple spatial levels, for example, using a "chain of explanation" (Box 5.2).

Second, political ecology has asked how and why conflicts over the environment occur, paying close attention to the ways in which access to natural resources is controlled and how the costs and benefits of resource use are distributed. Political ecology sees power as exercised in various ways, for example, through political and economic structures and institutions, but also through the narratives and discourses that shape conservation policy and practice.

Looking specifically at biodiversity conservation, political ecologists have studied the power relations of a wide range of conservation interventions. These include the impacts of protected areas on local communities (Neumann 1998; Brockington & Igoe 2006); payments for ecosystem services and the growing commodification of nature (Igoe & Brockington 2007; Brockington & Duffy 2010; Benjaminsen & Bryceson 2012; Fairhead et al. 2012); and the growing militarization and violence of antipoaching measures (Duffy 2014; Büscher & Ramutsindela 2016).

---

**Box 5.2   Methods: The chain of explanation**

The concept of a chain of explanation was used by Blaikie and Brookfield (1987) in developing their approach to political ecology. The chain of explanation seeks to integrate explanations of local phenomena (e.g. deforestation, habitat degradation, or soil erosion) across multiple spatial and temporal scales. It begins at the local level with an individual resource user (e.g. a farmer), then attempts to trace the factors that influence their resource-use decisions upward to the regional, national, and international levels. A chain of explanation stretches through space and time in order to understand the antecedents to the current conditions that shape the decisions of the resource user (e.g. the history of land ownership).

Thus, in the case of soil erosion, Blaikie and Brookfield argued that physical changes in soils and vegetation were linked to economic symptoms in particular places at particular times. In turn, these were linked to land-use practices in those places, especially the resources, skills, assets, time horizons, and technologies of land users. Farmers were also embedded in communities and interacted with other households, and were influenced by government policies, as well as the dynamics of the global economy (Figure 5.10).

---

*(Continued)*

---

**Box 5.2  (Continued)**

The chain of explanation shows that the links between environment, economy, and society are complex and multiscalar, and the process of development is both a *response* to and a *cause* of environmental change: "land degradation can undermine and frustrate economic development, while low levels of economic development can in turn have a strong causal impact on the incidence of land degradation" (Blaikie & Brookfield 1987, p. 13). More importantly, because the development process involves the transformation of social and economic relations, it relates to the ways in which individuals and groups within a society experience their environment, and the ways in which they use it.

While useful as a conceptual tool, Blaikie and Brookfield's chain of explanation has been criticized by some political ecologists for relying on abstract, arbitrary, and pre-given spatial levels (e.g. local, regional, national, international), as well as conceptualizing political and economic processes in an overly hierarchical fashion (Zimmerer & Bassett 2003), leading to explanations that inevitably end up attributing environmental change to global economic processes "up there." Furthermore, in attempting to understand the complexity of factors at multiple levels, a chain of explanation risks simply generating a list of everything that can influence the decisions of a person or household, rather than providing a theoretical basis for understanding the process of resource use (Black 1990).

---

As well as seeing human–environment relations as inherently political, political ecology pays explicit attention to the environment itself. However, its engagement with ecology is characterized by a skepticism about the equilibrium ideas of much mainstream ecology. Until recently, ecology has been dominated by the assumption that populations and communities of organisms exist under equilibrium conditions in habitats that are saturated with species in stable and balanced ecosystems (Rohde 2005). This dominant paradigm owes its roots to Darwinian ideas of competition for resources, as well as Malthusian ideas of limited resources and self-correcting mechanisms, and still reflects the Clementsian view of plant communities developing linearly and predictably through successional stages toward an end point: a climax vegetation (Clements 1928). Any disturbance (and especially human disturbance) is seen as external to the system and a move away from the stable climax. Following disturbances, ecosystems are assumed to return predictably to the equilibrium of the climax state (see also Box 3.3, Chapter 3, Anthropology, for a discussion of equilibrium and nonequilibrium models in anthropology).

Over the last 30 years, ecological research has challenged this equilibrium view and a "new ecology" has emerged that does not view ecosystems as inherently stable. Empirical evidence has come from arid and semi-arid ecosystems, where rainfall is the major limiting factor and is unpredictable in both space and time. The work of Caughley (1987) on domestic sheep and kangaroos in semi-arid ecosystems in Australia suggested that rainfall was the predominant factor affecting population size and that unpredictable droughts resulted in nonequilibrium conditions. This work was subsequently supported by research

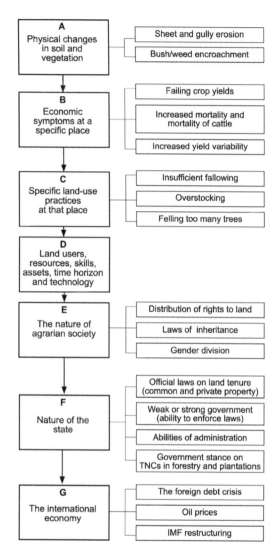

**Figure 5.10** A chain of explanation for land degradation. The chain of explanation seeks to explain environmental change (e.g. deforestation, habitat degradation, or soil erosion) by connecting local land use decisions to social, economic and political factors operating across multiple spatial and temporal scales

on desertification in the Sahel (Behnke et al. 1993) and overgrazing by East African pastoralists (J. Ellis & Swift 1988).

In such unpredictable ecosystems, concepts of stability and equilibrium are misleading and nonequilibrium is the norm (Bartels et al. 1993; Behnke et al. 1993). Non-equilibrial behavior is not limited to arid and semi-arid ecosystems, and has important implications for conservation planning, since landscapes and species cannot be assumed to be static and stable (Holling & Meffe 1996). Conservation planning therefore needs to be based around the idea of dynamic landscapes and species. Conservation policy might usefully turn to the field of resilience science, which looks at the capacity of ecosystems to tolerate disturbance. According to the resilience approach, a resilient ecosystem, while not static or necessarily stable, is able to withstand shocks and rebuild itself (Berkes & Folke 1998; Gunderson & Holling 2002). For conservation policy, this means a move away from focusing on narrow ranges of species and fixed protected areas to consider (dynamic) ecosystems and landscapes more broadly and how resilience might be increased through management practices.

### 5.3.2.4  Hybridity and Actor-network Theory

So far, the relationship between society and the environment has been discussed as if the two are essentially separate. Whether the environment shapes human action, human actions shape the environment, or human minds socially construct "nature," the world is divided into two distinct groups of objects and processes. The things that make up the world are thought about as *either* "social or natural, active or passive, agent or acted upon ... nature is separate from humanity and humans have the monopoly on knowledge, agency and morality" (Dyer 2008, p. 209).

Some human geographers have challenged this approach, rejecting binary divisions between people and things, as well as between nature and culture. In particular, they have drawn on the posthumanist idea of hybrids, developed by sociologists such as John Law, Bruno Latour, and Michel Callon. The actor-network school of thought is built on a radically different **ontology**. First, it does not distinguish between people and objects, or between biotic and abiotic entities. Instead, the world is made up of diverse materials. Second, it sees the world as made up not only of physical entities but also of concepts, ideas, and languages. Ontologically, no distinction is made between the world of material objects and ideas. Instead, the world is made up of diverse and heterogenous *actants*. An actant can be an object, a person, an idea, or a word. These actants come together in networks, and each actant in a network has a role to play—it has **agency**. In this hybrid view, the world around us is seen as complex and messy.

Posthumanist ideas have influenced many branches of human geography, including political ecology (N. Watts & Scales 2015), but have been strongest in an area now referred to as more-than-human geographies (Whatmore 2002). Insights from posthumanism have important implications for thinking about our relationship to non-human life, since they challenge categories that conservationists often take for granted: human versus non-human, urban versus rural, synthetic versus organic, managed versus wild (Marris 2011; Barua 2014; Lorimer 2015). Conservation has conventionally been seen as the result of the work of human actors: individuals, institutions, or collectives. When thinking about conservation, Jepson et al. (2011) argue for the need to include a much wider range of non-human actants (especially other species) and devices (especially prioritization devices like lists of rare species). All these play a role in shaping where and how conservation takes place. As shown in Box 5.3, thinking has the scope to broaden our understanding of what conservation means and where it should take place.

---

**Box 5.3    Crossing boundaries: Actor-networks and hybrids in conservation: The case of the urban water vole**

Research on water vole (*Arvicola amphibius*) conservation (Hinchliffe et al. 2005) shows how some concepts of actor-network theory and hybridity can be applied to conservation, and how this reveals complex relationships. The case study centers around a small former industrial site (less than a square kilometer in size) at the heart of the city of Birmingham, UK. The site is a mosaic of habitats, and a variety of species live there or pass through it, including European badgers (*Meles meles*) and water voles. Water voles have experienced dramatic declines in population in the UK, mostly due to predation by the introduced American mink (*Neovison vison*) and habitat destruction through farming and watercourse management. As a result, the industrial site was identified as a potential location for the city's first urban nature reserve. However, it had also been earmarked as a potential area for urban renewal, through the construction of a hospital and shopping complex. An environmental impact assessment (EIA) carried out on behalf of the developers concluded that there was nothing of value on the site. This verdict was challenged by local conservationists, who set about trying to prove the conservation value of the land.

This urban, postindustrial site of potentially high conservation importance challenges the binaries that have traditionally underpinned conservation (e.g. urban versus rural, managed landscapes versus wilderness). Further investigation reveals that the hybridity of the site goes even deeper. One of the reasons the EIA had missed the water voles is that they are very difficult to spot, especially in urban settings where they tend to reduce their foraging time. They are less visible to people in urban sites. Revealing the presence of water voles involves a considerable amount of human effort, technology, and skill. It is only through the use of particular tools and experience, built up over time, that water voles are made visible. Experience and technology can therefore be added to the hybrid network, which is now made up of (i) the non-rural, non-"wild" and anthropogenic fabric of the site, (ii) the (urban) water voles, and (iii) the tools and knowledge necessary to make them visible.

The deeper we go, the less familiar and comfortable this hybrid becomes. As more research was carried out, some startling findings emerged. Rather than simply revealing the presence or absence of water voles, the studies revealed ways in which this particular hybrid was novel. The research suggested that water voles at the site were cohabiting with brown rats (*Rattus norvegicus*). Prior to this, biologists thought that the two species kept to separate areas due to predation and competition for food and habitat. However, it seems that the complexity of the habitat in this urban site allowed the species to cohabit. This suggested that the very non-"naturalness" of this habitat allowed the endangered water vole to thrive, where elsewhere it might not have done so due to threats from mink (which had not managed to spread into urban areas) and rats. Thus, the hybrid actor-network in the middle of Birmingham challenges "the universalism of water vole ecology" and suggests that "the urban stream and habitat [are] becoming more interesting than pale imitations of the rural idyll" (Hinchliffe 2007, p. 134).

This case study shows how a hybrid approach moves us away from simple binaries, to reveal a network made up of a diverse range of actants: conservation biologists, water

*(Continued)*

---

**Box 5.3 (Continued)**

voles, urban habitats, and research tools. Each of these actants has agency. Without them, the network changes or even collapses. Without the tools of the researchers, the voles can't be made visible and the site loses its value. However, through the same tools, the water voles are not only made visible but new and challenging ecologies are revealed that further dissolve preconceived boundaries.

---

### 5.3.3 Cartography and Power

Maps are not just two-dimensional representations of the Earth's surface. Every map is made for a purpose and involves choices of what to map and how to represent it:

> Maps are never value-free images; except in the narrowest Euclidean sense they are not in themselves true or false. Both in the selectivity of their content and in their signs and styles of representation maps are a way of conceiving, articulating, and structuring the human world which is biased towards, promoted by, and exerts influence upon particular sets of social relations. (Harley 1988, p. 277)

The power of maps is most obvious in the historical relationship between exploration and cartography, which underlay the knowledge revolution of European imperialism. For example, the British Royal Navy's charts of oceans and distant shores enabled both the extension of the British Empire and the application of Western culture, economy, and science to the task of subduing and exploiting colonized lands (Livingstone & Withers 1999; Drayton 2000). Maps of annexed and conquered territories in the Americas, Australasia, Asia, and Africa demarcated territories, and in many instances created the evidence that allowed the legal view to prevail that indigenous lands were empty (*terra nullius*), effectively unused and unclaimed (Crosby 1986; Adams 2003). The hegemony of the maps created as part of imperialism have been challenged repeatedly by indigenous peoples in struggles to reclaim disputed territories against the demands of conservation (e.g. to establish protected areas) and economic growth (e.g. for mining) (Whyte 2017).

#### 5.3.3.1 Maps and the Control of People and Nature

As European imperialism tightened its grip on tropical possessions, knowledge and intervention together comprised a tightening "government" of nature (Drayton 2000). Maps, which guided colonial annexation and divided land for settlement, made the rational efficiency of imperial governance possible (Drayton 2000). Thus, in the USA, Thomas Jefferson's survey of the lands West of the Appalachians in the 1780s and 1790s divided the continent into a grid of six-mile-square townships, and 640-acre sections (Meine 2004). This geodesy brought administrative order to the barely explored West, confining indigenous people within reservations, filling the cleared land with settler farmers, and setting aside vast federal lands for timber exploitation, grazing, and (eventually) conservation. Under Progressive Era conservation at the turn of the nineteenth century to the twentieth century, forest, soil, and water would be harnessed by science to serve the common good (Hays 1959).

Modern state governance was built on the idea that nature could be understood, manipulated, and controlled for social benefit through the development of schematic knowledge (Scott 1998). Thus, scientific forestry, developed in eighteenth-century Prussia, was adopted through the nineteenth century in France, in British colonial possessions (notably India, where imperialism, science, and environmentalism became inextricably interlinked (Barton 2002)), and in the USA (Demeritt 2001). In the twentieth century, this type of scientific management became the standard global approach to renewable resources such as forests and management of ecosystems for conservation (Adams 2020).

Contemporary geographers have drawn extensively on the ideas of Foucault to explore issues of **governmentality** (the process by which states use various administrative procedures to control populations), and biopolitics (the development of schematic knowledge about the human body to enable control and manipulation). Science and cartography have allowed nature to be classified, counted, mapped, and (at least in theory) controlled by government bureaucracies (Willems-Braun 1997; Demeritt 2001). The same reductionist approach was applied to the people in colonial territories. In Kenya, for example, the colonial state annexed land for white settlement in the highlands, confining Africans to prescribed native reserves and, later, attempting to control the way they used land and enforce the construction of soil conservation terraces (Mackenzie 1998). In Madagascar, the colonial government decided to classify the island's population into 18 official "tribes" and drew maps to delineate their boundaries, unaware of both the flexibility of ethnic identity and the mobility of rural Malagasy (Randrianja & Ellis 2009). The government thus turned a fluid, complex, and mobile indigenous population into discrete and geographically fixed units which it could work with and govern.

### 5.3.3.2 Maps and the Segregation of Nature

Geographers have paid particular attention to one feature of the ordering of space by colonial and postcolonial states, namely the segregation of nature and people through the establishment of protected areas. Ironically, "fortress conservation" (Brockington 2002) is built on exactly the same conceptual distinction between "nature" and "human" as the destructive process of development that it opposes. Arguably, this conceptual separation is essential to the creation of conservation as a practical project. In particular, the idea of nature as pristine, with complexes of species existing in a natural state, matched a view of humanity as a destructive force analytically external to the natural world. Strictly protected areas achieve the physical separation of natural and human-transformed landscapes. Historically, conservation must therefore be understood as a part of the project of ordering and simplification undertaken by the modern state (Neumann 2004).

Conservation action has a very specific geography, in that it involves the exercise of power to classify areas of land (or sea) to be managed in particular ways, and the exercise of power to establish rules about who may enter such areas, and what may be done within them (Schroeder 1999). As Hingston (1931, p. 406), promoting national parks in East Africa, commented "human life and the wildlife must be separated permanently and completely. So long as man and animals live together there will always be trouble."

In the 1930s, for example, the colonial state in Liwale District in Tanganyika (now Tanzania) began a process that eventually gave rise to the Selous Game Reserve. At the end of the nineteenth century, economic and environmental conditions were severely disrupted in this part of Africa following annexation by Germany. By the 1920s, under British administration, tsetse fly and sleeping sickness were widespread, human populations were at low-density, and elephants were a serious threat because of crop raiding. In 1933, the government proposed a scheme to separate people and wildlife, driving elephants west into the land that became Selous and moving farmers east, toward the coast where more intensive farming would discourage sleeping sickness, and the administration could more effectively reach rural communities. They did this by intensifying shooting of elephants in the east and abandoning it in the west "to try and force the natives in the West to come into country where they could be protected" (Neumann 2004, p. 204). In total, 40,000 people were relocated. Neumann (2004, p. 212) argues that the creation of the vast Selous Game Reserve has "all the hubris and faith in science and progress as any high modernist project," and is "as much an expression of modernism as skyscrapers." For Neumann (2004, p. 212), the establishment of protected areas like the Selous Reserve "represents an historic transition under modernity wherein human civilization becomes the caretaker of a wild nature that poses no threat beyond the threat of disappearing. Nature becomes a ward of the state."

No idea has had more power in the bounding of nature for conservation than that of wilderness. When biologist Edward Wilson (1992, p. 335) argued that "wilderness settles peace on the soul because it needs no help; it is beyond human contrivance," he captured the dominant view of wilderness in the Western conservation imagination. Western thinking about wilderness has evolved from a zone of fear and danger beyond civilization to a wonderful and sublime place free from the threat of human despoliation (Cronon 1995). This Anglo-American nature aesthetic (Neumann 1998), comprising both wilderness and the picturesque, has had huge influence on Africa, whose large mammals and open savannas were seen by colonial hunters and administrators as a kind of "lost Eden in need of protection and preservation" (Neumann 1998, p. 80). Africans were perceived as intruders into pristine or natural landscapes, denied both history and rights. This idea of "wild" nature continues to exert a powerful influence in conservation thinking.

### 5.3.3.3 Geographical Information Science and New Mapping Technologies

The making of maps and their use to delineate spaces for people and nature goes on, although the technologies have changed. GISs store, analyze, and display information with a fixed location in space or, to put it another way, "information systems that keep track not only of events, activities, and things, but also of *where* these events, activities, and things happen or exist" (Longley et al. 2001, p. 2). Geographical information science has transformed practices of cartography, massively extending its analytical possibilities through digital analysis, with important political implications.

A GIS has four key functions (Fischer et al. 1996): (i) the input of geographically referenced data; (ii) the storage and management on these data in a database; (iii) the analysis of data through exploration, manipulation, and statistical tests; and (iv) the visual display of data and the results of analysis. Users are able to create separate layers of geographically referenced data that can be used to explore and analyze the relationships between different entities or processes (Figure 5.11).

The mapping of biodiversity can help policy makers prioritize conservation action (Box 5.4). GISs also can be used to investigate the environmental impacts of human activities. For example, a GIS might be used to shed light on the drivers of bushmeat hunting in tropical

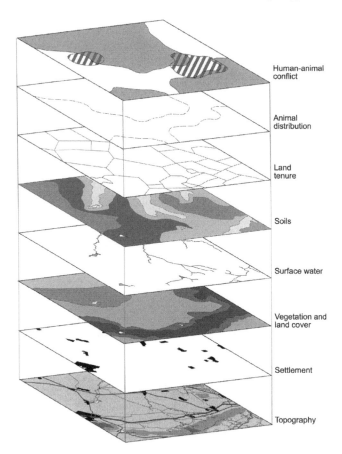

**Figure 5.11** Layers in a Geographical Information System (GIS). A GIS enables users to create separate layers of geographically referenced data. These can be used to explore and analyze the relationships between different entities or processes.

forests by creating a database with layers corresponding to the population densities of different primate species and a range of human factors (e.g. the location and size of human settlements, the location of markets for bushmeat, the location of roads) to test which shows the strongest spatial correlation. Such GIS applications are increasingly common in conservation planning and assessment (e.g. Ribeiro & Atadeu 2019; Börner et al. 2020).

A wealth of guides and manuals explain how to use GIS and remote sensing tools, but the social implications of these technologies are perhaps less well understood. With the increasing power of hardware and sophistication of software (as well as their increasing accessibility as prices have fallen and computers have become more widely available globally), GIS has become widely used in both conservation research and policy making. However, "doing GIS" involves more than simply interacting with a particular piece of software (Wright et al. 1997). Rather, it raises important questions about "the message it sends, whom it empowers, and the responsibility its developers should bear for its eventual use" in the process of biodiversity conservation (Wright et al. 1997, p. 347).

---

**Box 5.4    Applications: Hotspots, ecoregions, and mapping for conservation**

In order to prioritize conservation policy, conservationists have often turned to spatial tools. Examples include the World Wildlife Fund's Global 200 list of important ecoregions, which are defined as large units of land or water containing a geographically distinct assemblage of species (Olson & Dinerstein 2002). Another example of spatially explicit prioritizing is Conservation International's (CI's) use of biodiversity hotspots (Figure 5.12). These are areas with exceptional species richness and concentration of endemic species experiencing severe habitat loss (Myers et al. 2000).

Norman Myers first identified ten tropical forest hotspots in 1988, which were characterized by exceptional levels of plant endemism and serious levels of habitat loss. CI adopted Myers' hotspots in 1989. Three years later, an extensive global review was carried out by CI, based on quantitative thresholds for the designation of biodiversity hotspots.

To qualify as a hotspot a region must meet two criteria. First, it must have a minimum of 1,500 species of endemic vascular plants (> 0.5% of the world's total) and must have lost at least 70% of its original habitat. In a subsequent analysis carried out in 1999 (Myers et al. 2000), 25 biodiversity hotspots were identified. In total, these areas contained as endemics at least 44% of the world's plants and 35% of terrestrial vertebrates, in an area covering11.8% of the planet's land surface.

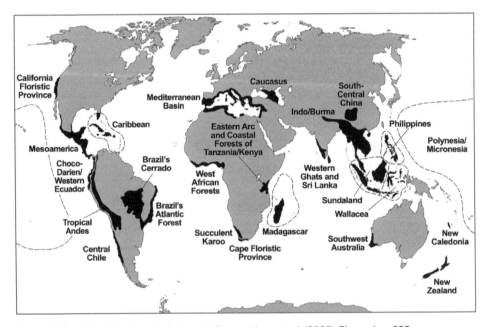

**Figure 5.12**    Global biodiversity hotspots. *Source:* Myers et al. (2000), Figure 1, p. 853

The science-based planning approaches to the selection of protected areas developed in the 1990s (e.g. Margules & Pressey 2000) used remote sensing and algorithms to acquire, store, present, and exchange data in a global system of conservation planning (Ladle & Whittaker 2011). Such strategic conservation planning by international conservation organizations has become the norm, and as Bryant (2002) discusses in the context of the Philippines, largely external perceptions of biodiversity have been powerful in framing government policy. Local people have played little or no part in the planning process and local uses of nature had little or no place in this analysis.

As noted previously, knowledge and power are closely related, and science has an important role in shaping environmental discourses. Spatial data are particularly powerful. Maps are easily understood by non-specialists and give us an instant idea of *what* is *where*. If one wants to get an idea of the how forest cover has changed over time, for example, a map derived from satellite imagery can help to understand both the extent and pattern of change in a single image. However, it is these very characteristics that should make us stop for a moment to consider how these tools are being used and for what purpose. Something as seemingly straightforward as measuring and mapping forest loss can reveal different perceptions of landscapes and the politics of natural resource use (Box 5.5).

---

**Box 5.5 Methods: The causes and consequences of land-cover categorizations**

The landscape of the Godwar region of Rajasthan in India is a complex mosaic of deciduous forest, farmland, savanna, and shrubland dominated by Mexican mesquite (*Prosopis juliflora*), an invasive and highly aggressive species. Robbins' (2003) research focused on how different people perceived and classified this landscape (Figure 5.13). In order to explore the different ways in which herders, farmers, and forestry officials saw their surrounding landscape, Robbins asked them to identify and describe different land-cover types in photographs. He then used these definitions to classify satellite images. This revealed that the different stakeholders saw the landscape in very different ways, and that their differing definitions of forest cover led to varying estimates of forest cover. The biggest contrast was in how the stakeholders saw the expanding shrubland. Foresters saw it as forest, whereas local herders and farmers saw it as degraded land cover, referring to it as *banjar* (wasteland) due to its low fodder value and inferior quality as wood fuel. In contrast, the foresters saw it as a legitimate form of land cover that was dominated by woody vegetation and therefore counted as forest.

This example shows that behind the supposedly objective façade of land-cover maps, there are often complex debates over land-cover categories. While remote sensing is a powerful tool for measuring habitat loss and degradation, it is worth noting that there is still a level of subjectivity in the process of land-cover classification. Take forests, for example. It might seem obvious what a forest is, and yet in order to classify a satellite image, it is necessary to make operational decisions such as when a shrub becomes a tree and how many trees it takes to make a forest. There is no single definition of what a forest is, nor what constitutes deforestation. Definitions of deforestation range from the total removal of forest cover to small changes in forest composition and structure (e.g. by selective logging), and there is often no distinction made between

---

*(Continued)*

**Box 5.5 (Continued)**

permanent and temporary conversions, between conversion and alteration, and between forest loss and degradation (Angelsen 1995). While satellite imagery and other types of remotely sensed data are often used to provide "hard facts" about environmental change, and thereby settle debates, "such imagery, rather than reducing the contentiousness of landscape change claims, actually reinforces it" (Robbins 2003, p. 181). Rather than being used as a means to close debates about landscapes—to establish hard, incontestable facts—maps should be used to create dialogue between stakeholders.

**Figure 5.13** Woodland (or wasteland?) in Rajasthan, India. Government foresters see this landscape, dominated by the invasive *Prosopis juliflora* (Mexican mesquite) as forest. Local herders and farmers refer to it as *banjar* (wasteland) due to its low fodder value and inferior quality as wood fuel. Photo courtesy of Sushil Saigal.

Because maps are powerful, they also can be used to challenge received wisdom and empower local communities through a form of "counter-mapping." Participatory mapping, the creation of maps by local communities, can provide a visual representation of how local people perceive their surrounding environment. The use of participatory maps has its roots in participatory rural appraisal, a set of techniques aimed at incorporating the knowledge of people in the planning of development projects (Chambers 1994). Over the last decade, participatory techniques have increasingly relied on participatory GIS (PGIS). This has been made possible through the drop in prices in global positioning systems (GPS), imagery from satellites, and computer hardware. PGIS can be used to highlight divergent knowledge and multiple perceptions and classifications of landscapes.

The use of PGIS has the potential to redress the power and knowledge imbalance in conservation, as well as reveal a different "view from below." However, research has shown that it may also lead to unintended effects, such as creating conflict within and across communities when individuals and groups use mapping to make claims on resources (Fox et al. 2008). Technologies such as GIS must be used with care and should be thought of as one of multiple strands of analysis to inform research rather than dominate it (Rindfuss et al. 2003).

## 5.4 Future Directions

Conservation clearly has a geography. It occurs in particular places and spaces, which are both material and socially constructed. Moreover, geography has a long history of researching the relationship between people and the environment, and geographers have approached the subject by drawing on a wide range of epistemologies, theories, and methodologies. To try to paint a unified picture of geography as a discipline would be impossible. Tensions lie at the heart of the discipline: between physical and human geography, between those seeking to find universal laws and those seeking to describe the unique, between quantitative and qualitative approaches, and between scientific positivists and social constructivists. However, the discipline's diversity is something to be celebrated. Its unique history and experiences offer important lessons for conservation social science and the future directions it might take.

Geographers, through their exploration of space and place, have shown that *where things happen* is crucial to understanding how and why they happen. Furthermore, the discipline's focus on regions has shown that human–environment interactions shape and are shaped by biophysical and social processes coming together in particular places. Threats to biodiversity are rarely universal—they touch down in specific areas and contexts. This suggests that a large component of any future conservation social science will be based on case studies and fieldwork. Policy makers should understand that a universal win-win solution—the elusive magic bullet—is unlikely. A question for future researchers is, to what extent are the various processes of biodiversity loss generalizable or site specific?

Another key lesson from human geography, and more specifically political ecology, is the importance of scale. Conservation science has too often looked at proximate causes of biodiversity loss without understanding deeper underlying drivers. The chain of explanation approach has shown that local practices are often nested in broader political and economic processes. When seeking to understand complex human–environment interactions, it is important to avoid the problem of ecological inference. In other words, we cannot deduce the behavior of individuals from patterns observed in groups of individuals (or vice versa). Relationships between factors depend on the level at which we observe them. This strengthens the need for research on human–environment interactions to take into consideration multiple levels or, at the very least, to avoid scaling up or scaling down relationships unless there is direct empirical evidence for doing so.

Natural resource use often involves contests between multiple and diverse stakeholders with different perceptions and priorities. Moreover, the costs and benefits of natural resource use are often unevenly distributed. The question of power should therefore be at the core of conservation social science. However, power is not only exercised through rules and force (political power), or through money (economic power), but also through the production and reproduction of certain forms of knowledge. Knowledge is always situated (i.e. shaped by the social context it inhabits). Conservation science can be particularly powerful since decisions to set up protected areas and limit certain human activities can affect the lives of thousands, if not millions, of human beings. Given the fact that much of the planet's biodiversity is located where some of its poorest inhabitants are also found, conservation inevitably comes up against questions of development and justice. However, it should not be assumed unquestioningly that biodiversity loss is primarily a function of poverty and there is more research to be done on the precise relationship between economic growth, poverty, and natural resource use.

Human geographers have found that conservation is often driven by powerful narratives. While these narratives help to frame complex problems and simplify the world, they can lead to researchers' and policy makers' adopting untested (and often neo-Malthusian) assumptions. Conservation social science therefore needs to do two things. First, researchers need to be more reflexive and look carefully at their **positionality**, thinking about how their culture, political views, education, age, and gender influence their research. Second, there is more work to be carried out on conservation as a social process, looking carefully at how the values held by conservationists, as well as how different theories, assumptions, narratives, politics, and economic realities affect the way conservation is carried out.

## For Further Reading

1 *Green Development: Environment and Sustainability in a Developing World*, 4th ed. (Adams 2020, Routledge, London).
2 *Nature Unbound: Conservation, Capitalism and the Future of Protected Areas* (Brockington, Duffy & Igoe 2008, Earthscan, London).
3 *The Conservation Revolution: Radical Ideas for Saving Nature beyond the Anthropocene* (Büscher & Fletcher 2020, Verso Books, London).
4 *Nature* (Castree 2005, Routledge, London).
5 *Green Grabbing: A New Appropriation of Nature* (Fairhead, Leach & Scoones, eds. 2013, Routledge, London).
6 *Conservation Biogeography* (Ladle & Whittaker, eds. 2011, Wiley-Blackwell, Oxford).
7 *The Lie of the Land* (Leach & Mearns, eds. 1996, James Currey, Oxford).
8 *Wildlife in the Anthropocene: Conservation after Nature* (Lorimer 2015, Minnesota University Press, Minneapolis, MN).
9 *Rambunctious Garden: Saving Nature in a Post-Wild World* (Marris 2011, Bloomsbury, London).
10 *Imposing Wilderness: Struggles over Livelihood and Nature Preservation in Africa* (Neumann 1998, University of California Press, Berkeley, CA).

## References

Adams, W.M. (2003) Nature and the colonial mind. In: *Decolonizing Nature: Strategies for Conservation in a Post-Colonial Era* (ed. W.M. Adams & M. Mulligan), 16–50. London: Earthscan.

Adams, W.M. (2020) *Green Development: Environment and Sustainability in a Developing World*, 4th ed. London: Routledge.

Angelsen, A. (1995) Shifting cultivation and "deforestation": a study from Indonesia. *World Development* 23: 1713–1729.

Bartels, G.B., Norton, B.E. & Perrier, G.K. (1993) An examination of the carrying capacity concept. In: *Range Ecology at Disequilibrium* (ed. R.H. Behnke, I. Scoones & C. Kerven), 89–103. London: Overseas Development Institute.

Barton, G.A. (2002) *Empire Forestry and the Origins of Environmentalism*. Cambridge, UK: Cambridge University Press.

Barua, M. (2014) Circulating elephants: unpacking the geographies of a cosmopolitan animal. *Transactions of the Institute of British Geographers* 39: 559–573.

Behnke, R.H., Scoones, I. & Kerven, C. (1993) *Range Ecology at Disequilibrium*. London: Overseas Development Institute.

Bell, S.E. (2016) *Fighting King Coal: The Challenges to Micromobilization in Central Appalachia*. Cambridge, MA: MIT Press.

Benjaminsen, T.A. & Bryceson, I. (2012) Conservation, green/blue grabbing and accumulation by dispossession in Tanzania. *Journal of Peasant Studies* 39: 335–355.

Berkes, F. & Folke, C. eds. (1998) *Linking Social and Ecological Systems: Management Practices and Social Mechanisms for Building Resilience*. Cambridge, UK: Cambridge University Press.

Bernstein, H. (2010) *Class Dynamics of Agrarian Change*. Hartford, CT: Kumarian Press.

Black, R. (1990) "Regional political ecology" in theory and practice: a case study from northern Portugal. *Transactions of the Institute of British Geographers* 15: 35–47.

Blaikie, P. & Brookfield, H. (1987) *Land Degradation and Society*. London: Methuen.

Bond, W.J., Silander, J.A., Ranaivonasy, J. et al. (2008) The antiquity of Madagascar's grasslands and the rise of C4 grassy biomes. *Journal of Biogeography* 35: 1743–1758.

Börner, J., Schulz, D., Wunder, S. et al. (2020) The effectiveness of forest conservation policies and programs. *Annual Review of Resource Economics* 12: 45–64.

Bourgeat, F. & Aubert, G. (1972) Les sols ferralitiques à Madagascar. *Revue de Géographie* 20: 1–23.

Bradnock, R.W. & Saunders, P.L. (2000) Sea-level rise, subsidence and emergence: the political ecology of environmental change in the Bengal delta. In: *Political Ecology: Science, Myth and Power* (ed. P. Stott & S. Sullivan), 66–90. London: Arnold.

Brockington, D. (2002) *Fortress Conservation: The Preservation of the Mkomazi Game Reserve, Tanzania*. Oxford: James Currey.

Brockington, D. & Duffy, R. (2010) Capitalism and conservation: the production and reproduction of biodiversity conservation. *Antipode* 42: 469–484.

Brockington, D. & Igoe, J. (2006) Evictions for conservation: a global overview. *Conservation and Society* 4: 424–470.

Bryant, R.L. (2002) Non-governmental organizations and governmentality: "consuming" biodiversity and indigenous people in the Philippines. *Political Studies* 50: 268–292.

Bullard, R.D. (1990) *Dumping in Dixie: Race, Class and Environmental Quality*. Boulder, CO: Westview.

Burney, D.A. (1987) Late Holocene vegetational change in central Madagascar. *Quaternary Research* 20: 130–143.

Burney, D.A. (2003) Madagascar's prehistoric ecosystems. In: *The Natural History of Madagascar* (ed. S.M. Goodman & J.P. Benstead), 47–51. Chicago, IL: University of Chicago Press.

Büscher, B. & Ramutsindela, M. (2016) Green violence: rhino poaching and the war to save Southern Africa's peace parks. *African Affairs* 115: 1–22.

Carswell, G. (2007) *Cultivating Success: Kigezi Farmers and Colonial Policies*. Oxford: James Currey.

Caughley, G. (1987) Ecological relationships. In: *Kangaroos: Their Ecology and Management in the Sheep Rangelands of Australia* (ed. G. Caughley, N. Shepherd & J. Short), 159–187. Cambridge, UK: Cambridge University Press.

Chambers, R. (1994) Paradigm shifts and the practice of participatory research and development. IDS Working Paper No. 2. Brighton, UK: Institute of Development Studies.

Chorley, R.J. & Kennedy, B.A. (1971) *Physical Geography: A Systems Approach*. London: Prentice Hall.

Christaller, W. (1933) *Die Zentralen Orte in Süddeutschland*. Jena, Germany: Gustav Fischer.

Clements, F.E. (1928) *Plant Succession and Indicators*. New York: H.W. Wilson.

Cloke, P., Crang, P. & Goodwin, M. (2005) *Introducing Human Geographies*, 2nd ed. London: Hodder Arnold.

Corson, C.A. (2016) *Corridors of Power: The Politics of Environmental Aid to Madagascar*. New Haven, CT: Yale University Press.

Cowen, M.P. & Shenton, R.W. (1996) *Doctrines of Development*. London: Routledge.

Cronon, W. (1995) *Uncommon Ground: Toward Reinventing Nature*. New York: W.W. Norton.

Crosby, A.W. (1986) *Ecological Imperialism: The Ecological Expansion of Europe, 1600–1900*. Cambridge, UK: Cambridge University Press.

Crush, J. (1995) *Power of Development*. London: Routledge.

Crutzen, P.J. & Stoermer, E.F. (2000) The "Anthropocene". *Global Change Newsletter* 41: 17–18.

Cutter, S.L. (2020) The changing nature of hazard and disaster risk in the Anthropocene. *Annals of the American Association of Geographers* 111: 819–827.

Darby, H.C. (1977) *Domesday England*. Cambridge, UK: Cambridge University Press.

Davis, M. (2006) *Planet of Slums*. London: Verso.

Demeritt, D. (2001) Scientific forest conservation and the statistical picturing of nature's limits in the Progressive Era United States. *Environment and Planning D: Society and Space* 19: 431–459.

Dewar, R.E. (1984) Extinctions in Madagascar: the loss of the subfossil fauna. In: *Quaternary Extinctions: A Prehistoric Revolution* (ed. P.S. Martin & R.G. Klein), 574–593. Tucson, AZ: University of Arizona Press.

Diamond, J. (1997) *Guns, Germs, and Steel: The Fates of Human Societies*. New York: W.W. Norton.

Diamond, J. (2005) *Collapse: How Societies Choose to Fail or Survive*. London: Allen Lane.

Drayton, R. (2000) *Nature's Government: Science, Imperial Britain and the "Improvement" of the World*. New Haven, CT: Yale University Press.

Driver, F. (2001) *Geography Militant: Cultures of Exploration and Empire*. Oxford: Blackwell.

Duffy, R. (2014) Waging a war to save biodiversity: the rise of militarized conservation. *International Affairs* 90: 819–834.

Dyer, S. (2008) Hybrid geographies (2002): Sarah Whatmore. In: *Key Texts in Human Geography* (ed. P. Hubbard, R. Kitchin & G. Valentine), 207–213. London: Sage.

Ehrlich, P. (1968) *The Population Bomb*. London: Ballantine.

Ellis, E.C., Goldewijk, K.K., Siebert, S. et al. (2010) Anthropogenic transformation of the biomes, 1700 to 2000. *Global Ecology and Biogeography* 19: 589–606.

Ellis, E.C. & Ramankutty, N. (2008) Putting people in the map: anthropogenic biomes of the world. *Frontiers in Ecology and the Environment* 6: 439–447.

Ellis, J.E. & Swift, D.M. (1988) Stability of African pastoral ecosystems: alternate paradigms and implications for development. *Journal of Range Management* 41: 450–459.

Escobar, A. (1995) *Encountering Development: The Making and Unmaking of the Third World*. Princeton, NJ: Princeton University Press.

Escobar, A. (2004) Beyond the Third World: imperial globality, global coloniality and anti-globalisation social movements. *Third World Quarterly* 25: 207–230.

Fairhead, J. & Leach, M. (2003) *Science, Society and Power: Environmental Knowledge and Policy in West Africa and the Caribbean*. Cambridge, UK: Cambridge University Press.

Fairhead, J., Leach, M. & Scoones, I. (2012) Green grabbing: a new appropriation of nature? *Journal of Peasant Studies* 39 (2): 237–261.

Fischer, M.M., Henk, J.S. & Unwin, D. (1996) Geographic information systems, spatial data analysis and spatial modelling: an introduction. In: *Spatial Analytical Perspectives on GIS* (ed. M.M. Fischer, H.J. Scholten & D. Unwin), 3–19. London: Taylor & Francis.

Forsyth, T. ed. (2005) *Encyclopedia of International Development*. Oxon, UK: Routledge.

Foucault, M. (1980) *Power/Knowledge: Selected Interviews and Other Writings 1972–1977* (ed. C. Gordon; trans. C. Gordon, L. Marshall, J. Mepham et al.). New York: Pantheon Books.

Fox, J., Suryanata, K., Hershock, P. et al. (2008) Mapping boundaries, shifting power: the socio-ethical dimensions of participatory mapping. In: *Contentious Geographies: Environmental Knowledge, Meaning, Scale* (ed. M.K. Goodman, M.T. Boykoff & K.T. Evered), 203–217. Aldershot, UK: Ashgate Publishing.

Ganzhorn, J.U., Lowry II, P.P., Shatz, G.E. et al. (2001) The biodiversity of Madagascar: one of the world's hottest hotspots on its way out. *Oryx* 35: 346–348.

Gregory, D. (1978) *Ideology, Science and Human Geography*. New York: Harper Collins.

Gregory, D. (1994) *Geographical Imaginations*. Oxford, UK: Blackwell.

Gregory, D., Johnston, R., Pratt, G. et al. eds. (2009) *The Dictionary of Human Geography*. London: Wiley-Blackwell.

Gunderson, L.H. & Holling, C.S. eds. (2002) *Panarchy: Understanding Transformations in Human and Natural Systems*. Washington, DC: Island Press.

Harley, J.B. (1988) Maps, knowledge, and power. In: *The Iconography of Landscape* (ed. D. Cosgrove & S. Daniels), 277–312. Cambridge, UK: Cambridge University Press.

Hartshorne, R. (1959) *Perspective on the Nature of Geography*. Chicago, IL: Rand McNally.

Harvey, D. (1973) *Social Justice and the City*. Baltimore, MD: Johns Hopkins University Press.

Harvey, D. (1996) *Justice, Nature and the Geography of Difference*. Oxford, UK: Blackwell.

Hays, S.P. (1959) *Conservation and the Gospel of Efficiency: The Progressive Conservation Movement, 1890–1920*. Cambridge, MA: Harvard University Press.

Hewitt, K. ed. (1983) *Interpretation of Calamity*. London: Allen and Unwin.

Hinchliffe, S. (2007) *Geographies of Nature: Societies, Environments, Ecologies*. London: Sage.

Hinchliffe, S., Kearnes, M.B., Degen, M. et al. (2005) Urban wild things: a cosmopolitical experiment. *Environment and Planning D: Society & Space* 23: 643–658.

Hingston, R.W.G. (1931) Proposed British national parks for Africa. *Geographical Journal* 77: 401–428.

Hofer, T. & Messerli, B. (2006) *Floods in Bangladesh: History, Dynamics and Rethinking the Role for the Himalayas.* New York: United Nations University Press.

Holling, C.S. & Meffe, G.K. (1996) Command and control and the pathology of natural resource management. *Conservation Biology* 10: 328–337.

Hudson, B. (1977) The new geography and the new imperialism: 1870–1918. *Antipode* 9: 12–19.

Hulme, D. & Murphree, M. (2001) *African Wildlife and Livelihoods: The Promise and Performance of Community Conservation.* Oxford, UK: James Currey.

Humbert, H. (1927) Principaux aspects de la végétation à Madagascar. *Memoires de l'Academie Malgache* 5: 1–89.

Huntington, E. (1915) *Civilization and Climate.* New Haven, CT: Yale University Press.

Huxley, T. (1877) *Physiography: An Introduction to the Study of Nature.* London: Macmillan.

Igoe, J. & Brockington, D. (2007) Neoliberal conservation: a brief introduction. *Conservation and Society* 5 (4): 432–449.

IPBES (Intergovernmental Science-Policy Platform on Biodiversity and Ecosystem Services) (2019) *Summary for Policymakers of the Global Assessment Report on Biodiversity and Ecosystem Services of the Intergovernmental Science-Policy Platform on Biodiversity and Ecosystem Services.* Bonn, Germany: IPBES.

Ives, J. & Messerli, B. (1989) *The Himalayan Dilemma: Reconciling Development and Conservation.* London: Routledge.

Jepson, P., Barua, M. & Buckingham, K. (2011) What is a conservation actor? *Conservation and Society* 9 (3): 229–235.

Knox, P. & Marston, S. (2004) *Places and Regions in Global Context: Human Geography.* Upper Saddle River, NJ: Prentice Hall.

Kull, C.A. (2000) Deforestation, erosion, and fire: degradation myths in the environmental history of Madagascar. *Environment and History* 6: 423–450.

Ladle, R.J. & Whittaker, R.J. eds. (2011) *Conservation Biogeography.* Oxford, UK: Wiley-Blackwell.

Leach, M. & Fairhead, J. (2002) Fashioned forest pasts, occluded histories? International environmental analysis in West African locales. *Development and Change* 31: 35–59.

Lefebvre, H. (1991) *The Production of Space* (trans. D. Nicholson-Smith). Malden, MA: Blackwell Publishing.

Livingstone, D.N. (1992) *The Geographical Tradition.* Malden, MA: Blackwell Publishing.

Livingstone, D.N. & Withers, C.W.J. (1999) *Geography and Enlightenment.* Chicago, IL: Chicago University Press.

Longley, P.A., Goodchild, M.F., Maguire, D.J. et al. (2001) *Geographic Information Systems and Science.* Chichester, UK: John Wiley and Sons.

Lorimer, J. (2015) *Wildlife in the Anthropocene: Conservation after Nature.* Minneapolis, MN: University of Minnesota Press.

Low, N. & Gleeson, B. (1998) *Justice, Society, and Nature: An Exploration of Political Ecology.* Oxon, UK: Routledge.

Mackenzie, A.F.D. (1998) *Land, Ecology and Resistance in Kenya 1880–1952*. London: Heinemann.

Malhi, Y. (2017) The concept of the Anthropocene. *Annual Review of Environment and Resources* 42: 77–104.

Malthus, T. (1798) *An Essay on the Principle of Population, as It Affects the Improvement of Society with Remarks on the Speculations of Mr. Godwin, M. Condorcet, and Other Writers*. London: Printed for J. Johnson in St Paul's Churchyard.

Margules, C.R. & Pressey, R.L. (2000) Systematic conservation planning. *Nature* 405: 243–253.

Marris, E. (2011) *Rambunctious Garden: Saving Nature in a Post-Wild World*. London: Bloomsbury.

Marsh, G.P. (1864) *Man and Nature; or, Physical Geography as Modified by Human Action*. New York: Charles Scribner.

McCarthy, J. & Prudham, S. (2004) Neoliberal nature and the nature of neoliberalism. *Geoforum* 35: 275–283.

Meine, C. (2004) *Correction Lines: Essays on Land, Leopold and Conservation*. Washington, DC: Island Press.

Meine, C. & Meffe, G.K. (1996) Conservation values, conservation science: a healthy tension. *Conservation Biology* 10: 916–917.

Mills, S. (1997) *Discourse*. London: Routledge.

Morris, J. (2005) Confuse: how Jared Diamond fails to convince. *Energy & Environment* 16: 395–421.

Mortimore, M. (1998) *Roots in the African Dust: Sustaining the Drylands*. Cambridge, UK: Cambridge University Press.

Mortimore, M. & Adams, W.M. (1999) *Working the Sahel: Environment and Society in Northern Nigeria*. London: Routledge.

Myers, N., Mittermeier, R.A., Mittermeier, C.G. et al. (2000) Biodiversity hotspots for conservation priorities. *Nature* 403: 853–858.

Neumann, R.P. (1998) *Imposing Wilderness: Struggles over Livelihood and Nature Preservation in Africa*. Berkeley, CA: University of California Press.

Neumann, R.P. (2004) Nature-state-territory: towards a critical theorization of conservation enclosures. In: *Liberation Ecologies: Environment, Development, Social Movements* (ed. R. Peet & M. Watts), 195–217. London: Routledge.

Nystuen, J.D. (1963) Identification of some fundamental spatial concepts. *Papers of the Michigan Academy of Science, Arts and Letters* 48: 373–384.

Olson, D.M. & Dinerstein, E. (2002) The Global 200: priority ecoregions for global conservation. *Annals of the Missouri Botanical Gardens* 89: 199–224.

Peet, R. & Watts, M. (1996) *Liberation Ecologies: Environment, Development, Social Movements*. London: Routledge.

Perrier de la Bâthie, H. (1921) La végétation Malgache. *Annals du Musee Colonial de Marseille* 3: 1–266.

Potschin, M.B. & Haines-Young, R.H. (2011) Ecosystem services: exploring a geographical perspective. *Progress in Physical Geography* 35: 575–594.

Potter, R.B., Binns, T., Elliot, J.A. et al. eds. (2008) *Geographies of Development: An Introduction to Development Studies*. London: Pearson.

Radcliffe, S.A., Watson, E.E., Simmons, I. et al. (2010) Environmentalist thinking and/in geography. *Progress in Human Geography* 34: 98–116.

Randrianja, S. & Ellis, S. (2009) *Madagascar: A Short History*. London: Hurst.

Ribeiro, B.R. & Atadeu, M. (2019) Systematic conservation planning: trends and patterns among highly cited papers. *Journal for Nature Conservation* 50: art. 125714.

Richardson, D., Castree, N., Goodchild, M.F. et al. (2016) *International Encyclopedia of Geography*. Oxford, UK: Wiley-Blackwell.

Rindfuss, R.R., Walsh, S.J., Mishra, V. et al. (2003) Linking household and remotely sensed data. In: *People and the Environment: Approaches for Linking Household and Community Surveys to Remote Sensing and GIS* (ed. J. Fox, R.R. Rindfuss & S.J. Walsh et al.), 1–29. Boston, MA: Kluwer Academic Publishers.

Robbins, P. (2003) Fixed categories in a portable landscape: the causes and consequences of land cover categorization. In: *Political Ecology: An Integrative Approach to Geography and Environment-Development Studies* (ed. K.S. Zimmerer & T. Bassett), 181–200. New York: Guilford Press.

Robbins, P. (2019) *Political Ecology: A Critical Introduction*, 3rd ed. Oxford, UK: Wiley-Blackwell.

Robinson, M. (1994) Governance, democracy and conditionality: NGOs and the new policy agenda. In: *Governance, Democracy & Conditionality: What Role for NGOs?* (ed. A. Clayton), 35–52. Oxford, UK: INTRAC.

Roe, E.M. (1991) Development narratives, or making the best of blueprint development. *World Development* 19: 287–300.

Rohde, K. (2005) *Nonequilbrium Ecology*. Cambridge, UK: Cambridge University Press.

Rostow, W.W. (1960) *The Stages of Economic Growth: A Non-Communist Manifesto*. Cambridge, UK: Cambridge University Press.

Said, E. (1979) *Orientalism*. New York: Vintage Books.

Sandbrook, C., Nelson, F., Adams, W.M. et al. (2010) Carbon, forests and the REDD paradox. *Oryx* 44: 330–334.

Scales, I.R. (2011) Farming at the forest frontier: land use and landscape change in western Madagascar, 1896 to 2005. *Environment and History* 17: 499–524.

Scales, I.R. (2012) Lost in translation: conflicting views of deforestation, land use and identity in western Madagascar. *The Geographical Journal* 178: 67–79.

Scales, I.R. (2014) The future of conservation and development in Madagascar: time for a new paradigm? *Madagascar Conservation and Development* 9: 5–12.

Scales, I.R. (2015) Paying for nature: what every conservationist should know about political economy. *Oryx* 49: 226–231.

Schaefer, F.K. (1953) Exceptionalism in geography: a methodological examination. *Annals of the Association of American Geographers* 43: 226–249.

Schroeder, R.A. (1999) Geographies of environmental intervention in Africa. *Progress in Human Geography* 23: 359–378.

Scott, J.C. (1998) *Seeing Like a State: How Certain Schemes to Improve the Human Condition Have Failed*. New Haven, CT: Yale University Press.

Sen, A. (1999) *Development As Freedom*. Oxford, UK: Oxford University Press.

Stoddart, D.R. (1986) *On Geography and Its History*. Oxford, UK: Blackwell Publishers.

Swift, J. (1996) Desertification narratives; winners and losers. In: *The Lie of the Land: Challenging Received Wisdom on the African Environment* (ed. M. Leach & R. Mearns), 73–90. London: James Currey/Heinemann.

Swyngedouw, E. (2004) *Social Power and the Urbanisation of Water: Flows of Power*. Oxford, UK: Oxford University Press.

Thomas, W.L., Jr. ed. (1956) *Man's Role in Changing the Face of the Earth*, vol. 1. Chicago, IL: University of Chicago Press.

Tiffen, M., Mortimore, M. & Gichuki, F. (1994) *More People, Less Erosion: Environmental Recovery in Kenya*. Chichester, UK: Wiley.

Turner, M.D. (2003) Methodological reflections on the use of remote sensing and geographic information science in human ecological research. *Human Ecology* 31: 255–279.

UN DESA (United Nations Department of Economic and Social Affairs). (2019) *World Urbanization Prospects: The 2018 Revision*. New York: United Nations.

UNDP (United Nations Development Programme) (2020) *Human Development Report 2020: Technical Notes*. New York: United Nations.

Von Thünen, J.H. ([1826]1966) *Der Isolierte Staat in Beziehung auf Landwirtschaft und Nationalökonomie (Hamburg)*, 2nd ed. Reprint. Stuttgart, Germany: Gustav Fischer.

Waldron, A., Mooers, A.O., Miller, D.C. et al. (2013) Targeting global conservation funding to limit immediate biodiversity declines. *Proceedings of the National Academy of Sciences of the United States of America* 110: 12144–12148.

Watts, M. (1983) *Silent Violence: Food, Famine and Peasantry in Northern Nigeria*. Berkeley, CA: University of California Press.

Watts, N. & Scales, I.R. (2015) Seeds, agricultural systems and socio-natures: towards an actor–network theory informed political ecology of agriculture. *Geography Compass* 9: 225–236.

Whatmore, S. (2002) *Hybrid Geographies: Natures, Cultures, Spaces*. London: SAGE Publications Ltd.

Whyte, K.P. (2017) Our ancestors' dystopia now: indigenous conservation and the Anthropocene. In: *The Routledge Companion to the Environmental Humanities* (ed. U. Heise, J. Christensen & M. Niemann), 206–215. London: Routledge.

Willems-Braun, B. (1997) Buried epistemologies: the politics of nature in (post) colonial British Columbia. *Annals of the Association of American Geographers* 87: 3–31.

Williams, M. (2003) *Deforesting the Earth*. Chicago, IL: University of Chicago Press.

Wilson, E.O. (1992) *The Diversity of Life*. Cambridge, MA: Harvard University Press.

Wright, D.J., Goodchild, M.F. & Proctor, J.D. (1997) GIS: tool or science? Demystifying the persistent ambiguity of GIS as "tool" versus "science". *Annals of the Association of American Geographers* 87: 346–362.

Zimmerer, K.S. & Bassett, T. (2003) *Political Ecology: An Integrative Approach to Geography and Environment-Development Studies*. New York: The Guilford Press.

# 6

# Political Science and Conservation

*Daniel C. Miller and Arun Agrawal*

## 6.1 Defining Political Science

Political science is the study of political processes, systems, and behavior. This long-standing social science discipline is, in short, concerned with politics. Ubiquitous in social life, politics suffuse human relationships and our relationship with the natural world.

Political scientists have defined politics as "who gets what, when, and how" (Lasswell 1936, p. 345), the "authoritative allocation of values" (Easton 1953, p. 129), and the "conciliation of differing interests" (Crick 1964, p. 148). As these classic definitions indicate, politics concerns the making of decisions that affect people and groups of people. These definitions also highlight the diverse and competing interests of different groups in collective decision-making. The tensions that mark collective decision-making mean that politics is fundamentally about the exercise of power in the allocation of resources; is based on potentially competing values; requires institutional or other means to manage differences; and results in the distribution of rights, resources, and responsibilities.

Patterns of social behavior involving any of the above elements can fall under the purview of political science. Because **biodiversity conservation** requires making science- and value-based decisions, and because it inevitably results in the allocation of rights, resources, and responsibilities, it is inherently political. It is therefore well suited to analysis using concepts, theories, and tools from political science.

Political scientists study the "capital P" politics that unfolds and takes shape in Beijing, Brussels, Washington, DC, and other places where official political activities are conducted. They also study the "small p" politics of households, neighborhood associations, religious institutions, corporate boardrooms, and less formal group settings (Shepsle 2010). Studies of politics span social, temporal, and geographic contexts stretching around the globe and throughout history. They build from theories and tools used in multiple disciplines— including anthropology, economics, sociology, and, of course, political science itself—to describe the structure of the political world, analyze how it works, and explain why it works the way it does.

In this chapter, we provide a portrait of some of the variety that constitutes political science, with an eye to those features that are most relevant to conservation. We describe the main areas of inquiry, concepts, and approaches in the discipline. In so doing, we discuss

*Conservation Social Science: Understanding People, Conserving Biodiversity*, First Edition.
Edited by Daniel C. Miller, Ivan R. Scales, and Michael B. Mascia.

the engagement between political science and conservation, which, despite some notable exceptions, has remained limited. The core of the chapter then focuses on four concepts central to conservation policy and practice: power, participation, institutions, and governance. We conclude with future directions, including suggestions for enhancing the conversation between those interested in conservation *and* politics.

### 6.1.1 Subfields in Political Science

Political science is structured into subfields, including comparative politics, international relations, American politics, political theory (or political philosophy), **political economy**, public policy, and political **methodology**. These fields of inquiry inform and blend into one another through the use of common methods or by addressing questions that cut across multiple subfields (Goodin 2011). Nevertheless, significant differences among them remain, including the distinction between **normative** and empirical approaches (Box 6.1).

---

**Box 6.1   Debates: Political science, values, and public policy**

Political science has had a complex relationship with public policy. As a field, public policy examines what government officials choose to do—or not do—about issues of importance to the public. Given its more practical focus on policy, the subfield of public policy may be of special interest to those who care about conservation issues. Public policy is sometimes part of political science department. It also exists in its own departments or schools.

Debate about the proper role and place of public policy research remains lively within political science. Perhaps the most important disagreement stems from the normative orientation public policy can have and the ambition of mainstream political analysis to develop a science of politics. Many scholars of policy have some interest in actually influencing policy in their areas of specialization. Policy studies in general "embody a bias toward acts, outputs, and outcomes—a concern with consequences" (Goodin et al. 2006, p. 28). This interest contrasts to much of the rest of political science, which focuses on political behavior, processes, and systems rather than on the implementation and outcomes of specific policies. Many political scientists thus have little interest in engagement with specific policy debates, seeking instead to advance a general understanding of politics.

One can identify two broad types of political science research on environmental and conservation issues: (i) that for which the environment holds interest as a specific context in which to explore broader questions of politics, and (ii) that motivated primarily by a normative commitment to improving environmental outcomes (Steinberg 2005). For researchers committed to the latter, political science has become more and more esoteric and less and less relevant to real-world issues. For the former, normative commitments on the part of policy analysts have sometimes led to work that fails to advance scientific knowledge of politics.

This divide mirrors the debate within conservation biology regarding the appropriate role of scientists in policy (Lackey 2007). As in the case of conservation biology, scope exists for normatively and empirically oriented political scientists to find common ground and effectively navigate the gap between science and policy advocacy. Both

---

**Box 6.1 (Continued)**

groups share a common interest in better understanding the policy process, which includes government policy decision-making and the outcomes of such decisions on society and the environment. The first part of this process has been the traditional focus of political science, and therefore developments in the wider discipline should be of interest to policy scholars. At the same time, contributions to knowledge about the functioning of government institutions by scholars in the policy subfield are relevant to all political scientists regardless of their interest in the effects of specific policies. In the environmental domain, both have an interest in better understanding how (biophysical) science affects the policy process. A growing interest in quantitative causal inference within political science suggests that studies estimating the effects of specific environmental policies and programs, and simultaneously testing theories of politics, will be of increasing interest to political scientists. Finally, the gravity and complexity of current environmental challenges may spur increased collaboration between normative- and theory-inspired political scientists in the service of greater understanding of the political dimensions of these pressing issues.

---

Much research in the subfields of political theory and at least some investigations in the fields of public policy and public administration are normative in orientation. That is, they are often more concerned with "what ought to be" rather than "what is." For example, a political philosopher might develop and defend an argument about the best form of government or a policy analyst may identify a particular policy option as preferable for achieving specific goals. In contrast, the majority of work in political science, including in comparative politics, American politics, and international relations, takes a **positivist** approach (see Chapter 2). This approach seeks to advance scientific understanding of political phenomena by accumulating knowledge that describes, analyzes, explains, and at times, seeks to predict empirical patterns based on clearly articulated premises in the belief that objective knowledge of political relationships is possible and desirable. Developing and testing hypotheses about the causes and consequences of patterns of political decision-making, policy implementation, and policy performance lie at the heart of this approach.

Political scientists use a variety of methods and approaches. These run the gamut from humanistic to scientific. They do so often in close, even if implicit, conversation with other disciplines, from anthropology and history to zoology and mathematics. They use **quantitative** methods such as survey research, statistical analysis, and field experiments, as well as **qualitative** methods like **ethnographic** research, case studies, and historical comparisons (see Chapter 2 for more on these methods and their **epistemological** underpinnings). Approaches based on deductive reasoning include modeling efforts such as game theory, agent-based modeling, and other mathematical models of strategic interactions to understand political behavior. Inductive approaches include experiments as well as observational studies that rely on specific cases or cross-case comparisons. In practice, political scientists often combine qualitative and quantitative methods, and deductive and inductive approaches. Political scientists also develop agent-based models (Box 6.2) to simulate the interactions of political decision makers and use spatial and time-series data to advance insights into political outcomes.

---

**Box 6.2    Methods: Agent-based modeling**

The world is complex and dynamic. Natural and social scientists have long sought to develop ways to understand this complexity. In recent years, agent-based modeling has attracted increased attention within a range of scientific disciplines, including political science. Part of the attraction of agent-based models (ABMs) stems from their ability to shed light on how various social and ecological phenomena emerge, are maintained, and sometimes disappear (Berry et al. 2002). ABMs derive outcomes from the dynamic interaction of rule-based "agents"—which could be landowners, resource users, or even animals—in specific environments and times. Unlike many other social science approaches that examine how social structure shapes behavior, agent-based modeling explores how social structures and patterns of behavior emerge from the interaction among individual agents. This bottom-up approach can thus yield insights that approaches focusing on single actors in isolation commonly miss (Berry et al. 2002). An ABM is highly flexible and capable of representing a wide variety of causal factors that influence social outcomes including differences among agents, non-linear relationships, and network interactions.

Computer simulation has greatly expanded the application of ABMs, making them more widely available to help stakeholders and decision makers understand the response of complex systems to a range of possible policy interventions or disturbances. For example, Baulenas et al. (2021) developed a model to simulate the effects of different payments for **ecosystem services** (PES) designs on the behavior of people and forest and water conditions in a river basin in Catalonia under different climate scenarios. Their results suggest that optimal PES policy design should include repeated payments to landholders, the presence of a local intermediary, and targeting one **ecosystem** goal (rather than more than one at once). The model showed that first-generation participants in the scheme would see benefits and thus have an incentive to remain in the program over time. A local intermediary between landowners and the government was key in building trust and ensuring targeted agents join the PES scheme voluntarily, thereby supporting its early success. Finally, focusing on one ecosystem goal helped avoid the ecosystem tradeoffs that can arise when two or more goals are targeted.

Agent-based modeling is especially useful when multiple, interactive forces influence the dynamics of land-use changes or management regimes for common-pool resources. This approach enables participants to see policies, laws, and institutions as complex, evolving pathways rather than fixed end points, thereby opening up possibilities for changing them.

---

Although political scientists often seek to develop scientific theories and explanations of politics, normative questions also motivate them (see Box 6.1). Political philosophy provides one important source of questions that concern empirical political scientists (Laitin 2004). Contemporary events, especially disasters, crises, and pressing social issues also give rise to important questions that occupy political science. For example, the September 11, 2001, attacks on the USA and growing awareness of human-induced climate change generated a new wave of studies within international relations (Keohane 2008;

Underdal 2017). Like scientists working in other domains such as medicine or conservation biology, political scientists may thus be motivated by normative interest, but seek a rigorous scientific approach to the study of politics.

### 6.1.2 Major Themes and Questions in Political Science

Contemporary political science research covers a wide variety of topics. There are many ways to parse major themes and questions in the discipline. We group them under four broad headings: (i) order and states, (ii) democracy and justice, (iii) political participation, and (iv) security. Together, these four broad topics illustrate some of the major orientations in the study of politics.

#### 6.1.2.1 Order and States

How do states arise and how do they break down? How do governing regimes within states change? How do cooperative or conflictual behaviors develop and change within and among states? What is the impact of **globalization** on states and those they would govern? These questions get to the heart of political organization and change. To answer them, political scientists use the concept of "the state," understood as "a complex apparatus of centralized and institutionalized power that concentrates violence, establishes property rights, and regulates society within a given territory while being formally recognized as a state by international forums" (Levi 2002, p. 40).

States are distinct, though related to "nations." A nation is an "imagined political community" (Anderson 1991) wherein a large group of people are bound together by commonalities such as language, ethnicity, or **culture**. Nations may coincide with state boundaries, but they often do not, and a state's territory may include many different nations. For example, Spain may be understood to include a number of different nations, such as Basque, Catalan, and others, which are located partly within the Spanish state and partly in other states (e.g. France in the case of Basques and Catalans). Mismatches between the boundaries claimed by particular nations and those of existing states can lead to conflict. States can also inspire devotion in the form of patriotism or nationalism. Understanding how modern states insinuate themselves into the core identities of their subjects is an active area of research in political science.

States are also not identical to government or the laws and bureaucracies they contain. For example, the USA exists as a state no matter which political party holds power. Governments set and administer public policies. They can collect taxes, provide public goods, administer justice, and wage war. There are many different types of governments, from democratic to totalitarian, and they are distinguished by arrangements for making and implementing law and legislation (Levi 2002). Many political scientists seek to understand the persistence and breakdown of states and political regimes (governments) within them. Others investigate the effects of globalization on the sovereignty of states, or authority within a given territory.

#### 6.1.2.2 Democracy and Justice

How do democracies emerge, survive, or fail? How do different institutional forms affect the incentives of citizens and the performance and endurance of democracies? What is the nature of citizenship? What are the best ways to distribute resources and opportunities among individuals in society? How do group identity, race, gender, and class affect political processes and outcomes? Why are inequalities growing in many countries around the globe and why does inequality persist?

Democracy has been a central concern of political science since the discipline's inception. With origins in the Greek words for "people" (*demos*) and "rule, strength" (*krátos*), democracy refers to "rule by the people." Democracy can be understood to include two basic principles: (i) political equality such that all members of a state or society have equal access to power and (ii) all members have universally recognized rights and freedoms (Dahl et al. 2003). Liberal democracies throughout the world are founded on these two principles.

Elections are the chief mechanism by which people exercise their "rule" in democracies. By choosing among alternative candidates, citizens determine the people charged with setting the state's policies for a designated period of time until the next election. Other key democratic institutions include representative bodies such as a parliament or congress, courts, bureaucracies, political parties, and interest groups. Together these institutions shape how preferences of individuals and groups are aggregated to determine outcomes ranging from the protection of the populace to the distribution of wealth. Here critical questions of justice arise. Does everyone in the society have equal say in the political process? What is the best way to ensure representation of different, often competing perspectives? What mix of freedom and responsibility does justice entail? How do past injustices affect current justice?

Political philosophers continue to debate the nature of justice while empirical political scientists seek to define and measure aspects of justice within and across polities. For example, research comparing the voting response of US senators to the opinions of constituents with high and low incomes has concluded that legislators are much more responsive to the wealthy (Bartels 2008). This disparity in political clout undermines democratic ideals of equal participation and fairness.

Political scientists often focus on the distributional aspects of justice. Distributional or social justice refers to how social policy and institutional arrangements affect the distribution of resources and opportunities within a society (Waldron 2002). Justice also includes recognitional and procedural aspects relating to which groups are acknowledged in political processes and the extent to which they are able to participate in such processes (Whyte 2011; Suiseeya & Caplow 2013).

### 6.1.2.3 Political Participation

Who participates in the political process and why? How does political participation affect policy making? How do the formal institutions of government translate individual and group interests into policy? What kinds of mobilization, organizations, and tactics of advocacy lead to what kinds of government responses? How do individual choices aggregate into social patterns of behavior and generate new institutions? How do institutions structure actions, aggregate individual choices, and shape collective and social outcomes?

Democracy and justice relate closely to political participation and a set of important concepts political scientists use to analyze it. **Civil society** designates the broad sphere of politics that is, in theory, distinct from the state, family, and market. Here we find a welter of organizations, ranging from non-government organizations (NGOs) concerned with conservation to business associations, from community groups and religious organizations to unions and social movements. Political scientists debate whether civil society is the preserve of groups in modern, democratic countries or encompasses many different kinds of associations such as the mafia, extremist Islamic groups, or far-right American militias and traditional groups based on inherited characteristics like religion and ethnicity (Edwards 2004). Scholars also look beyond national boundaries to explore the emergence of a "global civil society," understood as a sphere of voluntary societal associations located above the individual and below the state as well as across state boundaries (Keck & Sikkink 1998).

Political culture is a related concept, which denotes the set of all "attitudes towards the political system and its various parts, and attitudes toward the role of the self in the system" (Almond & Verba 1963, p. 13). These shared attitudes and values form the basis of political behavior. Because they vary across different societies, political behavior varies as well. This behavior, in turn, affects the form of political institutions and politics. Research in this tradition (e.g. Inglehart & Welzel 2005) underscores the importance of cultural factors in understanding the stability of democratic regimes.

Political scientists have also used and helped develop the related concept of **social capital**. Although its origins lie in urban studies (Jacobs 1961) and sociology (Bourdieu 1985), political science research has helped popularize the term (e.g. Putnam 1995). Social capital inheres in the features of social life—networks, **norms**, and trust—that enable participants to act together more effectively to pursue shared objectives (Putnam 1995, p. 67). It is seen as an important factor influencing the quality and stability of democracy. The erosion of social capital thus undermines participation in the political process and the functioning of political institutions.

### 6.1.2.4 Security

Why do wars start and how can peace be established and maintained? Why do states cooperate and what is the role of multilateral institutions? What triggers organized protest and how do social movements shape policy responses and institutional change? What are the implications of technological change for national and international peace and security? How is global environmental change affecting peace and security within and among countries? What policy action can effectively address the unprecedented stress that humans are placing on planetary systems?

Security has many meanings. It is a central subject in international relations and public policy, but also political theory as well as American politics. Security has traditionally focused on provision of order and a well-functioning society within states as well as the ability of states and international institutions, such as the United Nations, to ensure their mutual survival and safety. Research in this area examines the causes of war and other international conflict and the formation and enforcement of international agreements, among other topics.

Political scientists and others have broadened the concept of security considerably over the past several decades to encompass not only military aspects (e.g. avoiding war) but also general social, economic, and environmental well-being (Ullman 1983). Human security can be understood as freedom from both violent conflict and physical want. Political scientists have investigated the relationship between human security and environmental change broadly, as well as specific links between the environment and violent conflict. This research finds that climate anomalies and change—in the form of temperature and precipitation changes—have a positive association with both interpersonal and intergroup violence and that climate change likely amplifies the effects of other drivers of conflict (Burke et al. 2015; Koubi 2019). Scholarship on the subject has moved away from a focus on average effects of climate or other environmental variables to a focus on the mechanisms such as migration, agriculture, and economic shocks through which effects of environmental variable may translate into different types of conflict (Busby 2018; Koubi 2019). Other research has focused on the possible role of environmental conservation, cooperation, and collaboration in promoting peace, but understanding the consequences of peace or human security for the environment remains a research frontier (Khagram & Ali 2006; Busby 2018).

The cluster of concepts around security also encompasses the area of contentious politics. Contentious politics is non-routine politics that involves the use of disruptive measures such as general strikes, terrorism, and social movements to make a political point, change government policy, or more radically, undermine a government and change political systems. Such politics may start when individuals or organizational actors advance competing claims against existing powerholders. These claims may give rise to the coordination of efforts on behalf of shared interests or programs, in which governments are targets, the objects of claims, or third parties (Tilly & Tarrow 2006). Because governments are involved, contentious politics necessarily interact with non-contentious political processes such as routine public administration, elections, military conscription, tax collection, appointment of officials, and disbursement of funds (McAdam et al. 2009). Understanding the relationship between "normal" politics and contentious politics and what triggers organized protest, social movements, regime change, civil war, and revolution remain important topics in contemporary political science.

### 6.1.3 Political Science and Conservation

The major themes and questions described above help clarify the relevance of political science to conservation even if the specific domains in which political scientists study them may be distant from the concerns of conservationists. Some of these questions stem from mature research areas within the discipline, such as those relating to the four main concepts discussed in this chapter: power, institutions, participation, and governance. Other questions represent frontiers of inquiry beyond which much less is known. Several important, leading-edge questions at the intersection of research in political science and conservation can be grouped under the four broad themes of governance, equity, identity, and networks (Table 6.1).

Generally, biodiversity conservation has not been a major focus for the discipline of political science even as important differences exist across subfields in how much they have engaged with this topic. Engagement with conservation and other environmental issues is perhaps most developed in international relations and public policy. However, scholars in other subfields also have contributed substantially to analyses and concepts relevant to conservation.

**Table 6.1** Research questions at the frontier of political science and environmental conservation.

| Theme | Research questions |
|---|---|
| Governance | What leads to effective governance and institutions across scales? How are variations in forms and types of governance related to conservation outcomes? |
| Equity | How can policies be designed to address environmental problems like conservation or climate change that have diffuse, long-term benefits, but concentrated, short-term costs? How might these costs and benefits be distributed equitably across space and over time? How are attitudes about the future aggregated at the societal level? |
| Environmental identities | How do people come to care about and act on behalf of the environment? How does the formation of environmental identities affect governance? |
| Networks | How do policy networks affect policy change? What is the connection between science and policy? How does science travel through policy networks? How do politicians and policy makers use science? |

The connection between political theory and the environment stretches back to antiquity (e.g. Aristotle's [1992] efforts to classify political regimes based on a similar method of classifying the natural world), but the work of political theorists starting from Hobbes has been the staple of scholarship with insights relevant to conservation. Hobbes famously posited a "state of nature" in which life was "nasty, brutish and short," "a war of all against all," as a starting point to justify Leviathan, a powerful ruler capable of ensuring political order (Hobbes 1958 [1651]). His theories about the relationship between individual behavior and macro level institutions, such as a strong central government, continue to hold relevance for contemporary debates about the role of the state in facilitating collective action to address environmental concerns. By contrast, Thoreau (1993 [1849]) found inspiration in nature, which set the stage for critical awareness and political action like civil disobedience.

Empirical political science began to show interest in environmental conservation in the late 1960s. Debates about "overpopulation" and the "tragedy of the commons" (Box 6.3) as well as the 1972 United Nations Conference on the Human Environment in Stockholm spurred political science research on the environment. Much of this early work had a **neo-Malthusian** flavor, arguing that ecological limits inhibit cooperation and prevent politicians from making tradeoffs they ordinarily seek in politics, thereby requiring a powerful central government (Ophuls 1977). Subsequent work (e.g. Ostrom 1990) has countered such arguments by showing how and when coordination and collective action (Box 6.3) are possible. As discussed below, the development and modification of institutions can facilitate coordination, thereby solving problems of collective action.

---

**Box 6.3   Crossing boundaries: The importance and logic of collective action**

One of the more robust findings in political science holds key insights for effectively addressing environmental concerns like **biodiversity** loss: Individuals who do not contribute to the collective provision or maintenance of public goods (like roads, education, national defense, and healthy ecosystems) that benefit groups as a whole cannot be excluded from enjoying these goods. Therefore, each person is motivated not to contribute to the collective effort. Not contributing to the production or maintenance of a public good, but enjoying its benefits is known as *free-riding*. The risks of free-riding are ubiquitous in different social contexts. If many participants choose to free-ride, the collective benefit will not be produced (Ostrom 1990). The problem of collective action prevails when individuals pursuing their own narrow interests do not act to achieve their common or group interests (Olson 1971). This problem is especially difficult to address for large groups, such as when people around the globe face biodiversity loss (or climate change). The reason for this difficulty is that each individual's actions have only a minuscule influence on collective outcomes, so the incentive is even stronger for individuals not to contribute. The free-rider problem supplies the major argument for government, which can enforce individual contributions to the public good.

Environmental goods like biodiversity present another difficulty for collective action. Free-riders cannot be excluded from benefitting from a given resource, such as forests, pasture, or water, yet greater resource use leads to greater degradation. Excessive use can lead to collapse and disappearance of the good altogether. Goods that have these characteristics—difficulty of exclusion and reduction with use—are **common-pool resources**.

Ecologist Garret Hardin (1968) argued that commons suffer from the tragedy of overuse, but he mistook open access for communal control. Unfortunately, many concerned with conservation continue to see his argument as describing the use of commonly owned resources (Janssen et al. 2019) and are unaware of its racist underpinnings, including his views on "overbreeding" among certain human populations (Mildenberger 2019). Such oversimplified and problematic logic has led many environmental scientists, among others, to propose that there are only two solutions that can sustain the commons over the long run: a strong, centralized government and private property. These solutions may be appropriate in some cases, but they are not the only options. Elinor Ostrom's boundary-spanning work *Governing the Commons* (1990) demonstrated that resource users are not doomed to this grim scenario. Through systematic theoretical and empirical analyses, she demonstrated the viability of the creative institutional solutions many groups and communities have devised to solve collective-action problems and sustain their resources over time (see Section 6.3.2).

Political scientists are well aware of the dynamics and potential solutions to the free-rider problem. But the breadth and efficacy of possible solutions is less well understood outside the discipline. To effectively tackle the massive and urgent problems of biodiversity loss and climate change, it is critical that people ubiquitously understand and appreciate the problem of collective action and its solutions—as ubiquitously, for example, as is the case for the laws of supply and demand in economics (Mansbridge 2009).

In addition to political science studies examining conservation and natural resource issues directly, political science research in a few other domains is worth mentioning briefly, given its relevance to conservation. For example, pathbreaking work on resistance (Scott 1985) has influenced understanding of the importance of local people to the success of conservation policies. Similarly, the study of social capital (Putnam 1995) provides a theoretical and empirical basis for arguments that strong civil society organizations are necessary to create pressures favorable to positive policies for conservation. Scholarship in international relations (Keohane & Levy 1996; Young 2002) has demonstrated conditions under which international institutions, such as the Convention on Biological Diversity, the Montreal Protocol, and multilateral donors like The World Bank, are more or less effective in facilitating environmental protection.

Other conservation-relevant political science research that is important but not described in detail in this chapter includes studies of the **political economy** of wildlife policy (Gibson 1999), how interactions among government offices affect environmental management (Thomas 2003; Struthers et al. 2021), the impact of political decentralization and elections on forest condition (Agrawal 2005; Andersson & Gibson 2007; Sanford 2021), and how interest groups and private actors shape environmental policy from local to global scales (Green 2013; Stokes 2020).

## 6.2 A Brief History of Political Science

Political science (also known as "political studies," "government," and sometimes just "politics") has existed for more than a century as a separate social science discipline, but the study of politics has a longer history. The roots of political science lie in ancient China, Greece, India, and the Middle East. Political scientists today continue to draw inspiration and insight from canonical works by Confucius, Plato, Aristotle, Maimonides, Chanakya, and Ibn Khaldun, among others. Many consider Machiavelli (1950 [1532]) the first modern political theorist due to his keen analysis of political strategy and view of social order as created rather than natural or god-given. Major works by Hobbes (1958 [1651]), Locke (1980 [1689]), and Rousseau (1987 [1762]) offer competing views of enduring political questions relating to order, community, and distributive politics. Other scholars, such as Montesquieu (1989 [1748]), writing during the European Enlightenment sought to explain the development or decay of various forms of government based on environmental (e.g. climate, soil, natural resources) and social (e.g. religion, livelihoods, customs) conditions in different regions of the world, an approach presaging modern political science. Writings by Marx, Durkheim, and Weber continue to influence both the questions political scientists ask and the answers they generate.

The emergence of political science as an organized field of study can be traced to the mid-nineteenth century in Europe and the USA. The first university political science chair was named in 1857, and the first political science department was established in 1880, both at Columbia University (Farr & Seidelman 1993). The first political science professional society, the American Political Science Association, was founded in 1903. Since this early period, the discipline has grown markedly, with political scientists teaching in most of the large universities around the world and occupying positions in government, business, and non-profit sectors. The years following World War II saw the discipline spread more extensively outside the USA. The International Political Science Association was founded under the auspices of the United Nations Educational, Scientific and Cultural Organization in 1949, and a host of other national associations were created around the same time.

The postwar period also saw the beginning of a "behavioral revolution" in political science, which emphasized the systematic and rigorously scientific study of individual and group behavior (Farr 1995). This behavioral paradigm focused on the collection of large amounts of sociopolitical and economic data and analysis using statistical methods to develop generalized explanations of individual and group political behavior. Building from and reacting to behavioralism, institutional and rational choice approaches, which rely especially on game-theoretic and regression analyses, have been ascendant since the 1980s. Recent years have seen major growth in experimental research in political science (Druckman et al. 2011; John 2017). These approaches have spurred many political scientists to place greater emphasis on theorization and marry empirical verification with analytical models of politics.

The role of behavioral, experimental, and quantitative approaches more generally has been contentious within the discipline. Critics deride such approaches as being unrealistic (e.g. relying on models that bear little resemblance to the real world), practically irrelevant, and overly narrow in terms of issues studied and methodologies employed (Monroe 2005). They worry, for example, that privileging quantitative and formal modeling approaches (borrowed from economics and mathematics) over qualitative and interpretive approaches (drawn from anthropology and history) limits a rich and complex understanding of politics and distorts knowledge about how politics works. There is also concern that the increasingly vaunted status of sophisticated technical models and analyses of politics comes at the expense of substantive knowledge and simultaneously reduces the applicability of political analysis to policy and practice.

An apt metaphor for the discipline may be that of a Cubist painting, in which an image, fragmented and cacophonous, emerges on close inspection (Figure 6.1). Political science has now long been a discipline with recognizable content and boundaries (Goodin 2011). These boundaries may become more literal and cohesive over time, but the disciplinary tableau of political science remains a work in progress, reflecting a congeries of attempts to understand better the complex and dynamic world of politics.

**Figure 6.1** Juan Gris, 1913. "Carafe, Cups and Glasses." National Gallery of Ireland.

## 6.3   Concepts, Approaches, and Areas of Inquiry in Political Science

In this section, we provide in-depth treatment of four key political science concepts—power, institutions, governance, and participation—especially relevant to conservation policy and practice (Table 6.2).

### 6.3.1   Power

Power is arguably the central concept of political science. Indeed, some view political science as "the study of the organization of power" (Holden 2000, p. 3). Power connects to virtually all the concepts introduced above, as well as the other key concepts presented in this section. Power is also crucially relevant to understanding the causes and effects of conservation.

One political science approach to studying power views it as having three "faces" (Lukes 1974). The first face of power is perhaps the one that most readily comes to mind. It refers to the ability of an individual or group to get another to do something the latter would not otherwise do. This conception of power has dominated much of political science. Power in this sense is about influence exerted on one by another through resource disparities or coercion, including violence (see Chapter 7). Political analysts taking this view have sought to gauge the presence and distribution of power based on changes in observable behavior (e.g. Dahl 1961).

Consider, as an example, the more than 40-year policy conflict between proponents of drilling for oil in Alaska's Arctic National Wildlife Refuge (Arctic Refuge) and environmentalists wishing to retain the area's protected status (Figure 6.2). The analyst establishes the contrasting demands of each group and its initial position relative to one another and the government. The analyst also assesses the resources each brings to the conflict: money, volunteers, well-placed allies, and skill in deploying these resources. Based on an analysis of policy outcomes over time, it is possible to deduce which group achieved more of their objectives than the other. In this way, the distribution of power is revealed and analysis of the role of power in the dispute and its resolution becomes possible (Doyle et al. 2015). In the case of the Arctic Refuge, environmentalists have been able to exercise power more effectively than their opponents to prevent drilling, though the struggle to protect the refuge continues.

**Table 6.2**   Four key political science concepts and associated approaches.

| Concepts | Approaches | | |
|---|---|---|---|
| 1. Power | Game theory | Principal-agent theory | Legibility |
| 2. Institutions | Rational-choice institutionalism | Historical institutionalism | Sociological institutionalism |
| 3. Governance | Advocacy coalitions | Structural choice | Veto player |
| 4. Participation | Direct | Indirect | |

**Figure 6.2** Rally for the protection of the Arctic National Wildlife Refuge in Fairbanks, Alaska, USA, March 2018. Photograph courtesy of Pamela A. Miller.

This approach has the benefit of being straightforward, but it misses important aspects of power, including the second face of power: "non-decision-making" power (Bachrach & Baratz 1962). This form of power relates to the ability to set the agenda for issues to be debated and acted upon in the policy process. A focus on overt conflict and behavioral causality neglects this kind of power and can thereby shroud inequality and privilege. The context of democratic politics is not neutral or even fair. Rather those who have been able to set up and shape "the rules of the game" have an advantage (Junn 2009, p. 27). Research on environmental justice (see Chapter 8), for example, has shown how certain issues such as toxic waste in minority communities in US cities do not appear on the policy-making agenda, while other issues, such as setting aside wilderness or open-space areas, do (Taylor 2009). A major reason for this disparity is that entities with powerful, vested interests, such as the chemical, oil, and gas industries, have been able to exert more influence (e.g. through lobbying and campaign contributions) in policy-making processes so their interest in maximizing profits outweighs any other concern, such as local public health.

To understand the operation of power, one must also consider a more hidden, third face of power (Lukes 1974). This aspect of power is subtle but pervasive. It extends the idea behind power's second face to include the ways in which some people's interests may be denied or harmed without even the consciousness of such harm being apparent to those being harmed. This lack of awareness prevents the articulation of grievances around which political action can be mobilized. Power in this sense is so effective that it forms the desires and interests of those who are subject to it to serve the ends of others. This third form—control over an individual's preferences—is thus critical to understanding power, complementing the two other ways power operates: through economic or coercive force and by setting agendas.

Scholarly studies of power influenced by the social theorist Michel Foucault go beyond the third face of power. Such studies understand power as operating to turn people into accomplices in regulating their own behavior and that of others (e.g. Foucault 1979). Power in this sense is a form of social control working through institutions such as schools, hospitals, and prisons, and also through the production of knowledge (see Chapter 2). Foucault developed the concept of **governmentality** to analyze this aspect of power through knowledge. He described how individuals internalize certain knowledges and related **discourses** of power such that their behaviors conform to guidance from powerful institutions and agents. Rational calculations demonstrate appropriate forms of conduct so that individuals "behave as they ought" to behave (Foucault 1991 [1978]). In this conception of power, consent of the governed is self-generated through their formation in social institutions like schools, the census, norms of behavior inculcated by families, and the panoply of institutional forms that govern everyday life. Narratives and discourses help make prevalent social relations and ideas seem natural because individuals in these institutions come to see institutional rules as serving not just the institutions' but also the individual's interests and thereby seem reasonable to follow (see Chapter 5 for more on environmental narratives and discourses). The result can be a subtle and pervasive form of social control.

Research on community forest management in the Indian Himalaya illustrates how this conception of power has been applied in a conservation context. Agrawal (2005) shows how the delegation of authority to communities to manage forests led residents to embrace forest conservation, whereas they had previously resisted such protection efforts. Power worked through community-based forestry institutions to develop new forms of collective consciousness about the environment, which in turn lead to its protection in ways encouraged, but not directly undertaken, by the central state. This study provides one of many available examples of how more subtle forms of power can be analyzed empirically. Research along these lines suggests the importance of investigating power as dynamic and context specific, less as a property of individuals and more as a means through which different social goals are accomplished.

In this section, we have focused primarily on power as exercised by and on individuals, or across fields of influence. But power is also critical to understanding relations between states and organizations, whose choices have important consequences for biodiversity. For example, wars between or within states can lead to a breakdown of law enforcement in protected areas and plundering of natural resources. Conversely, cooperation between organizations at different levels can lead to more effective solutions to biodiversity loss. For example, international donor support to countries where governance is more decentralized has been associated with better conservation outcomes in protected areas than in cases where governance is more centralized (Miller et al. 2015).

Scholars who study international relations have developed the concepts of "hard" and "soft" power to help understand the interactions in the interstate system. Hard power, which corresponds to the first face of power described above, refers to military and economic means a state uses to influence the behavior of another state. Soft power is the ability of states to get what they want through attraction rather than coercion (Nye 2004). It involves the use of diplomacy, economic assistance, cultural exchange, and other activities leading to a country's favorable public opinion and credibility abroad.

### 6.3.1.1 Approaches to Analyzing Power

Given the diversity of interpretations about the nature of power, it is not surprising that political scientists have developed a variety of frameworks for analyzing it. Here, we review three with particular relevance to biodiversity conservation: game theory, principal–agent theory, and legibility.

#### 6.3.1.1.1 Game Theory

Game theory relies on the development and application of models (simplified representations) of actors interacting strategically to produce outcomes. It is based on the assumption that the "players" are rational, that is, they have specific wants and beliefs and they act accordingly (Shepsle 2010). Their preferences in any given situation reflect these wants and beliefs. Preferences, in turn, are described as a player's utility, or the amount of "welfare" they derive from an object or an event. This welfare refers to an index of the actors' relative well-being. The formal reasoning employed by game theory requires that each player's utility be captured mathematically. This is done using a utility function, which maps ordered preferences onto real numbers. Actors are assumed to want to maximize their utility in their decision-making as they play in "games," which are understood as situations in which at least one actor can only act to maximize their utility through anticipating the responses by one or more other actors. Actors in a given game can choose among different strategies, which dictate the actions they should take in response to the strategies other players might use.

Game theory is particularly useful in analyzing the first face of power, but it also can be used to understand power in relation to setting an agenda. It has many potential applications to understanding conservation. For example, game theory has been used to help determine what protected-area monitoring strategies and incentives are likely to be most effective in a given situation (Box 6.4). It has been employed to yield conservation-relevant insights in situations as diverse as fishers coordinating to make use of common fishing grounds in Turkey (Ostrom 1990), international negotiations over environmental treaties (Hoel & Schneider 1997), and hunting and enforcement of community-based wildlife management policies in Africa (Gibson 1999).

---

**Box 6.4   Applications: Game theory and protected area management**

Game theory provides a useful tool for understanding the success of different management strategies for national parks and other protected areas in preventing unsustainable resource use. Walker (2009) applied a game-theoretic model of monitoring and rule enforcement developed within political science to analyze the strategic interaction between protected area managers and resource users under different scenarios. In this protected-area monitoring game, the resource user must decide whether to abide by the law or to illegally use a protected resource, and the manager must decide whether or not to monitor for illegal activity. Dilemmas arise when the use of a resource is disputed. This occurs when the resource—for example, wildlife or timber—is associated with a positive benefit for the user. In such situations, users are expected to continue using the resource unless there are negative consequences for

---

**Box 6.4 (Continued)**

doing so. If there is no monitoring to catch rule breakers and impose penalties, a user will choose to use the resource regardless of laws designed to protect it. Protected area managers also have incentives for monitoring – or not monitoring. For example, if managers do not monitor while users continue to use the resource, they suffer losses equivalent to the lost resource. If managers do monitor and catch a user not abiding by the rules, they gain both the corresponding benefits of the resource in the form consumed by the user (by confiscation) and any additional revenues from fines imposed on the offender.

By analyzing the incentives and potential strategic actions of the two players in this way, game theory yields three key variables useful in predicting the success of a protected area: (i) costs of monitoring for rule breakers; (ii) benefits of catching a rule breaker; and (iii) probability of catching a rule breaker if monitoring. Walker tested the predictions of the model using 116 cases sampled from peer-reviewed literature. The cases were from 35 countries spanning all continents except Antarctica. The model correctly predicted the outcome in more than 75% of the cases. It identified an important mismatch between typical protected-area circumstances and management policies. In situations where the costs of monitoring were greater than the product of the probability of catching a rule breaker and the benefit of doing so, conservation was unlikely to succeed. This is because monitoring will generally be sporadic and illegal resource use common. To curb illegal use, managers can turn to "carrot" (incentive) or "stick" (punishment) strategies to convince potential users to cooperate with rules.

This study contributes to ongoing debate in the conservation field regarding the relative effectiveness of positive or negative incentives in protected-area management. Which is the more effective conservation strategy: increasing the benefits of conservation to local communities or increasing punishment for rule breaking? The game theory model Walker developed suggests that *neither* tactic is effective in most typical protected-area situations. This is because the factors driving the monitor's decision are those that determine the outcome in most contexts. Rather than responding to "carrots" or "sticks," resource users make their decisions according to what they think the monitors will do.

The model does not prescribe a best management policy for conserving natural resources. Instead, it serves as a tool to help assess whether a proposed management policy will succeed in a given situation. With limited budgets and so much area worthy of protection, assessing the potential of different management strategies to succeed before investing too much money in any one strategy is critical. Walker's innovative application of game-theoretic insights from political science to protected area monitoring provides a tool to inform such assessment.

### 6.3.1.1.2 *Principal–agent Theory*

Principal–agent theory is a second approach to analyzing power. It is used to examine incentives and strategies that two players—a principal and an agent—engage in when interacting to accomplish a collective effort. In a principal–agent relationship, the principal is understood as the individual or organization that benefits from the outcomes achieved by the agent's actions, while the agent undertakes to act on behalf of the principal. The principal specifies what the agent must do or achieve, relying on the agent's concern for reputation, appropriate incentives, and other mechanisms to secure compliance (Shepsle 2010). Several problems can arise in this relationship, which is suffused with power. For example, the agent may have different preferences than the principal. Or, the principal may have limited information about the agent's actions. In general, careful design of initial agreements to capture the interests of both parties and ensure effective oversight can make principal–agent relationships more mutually beneficial.

The relationship between the US Congress and the Environmental Protection Agency (EPA) can be analyzed using a principal–agent framework. In this case, the US Congress is the principal. It passes laws intended to achieve some measure of environmental protection. However, Congress itself does not have all the scientific expertise and resources required to devise specific regulations appropriate for implementing the law. Thus, it delegates authority to an agent, the EPA, to perform this function. Here another dilemma arises: Congress does not have sufficient information about how and whether the EPA is implementing the law, nor does it have the required scientific expertise to know if EPA regulations are in line with the intent of the law. This is one of many classes of principal–agent problems that political scientists seek to understand in order to shed light on the different ways in which power is exercised.

Patron–client relationships (also known as "clientelism") represent another class of principal–agent problems relevant to conservation. Such relationships involve mutually beneficial but usually highly unequal relationships of power and exchange between principals (patrons) and agents (clients). Selective control over the allocation of resources lies at the heart of clientelism (Scott 1972). Those in control (patrons) supply goods and opportunities to those they favor but do so only to place themselves or their supporters in positions from which they can gain greater resources and services (Roniger 2004). Their partners (clients) are expected to return their benefactors' help politically or otherwise. For example, clients may work for their patrons to help them get elected or take other steps to boost their patrons' prestige and reputation. Patrons often know their clients personally, but this need not be the case. Some degree of charisma or other form of psychological attachment may also be involved in patron–client relations.

Patron–client relations can take different forms depending on the type of political regime in which they are embedded. In democratic settings, voting rights, party competition, and the option of exiting the relationship may strengthen a client's bargaining position with a patron. Electoral competition also promotes the scaling up of clientelistic networks from local politics with personalistic, face-to-face relations to the national level of hierarchical political machines (Kitschelt & Wilkinson 2007). In non-democratic regimes, clients typically have less opportunity to exit their relationships with patrons and patron–client relations are more often based on the patron's charisma or ethnic ties. These characteristics suggest that analysis of the effect of clientelism on conservation practice and outcomes should be attentive to the broader political context. The patron–client framework also helps us understand corruption, which has been hypothesized to negatively affect conservation by decreasing financial resources, law enforcement, and political support. Political scientists have used this framework to shed light on the complex relationship between corruption and conservation outcomes (Box 6.5).

---

**Box 6.5   Debates: Conservation and corruption**

A patron–client framework can facilitate understanding of the relationship between corruption and conservation, which has important policy implications. Political scientists have adapted the concept of rent-seeking from economics to build knowledge of the multiple dimensions of corruption. Rent-seeking arises when a public authority, such as a government bureaucracy, is able to impose rules that enable it to extract a profit (a "rent") greater than that which would be available in an open, competitive market. Natural resources such as minerals, timber, and wildlife are a common target for rent-seeking, as their extraction can lead to significant returns on investment for those who control access to them, typically government actors.

People who have been successful in the past in obtaining rents may become entrenched in a patron–client system in which the elite keep themselves in an advantaged position by distributing rents to clients in return for their support (Gibson et al. 2005). Like embezzlement of funds, acceptance of bribes, or overlooking illegal activities, rent-seeking is a form of corruption. It can impose substantial costs on others and society as a whole, as when, for example, government officials who control public or community land and natural resources allow their exploitation for personal gain.

Studies suggest that rent-seeking and other forms of corruption can negatively affect conservation, such as by reducing available financial resources, law enforcement, and political support (Smith & Walpole 2005). Researchers have sought to generalize an argument about the relationship between corruption and conservation outcomes by testing hypotheses with cross-national data. An influential study by Smith et al. (2003) found that that there is a significant and negative correlation between national-level indicators of corruption and changes in elephant and rhinoceros populations and in forest cover. However, despite the apparently commonsensical conclusion that corrupt central government officials in combination with patronage politics always leads to environmental destruction, Barrett and colleagues (2006) used insights and methods from political science to cast serious doubt on the results of that study. They demonstrate conceptual and statistical weaknesses in the links Smith and colleagues (2003) made between corruption and biodiversity loss to arrive at their generalization. For example, Barrett et al. show the complexities of the relationships, with corruption taking many different forms and operating on different levels with different effects, depending on the type of resource at issue (e.g. wildlife or forest resources). They also show how other confounding factors may have explained the relationship Smith et al. found. The study by Barrett et al. and the lively debate it engendered (see volume 21, issue 4 of *Conservation Biology*) provide a useful example of how political science can enrich understanding of the causes of biodiversity loss and suggest leverage points for more effective policy responses.

---

### 6.3.1.1.3   *Legibility*

Although the principal–agent framework is useful for analyzing more overt forms of power (e.g. its first and second faces), it has less utility for understanding more subtle ways in which power works in politics and natural resource management and policy. A "legibility" approach (Scott 1998) can be used to analyze such dimensions of power.

This approach sheds light on the means through which modern states have sought to control their territories and populations, particularly the invention and refinement of specific technologies, such as the census, maps, cadastral surveys, and standard units of measurement. These technologies simplify complexity and make society and nature legible to politicians and bureaucrats. This simplification, in turn, facilitates the state's ability to carry out its classic functions of taxation, conscription, and prevention of rebellion. However, when matched with an abiding faith in scientific intervention to improve every aspect of human life, a powerful authoritarian government, and a prostrate civil society, this legibility and its inevitable reductions can lead to "improvement" schemes with disastrous consequences for human and non-human populations alike (Scott 1998).

The Ujamaa village campaign in Tanzania during the 1970s illustrates this argument. This scheme comprised a massive effort by the Tanzanian government under the rule of Julius Nyerere to resettle most of the country's population in new, "modern" villages (Scott 1998). It sought to make the rural population "legible" by neatly arranging homes and land uses according to an idealized central plan incapable of adapting to local circumstances. It led to the relocation of at least five million Tanzanians and a profound reordering of the agricultural and forest landscape across the country. The plan failed. Peasant livelihoods were largely based on patterns of settlement and periodic movement adapted to an often sparse, but dynamic environment. State-mandated settlement "threatened to destroy the logic of this adaptation" (Scott 1998, p. 235). Indeed, the result was a disaster for agricultural production. The failure of this simplification scheme when confronted with local diversity provides a cautionary tale for would-be **development**—or conservation—projects planned from afar.

The legibility framework can be applied to understand how conservation organizations attempt to make land and sea areas legible as a means to recommend or implement conservation policy (most obviously through the establishment of protected areas). It also can reveal the extent to which such efforts undermine local livelihoods and support for conservation. Political ecologists have advanced this as the "conservation and control" thesis (Peluso 1993). This thesis emphasizes the degree to which many conservation efforts have failed to achieve their goals because they have marginalized traditional resource managers and buttressed the goals and desires of elites who have little or no interest in understanding ecosystem process, landscape, or local place (see Chapter 5 for more on **political ecology**).

Political science research using the legibility approach suggests several principles for conservation and development planners to adopt so as to minimize negative social impacts in their work (Table 6.3). Together, these principles suggest caution in relying too heavily on "expert" knowledge in the absence of local knowledge and support the need for adaptive governance of social–ecological systems (Folke et al. 2005; Ostrom 2007).

## 6.3.2 Institutions

Institutions are social entities that structure human actions and expectations. The term is used in two major ways. First, the term *institution* often refers to roles or organizations. This view of institutions is widely used in sociology and sometimes in public policy and administration. Political scientists tend to understand institutions a second way: as rules. This conception is distinct from the concept of organizations, which refers to concrete entities with resources and personnel that seek to achieve specific goals. Organizations are

**Table 6.3** Five principles for conservation and development planners to minimize negative social impacts.

1. *Take small steps.* Then evaluate the results and plan the next small step.

2. *Favor reversibility.* As well-known conservationist Aldo Leopold put it, "the first rule of intelligent tinkering is to keep all the parts" (Meine & Knight 1999).

3. *Plan on surprises.* Implementation will inevitably turn out quite different than anticipated.

4. *Anticipate human inventiveness.* Assume that "those who become involved in the project later will have or will develop the experience and insight to improve on the design" (Scott 1998, p. 345).

5. *There are no panaceas.* No single project model or governance approach exists that is applicable to all contexts (Ostrom 2007).

Note: Adapted from Scott (1998).

often seen as examples of broader institutions. Thus, for example, a stock exchange is an institution, but the New York Stock Exchange is an organization. Universities are institutions; University of Michigan is an organization.

The classic definition of institutions in economics and political science is "the rules of the game in a society" (North 1990, p. 3). Institutions are humanly devised constraints that shape human interaction with others and the natural world. These prescriptions organize all forms of repetitive and structured interactions, such as those within families, neighborhoods, markets, firms, sports leagues, churches, private associations, and governments at all scales (Ostrom 2005). They can be formal, detailed in writing and officially conveyed and enforced, or informal, created, communicated, and enforced outside officially sanctioned channels (Helmke & Levitsky 2004). Informal institutions include social sanctions, taboos, codes of conduct, and networks. Norms—shared expectations about behavior within a society or group—are an important class of informal institutions (Axelrod 1986). Constitutions, laws, and property rights represent prominent examples of formal institutions. Protected areas, biosphere reserves, the Convention on Biological Diversity or the Convention on International Trade in Endangered Species (CITES) represent some key international conservation institutions.

Institutions influence individual behaviors and expectations, the crucial links between institutions and outcomes. By constraining certain behaviors while enabling others, and by undermining some expectations and promoting others, institutions increase the predictability of human interactions and facilitate activities that may not be possible in their absence. Institutional change affects the way societies evolve through time, and is key to the understanding of historical change more generally. Institutions are powerful. They "reflect the resources and power of those that made them and, in turn, affect the distribution of resources and power in society" (Campbell 2004, p. 1). Given the importance of institutions, their analysis features centrally in much of the theoretical science of politics and has been applied to a multitude of substantive issues and empirical settings.

Institutions can arise to advance specific interests, but also to solve social and environmental problems. Institutions are critical to addressing two of the most important problems for biodiversity conservation and natural resource management: collective action (see Chapter 8) and negative externalities (see Chapter 4). Biodiversity has the characteristics of a common-pool resource (CPR) (Figure 6.3). This means that excluding people from

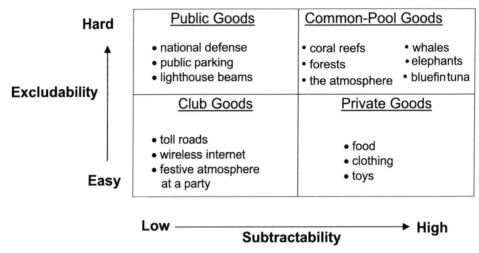

**Figure 6.3** A typology of goods with examples. Adapted from McKean 2000.

benefiting from it (e.g. to prevent exploitation of natural resources or use of ecosystem services) can only be done at great cost, while using it can lead to its eventual disappearance. Without institutional mechanisms to address these two characteristics of the good, which economists call "excludability" and "subtractability," respectively, it is effectively an open-access resource available to anyone (McKean 2000).

CPRs, like fisheries, pastures, or forests, are typically difficult to protect and easy to deplete. In such situations, individuals acting in isolation from one another may have no incentive to contribute to the joint benefit of environmental protection, leading to the rational overexploitation of a natural resource for private gain (see Box 6.3). This classic dilemma is especially difficult to overcome in low-income countries of the tropics, where the preponderance of the world's biodiversity is located, because the institutions necessary to address it are generally weak or lacking (Barrett et al. 2001).

Two prominent institutional solutions to the problem of collective action are (i) government with the power to enforce individual contributions to the supply and maintenance of CPRs or (ii) privatization of property rights for communally-owned resources. These solutions are diametrically opposed, though protection of property rights usually requires some state role. Scholars in the interdisciplinary field of common property have shown that these are not the only solutions to collective-action problems and that, under the right conditions, groups of people can sustainably manage their own resources without external intervention (Ostrom et al. 2002). Indeed, people have devised diverse institutions to manage resources in common for centuries (Ostrom 1990; McKean 2000). Political analysts, policy makers, and practitioners alike can learn a great deal from the ways in which these resource-user groups have done so.

In the context of natural resource conservation, it is important to distinguish among five major types of property rights (Table 6.4). These rights can be exercised at distinct levels (see Section 6.3.3). The exercise of access or withdrawal rights corresponds to an operational level whereas management, exclusion, and alienation rights require that the right holders operate at a collective-choice level. Decisions at this level affect future operational decisions and action. Furthermore, for communities to possess collective choice-making capabilities,

**Table 6.4**  Types of property rights.

| Right | Description |
| --- | --- |
| Access | The right to enter a defined physical area or property. |
| Withdrawal | The right to obtain resource units or products of a resource system (e.g. cutting firewood or timber, harvesting mushrooms, diverting water). |
| Management | The right to regulate internal use patterns and transform the resource by making improvements (e.g. planting seedlings and thinning trees). |
| Exclusion | The right to determine who will have right of withdrawal and how that right may be transferred. |
| Alienation | The right to sell or lease withdrawal, management, and exclusion rights. |

Note: Adapted from Schlager and Ostrom (1992). Refinements to this classic typology are presented in Galik and Jagger (2015) and Sikor et al. (2017).

some rules at a constitutional level (set locally or by a national government) must give them this authority. The analytical distinction between operational level rules, and collective and constitutional choice arenas should not create the impression that these correspond to three different formal levels of authority in a political or legislative system. A single political body can use operational rules, create them by deliberating at the collective-choice level, and may have powers in the constitutional-choice arena as well (Ostrom 2005).

As an example, consider forest CPRs in Kumaon in northern India. Local communities gained the rights to resource withdrawal, management, and exclusion from the colonial British government in 1931 (Agrawal 2005). They have continued to exercise these rights effectively for nearly 80 years now. In the process, they have improved forest cover in the areas they control (Somanathan et al. 2009) and at far lower costs than those incurred by the Forest Department. Their capacity to exercise different kinds of property rights in forests shows that even the lowest levels of administrative organization can effectively make operational, collective-choice, and constitutional decisions (Agrawal & Ostrom 2001).

### 6.3.2.1  Approaches to Analyzing Institutions
To understand the wide range of types of institutions and their effects on politics and policy, political scientists have developed three separate but related approaches: rational-choice, historical, and sociological institutionalism.

#### 6.3.2.1.1  Rational-choice Institutionalism
Rational-choice institutionalism focuses on (i) the effects of institutions, (ii) the reason for institutions in the first place, and (iii) the choice of particular kinds of institutions (Weingast 2002). Rational choice scholars work from a definition of institutions as humanly devised constraints on action (North 1990). Research in this vein investigates how institutions shape the sequence of interactions among relevant actors, the choices and information available to these actors, their beliefs, and the benefits ("payoffs") that accrue to individuals and groups operating within institutional constraints.

Building from a seminal study of majority rule in the US Congress (Riker 1980), rational-choice institutionalism has developed four characteristic features. First, it departs from specific assumptions about human behavior. Relevant actors are posited to have fixed preferences, to seek to maximize the attainment of these preferences, and to behave strategically in a manner that presumes extensive calculation (Shepsle 2010). Second, rational-choice institutionalism tends to see politics as a set of collective-action dilemmas. Third, this approach emphasizes the role of strategic interaction in determining political outcomes. Institutions constrain and coax actors by affecting the range and sequence of possible choices or by providing information and enforcement mechanisms that reduce uncertainty about the behavior of others. In this way, institutions can lead actors toward particular calculations and potentially better social outcomes. Finally, rational-choice institutionalism uses deductive reasoning to explain the existence and form of institutions. The existence of institutions on this account is usually attributed to voluntary agreement by interested parties to help them capture gains from cooperation (Weingast 2002).

Rational-choice institutionalism can yield insights for conservation. For instance, Agrawal and Goyal (2001) used this approach in combination with analysis of primary data on forests and institutions in Kumaon in the Indian Himalaya to analyze the relationship between group size and successful collective action. They show how forest resource users rely on calculations of personal gain to decide whether and how to monitor rule violations in their forests and the extent to which group characteristics such as its economic endowments affect the likelihood of effective enforcement of conservation goals. They found that medium-sized forest management groups were more likely than small or large groups to provide third-party monitoring, which can lead to more sustainable use of CPRs like forests. It is relatively easy for smaller groups to monitor forest resource use among themselves, but larger groups require greater effort and expense to do the same. It becomes more difficult to exclude those not contributing to maintenance of the resource as the number of people from whom the resource needs protection increases.

### 6.3.2.1.2 Historical Institutionalism

Three features distinguish historical institutionalism: (i) a focus on big, substantive questions important to broad publics; (ii) an emphasis on institutional change over time; and (iii) an interest in the effects of institutions and processes together rather than one institution or process at a time (Pierson & Skocpol 2002). In contrast to rational-choice institutionalists, who tend to use a set of common methodological principles, the historical institutionalist approach is more varied. What unites different approaches under the broad umbrella of historical institutionalism is that they take history seriously. They do not just look at the past, but trace processes over time. Institutions frequently persist over long periods and push historical developments along certain "paths."

The concept of path dependence is used in much contemporary work on institutions to refer to the dynamics of self-reinforcing or positive feedback processes in a political system (Pierson 2000; Page 2006). The logic is simple: outcomes at a specific historical moment trigger feedback mechanisms that reinforce the recurrence of a particular pattern in the future (Pierson & Skocpol 2002). Interestingly, relatively small events at early stages of an historical process can exert a disproportionate influence on subsequent events. A classic example is the QWERTY keyboard. The QWERTY layout on computer keyboards is a holdover from the typewriter era. More efficient keyboard arrangements have been developed, but QWERTY has persisted and predominates on computer keyboards around the world.

Importantly, unequal power relationships can also generate and solidify processes of path dependence that privilege certain groups over others (Moe 1990). Path dependence demonstrates the difficulty actors can have reversing course once they move down a particular institutional path. This "stickiness" means that alternatives that were once entertained may become irretrievably lost (Pierson & Skocpol 2002).

Recent efforts by communities to conserve wildlife in Kenya and Namibia illustrate the dynamics of path dependence. Sport hunting was banned in Kenya in 1977. Strong pressure by conservationists and a complex history of hunting by both blacks and whites ultimately led to the ban (Steinhart 2006). Thus, community conservation in Kenya must rely on tourism and forms of benefit generation that do not include revenues from sport hunting. By contrast, community-based and private conservation efforts in Namibia allow hunting as an incentive for local landowners to protect wildlife populations (Jones 2010). Namibia has long allowed hunting, and conservationist pressure has not changed it. An historical institutionalist approach emphasizing path dependence might trace these divergent outcomes to the critical juncture of the hunting ban in Kenya and its absence in Namibia.

### 6.3.2.1.3 Sociological Institutionalism

Sociological institutionalism emphasizes that institutions are more than freestanding entities through which autonomous individuals pursue their self-interests. Instead, they are embedded in specific social, economic, and political contexts (Granovetter 1985). Sociological institutionalism stresses the important role of values, norms, and symbols in defining an institution and in guiding the behavior of its members (Peters 1999). This emphasis distinguishes it from the other two institutionalisms discussed. It also draws this approach close to anthropology, which has well-developed understandings of "culture" (see Chapter 3). Another defining feature of this variant of institutionalism is its position that institutions do not simply affect the strategic calculations of individuals, but also their preferences and their very identity (Hall & Taylor 1996). Thus, the relationship between human action and institutions is understood as mutually constitutive. The claim that individuals and organizations develop or adopt new institutional practices not because it is more efficient, but because it conforms with existing social norms and enhances their social legitimacy is also characteristic (Hall & Taylor 1996). Institutions do not emerge fully formed from the imagination of some decision maker or leader but are usually variations on already existing institutions. This insight helps explain the presence of inefficient or ineffective social and political institutions.

Two important variants of institutional analysis have emerged in recent years from sociological institutionalism: network and discursive institutionalism (Table 6.5). These two approaches promise to shed new light on the formation of conservation policy and social responses to it in a variety of contexts.

Rational choice, historical, and sociological institutionalism are not mutually exclusive. In many ways, they complement one another, enriching our understanding of the multiple dimensions of politics. Together, these strands of political science research enhance our understanding of the importance of institutions to conservation and conservation scientists and practitioners might integrate richer analysis of the institutional landscape with analysis of the ecological landscape in order to design and implement better conservation measures (Table 6.6).

**Table 6.5** Emerging approaches to the study of institutions.

| Approach | Description | Conservation-relevant examples |
|---|---|---|
| Network institutionalism | This approach conceptualizes networks—sets of relationships between individuals, groups, or organizations—as institutions. To the extent that a network represents a stable or recurrent pattern of behavioral interaction or exchange between individuals or organizations, it can be thought of as an institution (Ansell 2006). The network approach in political science has been most developed in the study of policy networks. Policy networks are official government links with, and dependent on, other state and societal actors (Rhodes 2006). | Pioneering research has shown that networks of professionals with recognized knowledge and skill in a particular issue area, known as "epistemic communities," played an essential role in developing a coordinated plan for pollution control by states across the Mediterranean region (Haas 1989). Network institutionalism has also been used to study the management of estuaries in Florida (Schneider et al. 2003). This research suggests that national programs can facilitate the emergence of policy networks that effectively manage and conserve estuaries across political-administrative boundaries. |
| Discursive institutionalism | Discursive institutionalism focuses on ideas and discourse. An important concept in many social science disciplines, **discourse** refers to written or spoken language, stories, images, or terminology (see Chapters 3 and 5). Discourse "encompasses not only the substantive content of ideas but also the interactive processes by which ideas are conveyed" (Schmidt 2008, p. 305). Discursive institutionalism analyzes the discourse among policy actors, which helps them coordinate with one another. It also considers the discourse between political actors and the public. | Discourses are powerful. They play an important role in how we understand and respond to social problems. For example, "slash-and-burn" agriculture and "shifting cultivation" refer to the same activity—a rotational system of clearing forest land for farming—but have very different connotations. The former is more polemical, implying the practice is a grave environmental threat, while the latter is a more neutral term that includes the possibility that this is a sustainable system of land use. These discourses, in turn, can influence government policy and livelihood strategies. |

**Table 6.6** Four attributes of successful conservation institutions.

1. Authority, ability, and willingness to restrict access and use.

2. Wherewithal to offer incentives to use resources sustainably (including not to use resources at all).

3. Technical capacity to monitor ecological and social conditions.

4. Managerial flexibility to alter the array of incentives and the rules of access to cope with changes in the condition of the resource or its users.

Note: Adapted from Barrett et al. (2001). These characteristics are common to institutions whether they operate at the community, subnational, national, or international level or across several levels.

### 6.3.3   Governance

The third key political science concept we discuss in this chapter is governance. This concept has its roots in the Greek word *kybernan* (to steer). In political science, as in political life, governance has typically been conceived as the province of governments. Governments are the paramount official organizations within a given territory that perform controlling and regulating roles for society. Such organizations are indeed a central locus of governance, but they are not the only actors who govern. Civil society and private sector organizations can also govern—by themselves or in combination with government (Lemos & Agrawal 2006).

Governance can be defined as "the formal and informal institutions through which authority and power are conceived and exercised" (Larson & Soto 2008, p. 214) or simply as the creation and enforcement of socially binding agreements. While governments remain critical in conservation, businesses, communities, and NGOs have played an increasingly important role in environmental governance. The expansion of their role has generally corresponded with a diminished one for central state governments. In both domestic and international politics, the changing role of the state has led to theories of "governance without government" (Rosenau & Czempiel 1992). Around the globe, governments have increasingly partnered with private business and civil society organizations to implement policy. In the past three decades, governance generally, but particularly in relation to the environment (Figure 6.4), has thus taken on a hybrid character (Lemos & Agrawal 2006).

Governance occurs at multiple spatial scales defined by political jurisdictions (Table 6.7). These range from the household to international level with community, regional, and national levels in between. In addition to these spatially based scales, the three conceptual levels mentioned earlier are also involved in governance systems: constitutional, collective-choice, and operational (Ostrom 1990).

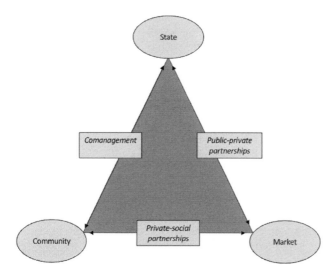

**Figure 6.4**   Actors, mechanisms, and strategies of environmental governance. Adapted from Lemos & Agrawal, 2006 / Annual Reviews.

**Table 6.7** The relationship between conceptual levels of human choice and geographic domains.

| Spatial levels of political jurisdictions | Conceptual levels of human choice | | |
|---|---|---|---|
| | Constitutional-choice | Collective-choice | Operational-choice |
| International | International treaties and their interpretation | Policy making by international agencies and multinational firms | Managing and supervising projects funded by international agencies |
| National | National constitutions and their interpretation; rules used by national legislatures and courts to organize their internal decision-making procedures | Policy making by national legislatures, executives, courts, commercial firms (who engage in interstate commerce), and NGOs | Buying and selling of land and forest products, managing property, building infrastructure, providing services, monitoring and sanctioning |
| Regional | State or provincial constitutions and charters of interstate bodies | Policy making by state or provincial legislatures, courts, executives, commercial firms, and NGOs with a regional focus | Buying and selling of land and forest products, managing property, building infrastructure, providing services, monitoring and sanctioning |
| Community | County, city, or village charters or organic state legislation | Policy making by county, city, or village authorities and local private firms and NGOs | Buying and selling of land and forest products, managing property, building infrastructure, providing services, monitoring and sanctioning |
| Household | Marriage/partnership contract embedded in a shared understanding of who is in a family and what the responsibilities and duties of members are | Policies made by different members of a family responsible for a sphere of action | Buying and selling of land and forest products, managing property, building infrastructure, providing services, monitoring and sanctioning |

*Source:* Adapted from Gibson et al., (2000) / With permission of Elsevier.

All the changes in governance described above raise important issues for democracy and accountability; that is, the exercise of counter power to balance arbitrary action, manifested in the ability to sanction (Larson & Soto 2008). Who is accountable in hybrid governance arrangements—the government or its partners? And what mechanisms might ensure that non-government actors can be held accountable? Changes in governance also raise questions about the extent to which contemporary governments are able to dictate the direction of the state and society. Although this capacity varies widely depending on regime type and other factors, official governments remain central to governance. The vital role of national governments in many countries across the world in response to the COVID-19 global pandemic provides but one stark indication of their continued importance. Nevertheless, the complexity and multiscale nature of most conservation problems mean that no one governance actor is likely to be able to address such problems on its own.

Political scientists have developed many measures to better understand variations in patterns of governance. There are hundreds of such indicators, nearly all of which focus at the national level. They are most often based on subjective data that reflect the views of a range of informed stakeholders from or knowledgeable about a given country. The Worldwide Governance Indicators (Table 6.8) represent one of the leading attempts to measure governance across countries over time (WGI 2021). Such indicators can be used to understand the relationship between governance and conservation, although attempts to do so must be grounded in theory and attend to complexities inherent in this relationship (see Box 6.5). The World Bank and other policy-related organizations use indicators such as these to make determinations about "good governance," which can also be applied to specific organizations like companies, universities, and NGOs. Whether governance is "good" or "bad" or results from the activities of governments or non-government actors, it is implemented through policy, defined as the mechanism by which officers of the state or other organizations attempt to rule, exercise control, and shape the world (Goodin et al. 2006).

**Table 6.8** Six Worldwide Governance Indicators

| Governance indicators | Description |
| --- | --- |
| 1. Voice and accountability | Perceptions of the extent to which a country's citizens are able to participate in selecting their government, as well as freedom of expression, freedom of association, and a free media |
| 2. Political stability and absence of violence | Perceptions of the likelihood that the government will be destabilized or overthrown by unconstitutional or violent means, including politically motivated violence and terrorism |
| 3. Government effectiveness | Perceptions of the quality of public services, the quality of the civil service and the degree of its independence from political pressures, the quality of policy formulation and implementation, and the credibility of the government's commitment to such policies |
| 4. Regulatory quality | Perceptions of the ability of the government to formulate and implement sound policies and regulations that permit and promote private-sector development |
| 5. Rule of law | Perceptions of the extent to which agents have confidence in and abide by the rules of society, and in particular the quality of contract enforcement, property rights, the police, and the courts, as well as the likelihood of crime and violence |
| 6. Control of corruption | Perceptions of the extent to which public power is exercised for private gain, including both petty and grand forms of corruption, as well as "capture" of the state by elites and private interests |

*Source:* Adapted from WGI (2021).

### 6.3.3.1 Approaches to Analyzing Governance

Along with the rational-choice approach discussed above, three prominent approaches political scientists use to understand governance and the policy process are advocacy coalitions, structural choice, and veto players. The goal of these frameworks is to explain how actors interested in particular substantive issues interact to create, implement, assess, and adjust public policies.

#### 6.3.3.1.1 Advocacy Coalitions

The advocacy coalition (AC) framework (Sabatier 1988) describes the multiple actors involved in policy change and focuses analysis on policy subsystems. A policy subsystem is comprised of "public and private organizations who are actively concerned with a policy problem" (Sabatier 1988, p. 131). Policy change is viewed as arising due to the interaction of competing ACs within a policy subsystem. This subsystem is, in turn, affected by changes outside the subsystem (e.g. economic change, an environmental disaster) and the effects of relatively stable system parameters (e.g. constitutional rules, basic social structure) (Schlager & Blomquist 1996). Individuals and organizations with shared values, causal assumptions, and perceptions of the problem that coordinate in some manner over time comprise ACs. Members of such coalitions act in concert on the basis of their belief systems in an effort to achieve shared goals. Tactics include advocacy and persuasion of decision makers, manipulation of the rules of relevant government institutions, and showing support to public officials who hold similar views.

Any given policy issue is likely to mobilize multiple, competing ACs. In the AC framework, major policy change can occur through several pathways, including (i) compromise among the coalitions; (ii) "external perturbations" like economic shocks, a shift in coalition resources, or perceptions of policy problems; (iii) iterative learning from the adoption, implementation, and evaluation of government programs, which alter belief systems; or (iv) one or more of the coalitions' belief systems changing substantially due to the accumulation of policy information (Schlager & Blomquist 1996). This framework places special emphasis on the role of learning and information in spurring the process of policy change. It also emphasizes the role of critical actors whose beliefs and preferences change over relatively long periods of time (decades) rather than the emergence of new actors.

The case of the ban on the international trade in ivory illustrates the utility of the AC framework for understanding conservation politics and policy. Beginning in the early 1980s, African elephant populations declined precipitously (Chasek et al. 2006). Illegal poaching of ivory for international markets was the main cause of this decline. The international framework designed to address the problem of threatened species, CITES, had proven ineffective in stemming the killing of elephants, and several countries resorted to drastic measures, such as shoot-to-kill anti-poaching campaigns in Kenya and Tanzania (Princen 1994). The publication of a study sponsored by the World Wildlife Fund and Conservation International showed that the rate of harvesting African elephants was unsustainable and marked a turning point in the policy process. Viewed through the AC framework's emphasis on the role of information and learning in policy change, the study can be seen to have altered the perceptions of many countries regarding the severity of the problem. The weight of evidence also shifted the position of policy advocates, with animal rights and conservation NGOs joining together in a coalition to advocate for a ban on the ivory trade (Princen 1994). This coalition successfully lobbied several governments to pass legislation to halt the ivory trade and accelerated the move toward a total trade ban on ivory, a policy approved by the parties of CITES in 1989 following often rancorous debate. As this case indicates, the AC framework could be used to analyze the politics surrounding contemporary conservation policy, such as that relating to an international agreement to mitigate climate change.

### 6.3.3.1.2  Structural Choice

The structural choice (SC) approach shares many of the assumptions of institutional rational choice (IRC). In short, IRC views public policy change as the result of actions by rational individuals pursuing their interests by affecting institutional arrangements (Schlager & Blomquist 1996). It assumes that the choices political actors make are guided by their perceptions of expected costs and benefits and are shaped by specific decision situations, or "action arenas" (Ostrom 1999). IRC conceives of policies as particular institutional arrangements that shape and are shaped by interaction among individuals at the three conceptual levels described in Section 6.3.2: operational, collective-choice, and constitutional. SC shares this conception of policies, but it explicitly emphasizes power. Politics and the policy-making process result in winners and losers. Thus, institutions are not merely solutions to collective-action problems. Rather, they are a means by which today's winners can impose structural outcomes on losers and insulate themselves should they fall from power.

Because of the competitive nature of the policy process, SC suggests that policy outcomes are *not* likely to be efficient or even effective. In well-established democracies, the people in control of government today know they may not be in control tomorrow. To ensure their preferred policies remain in force, they may, for example, specify in great detail how an agency must conduct its business, leaving as little discretion as possible to bureaucrats or future office holders (Moe 1990).

A second way government officials may attempt to exercise public authority is through compromise. But such compromise may lead to the imposition of conditions that cripple the institutions and policies they seek to advance or enable their political opponents to influence agencies and polices so they are more favorable to opponents' interests when they return to power (Moe 1990). Ironically, the SC framework thus predicts that political actors' preoccupation with political control leads them to make or accept choices in the policy-making process that divorce policy design from policy desire such that their policy goals are not achieved through the processes and structures they adopt.

Politicians and bureaucrats are assumed to try to minimize uncertainty where possible. Therefore, the SC framework analyzes potential sources of political uncertainty to anticipate the strategies different actors might adopt to protect their public agencies from political intervention. In this way, the SC framework can shed light on the bureaucratic politics underlying the ineffectiveness of conservation interventions in different political contexts. For example, a comparative study of two large, integrated conservation and development projects in Zambia showed how each was championed by a rival faction seeking to expand control over wildlife and place resources outside the reach of the powerful President Kaunda (Gibson 1999). The competition between these two projects led to detrimental short- and long-term effects on Zambian wildlife policy.

### 6.3.3.1.3 *Veto Players*

The theory of veto players (Tsebelis 2002) enables evaluation of the likelihood and direction of policy change based on information about the institutionalized process of government decision-making. This framework applies broadly to many different kinds of polities from parliamentary and presidential forms of democracy to authoritarian and even totalitarian governments.

The veto player (VP) framework begins from the premise that policy change requires agreement by a certain number of key individuals or collective actors (Tsebelis 2002). These actors are "veto players." Due to institutional arrangements in place, they are able to single-handedly stop a change from the status quo. VPs are specified by institutions such as a constitution (e.g. the President, House, and Senate in the USA) or by the political system (e.g. the different parties that are members of a governing coalition in Western Europe). These two types are "institutional" and "partisan" VPs, respectively. The VP framework includes rules for identifying the VPs in each political system. These rules form the basis for a specific configuration of VPs, meaning that there are a certain number of them, certain ideological distances among them, and a certain level of cohesion in each. These attributes affect the set of outcomes that can replace the status quo.

Using spatial models to quantify and visualize the preferences of relevant actors, the VP framework enables the identification of a "winset," that is, the set of possible outcomes preferred by a majority (Shepsle 2010). The size of the winset of the status quo has specific consequences for policy making. Significant departures from the status quo are impossible when the winset is small, i.e., when there are many VPs or when they have significant ideological differences among them (Tsebelis 2002). The difficulty of effecting significant change from the status quo, given the configuration of VPs, leads to policy stability within a given polity.

According to this logic, countries with many VPs, like Italy and the USA, will have high policy stability. The large number of VPs means the winset is small and it is harder to reach consensus on policy change. Countries like Greece and the UK have a single VP—the government in power—and, consequently, they may have high policy instability as the winset is relatively large with many possible options for policy, which can then be changed in similarly major ways when a new government comes into power. The VP framework suggests that the constellation of VPs determines policy stability. In turn, policy stability affects governance through a series of other policy and institutional characteristics. Thus, if the preferences of veto players, the position of the status quo, and the identity of the agenda setter are known, this approach enables "accurate predictions about policy outcomes" (Tsebelis 2002, p. 284).

The VP approach can be extended to understand many policy issues and choices, including conservation-related policies and governance arrangements at the subnational level. Thus, the fact that the USA has many VPs implies that passing comprehensive legislation to address climate change or indeed any other major shift in policy direction is likely to be difficult. Conversely, existing policies such as the Endangered Species Act have endured despite efforts at change because many officials and organizations interested in the legislation occupy VP positions.

### 6.3.4 Participation

Political science research on participation is of particular importance to those interested in conservation for two reasons. From a normative perspective, a persuasive case can be made that participation of people whose lives and livelihoods are affected by conservation interventions is important in its own right. People should have a say in decisions that affect them. On this perspective, participation is a good in itself. But participation is also important for instrumental reasons. Greater participation can lead to better conservation because participants view conservation as a legitimate social goal. In addition, the time- and place-specific information and resources that participants can bring to the decision-making process can improve the quality of decision-making and the strength of implementation and enforcement of decisions that are made collectively. Assessments of the impact of participation on conservation outcomes, as well as decisions about how to structure user participation in conservation initiatives, can and should be informed by existing research on the subject. Indeed, the entire field of community-based natural resource management can benefit substantially from political science scholarship on participation, because this management approach relies heavily on local participation to realize positive conservation outcomes (Box 6.6).

---

**Box 6.6   Applications: Participation in community forest management in India and Nepal**

Increased community participation in natural-resource governance has been adopted as a strategy for biodiversity conservation and human development across the world over the past two decades. Community forestry initiatives in Nepal and India exemplify this trend. In theory, increasing participation should provide incentives for more effective resource conservation and management and improve the allocation of benefits derived from local resources. But what happens when, as is often the case in rural areas of the developing world, communities are highly differentiated and stratified in terms of power, income and wealth, and social status? Adopting insights from political science research on democratic participation, recent studies examine several cases of community forest management in Nepal and India to answer this question.

Agrawal and Gupta (2005) surveyed 240 households adjacent to protected areas in the Parks and Peoples Program (PPP) in Nepal's Terai region. In keeping with literature on participation in political science (e.g. Verba et al. 1995), they found that wealthier and upper-caste households have a higher probability of participation in the forest user groups created by the PPP to decentralize environmental decision-making and distribute benefits from environmental resources. Another study, of 87 community forestry sites in India and Nepal (Agarwal 2001), found a similar pattern of "participatory exclusions" within ostensibly participatory institutions. Both of these studies highlight how women in particular are underrepresented, even though they may be directly affected by altered access regimes for resources such as grass, fodder, and firewood.

These findings have implications for conservation in that they undermine the assumption of many conservation and development projects that participation reduces

---

*(Continued)*

> **Box 6.6** **(Continued)**
>
> pressure on protected-area resources by improving participant livelihoods. If only a minority of mostly men with large landholdings, high incomes, and upper-caste status participate in user groups, other community members will have reduced incentive to participate. In turn, the latter group will have less incentive to conserve and effectively manage resources. The excluded community members may not comply with, or may actively resist management efforts deriving from new, ostensibly participatory institutions. Results from India and Nepal constitute a note of caution for the way many decentralization programs are implemented in relation to environmental policy: with minimal regard for the participation of poor and socially marginal groups. Fostering local-level institutions may not be enough to meet conservation and development aims. Instead, specific action must be taken to address inequalities and power relationships within communities. Studies of participation, institutions, and power within political science supply useful tools for designing such action.

Political scientists have long studied participation, but a broad spectrum of views on the nature, causes, and effects of participation on collective outcomes endures in the discipline. Participation can be direct or indirect, and can range from relatively surface to more intense involvement. Motivations for participation in different political processes, including conservation, can vary from more idealistic to more material. Motivations are linked to incentives. They can be intrinsic—part of what someone wants to do. Or they can be extrinsic—shaped by external factors. Research on motivations and incentives has begun to examine how they influence conservation behavior, and whether extrinsic incentives have the potential to undermine or crowd out intrinsic motivations (Agrawal et al. 2015; Rode et al. 2015).

Many theorists, advocates, and practitioners view broad-based participation in democratic decision-making as the foundation for more effective policies. Participation prompts leaders to act in the interests of their constituents. Others view direct participation with a more jaundiced eye, questioning both its relevance to democratic processes and its role in effective decision-making. According to this view, too much participation prevents considered and informed decision-making by leaders. Despite lasting debates on the nature, causes, and outcomes of participation within political science, most scholars of conservation who are interested in the impacts of conservation programs on humans tend to view greater participation as a means to greater responsiveness (Figure 6.5). In adopting this view, these scholars approximate ideas about polyarchy, which implies "the continued responsiveness of the government to the preferences of its citizens, considered as political equals" (Dahl 1971, p. 1).

Political science research has examined three broad sets of factors to explain why people participate: (i) the different incentives individuals might have, (ii) socioeconomic and structural factors, and (iii) normative or ideological forces. Some of the earliest research on the subject focused on political participation and suggested that an individual's social status, education, and organizational membership have strong effects on the propensity to participate in political activities (Dahl 1961; Almond & Verba 1963). For example, a

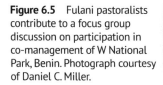

**Figure 6.5** Fulani pastoralists contribute to a focus group discussion on participation in co-management of W National Park, Benin. Photograph courtesy of Daniel C. Miller.

five-country study (Nie et al. 1969) found that as countries became more economically developed, people gained greater incomes, urbanization increased, and a greater density of economic and social organizations developed. These changes corresponded to higher levels of political participation. These findings suggest that, at the individual level, higher social status and increased organizational involvement tend to lead to higher levels of political participation.

Research on associations and interest groups has long addressed the issue of participation (Truman 1951). Olson's (1971) foundational focus on group size and selective incentives in explaining participation—or lack thereof—has strongly influenced subsequent research on the subject. For example, researchers Moe (1980) and Hansen (1985) have developed formal cost–benefit models to illustrate the factors that make an individual join a group, drawing from the insight that individuals in any group attempting collective action to provide public goods will have an incentive to free-ride on the efforts of others (see Box 6.3). This analysis has focused on non-economic factors such as **ideology** and social pressures, and individual-specific factors such as information, preferences, access to resources, and attitudes toward risk.

In contrast to general studies of political participation and the factors that affect it, studies of participation in conservation programs and natural resource management have been more concerned with questions of equity and distribution across gender, class, ethnicity, and other social distinctions (Agarwal 2001; Botchway 2001). A focus on distributive outcomes of conservation is particularly important because many conservation programs produce unequal effects. Protected areas may displace the poor and the livelihoods of the poor; their benefits may flow to those with the capacity to provide services to tourists and other visitors. A deeper and more comprehensive understanding of why people participate in different kinds of conservation efforts or program will likely require examination of socioeconomic characteristics of potential participants, their benefits from participation, and the nature of top-down efforts to secure higher levels of participation (Reed et al. 2018; Authelet et al. 2021).

If political scientists disagree about why people participate, their disagreements are as severe when it comes to examining the effects of participation on outcomes. There is a long and hallowed tradition in political science that views participation as beneficial for society in and of itself. But empirical study of the effects of participation suggests that evidence on the impact of political participation on policy making is sparse and mixed (Bartels 2008). Indeed, we know even less about how the impact of participation varies with the characteristics of the people who participate, institutions that structure participation, and resource systems that participants manage. The inconclusive nature of scientific evidence on the effect of participation on policy outcomes contrasts sharply with widespread rhetoric regarding the importance of participation in securing conservation success. This discrepancy suggests that political scientists might find important insights by investigating the relationship between participation and conservation processes more deeply, whether or not conservation rhetoric matches implementation.

### 6.3.4.1 Approaches to Analyzing Participation

One can understand participation in a variety of ways. One approach to participation focuses on direct vs indirect forms of participation. Direct participation implies that participants in a political process are contributing personal resources and are directly involved in decision-making related to the process. Indirect participation refers to a situation where citizens select representatives through some representative or electoral process, and their representatives make decisions on their behalf or as their agents. Although citizens in a democracy vote and members of civic associations take part in the activities of such associations, much participation in formal political processes in democratic societies is indirect. Voting itself is a form of indirect participation in the political processes responsible for the most important decisions in a democracy.

Forms of participation can be further subdivided based on the degree of direct engagement in decision-making. This engagement can range from passive to very active and intensive participation (Table 6.9). A related approach to classifying and analyzing participation focuses on the extent to which different forms of participation include violent or aggressive tactics. It should be clear that greater levels of participation do not necessarily mean that participation occurs in the absence of power, or that issues of political and social inequality are erased as a result of higher levels of participation. Indeed, various scholars, particularly those interested in development and conservation, have explored how reliance on greater participation and adoption of a participatory rhetoric in official conservation programs can mask continuing inequalities and political asymmetries (Cooke & Kothari 2001).

Contemporary political scientists have found it easy to treat participation as being both empirically and conceptually different from representation (i.e., speaking or acting on behalf of someone or having someone do so on your behalf) and accountability (see discussion in Section 6.3.3 on Governance). But scholars of conservation interested in participation have typically not made such distinctions, treating forms of organizational participation as a substitute for representation, and in some cases even for accountability.

More generally, existing research on participation in conservation initiatives remains somewhat limited in scope and ambition (Nzau et al. 2020). A deeper and more systematic understanding of the conditions under which greater participation can flourish is still

**Table 6.9** A typology of participation based on levels of decision-making involvement.

| Participation according to increasing involvement of people in making decisions | Characteristics of participation | Examples |
|---|---|---|
| Passive participation | People are told about what is going to happen, but their ideas and suggestions are not solicited and their responses are not taken into account | Community-based conservation in Tanzania (Songorwa 1999; Goldman 2003) |
| Participation by consultation | People affected by a political process are consulted, and their views are taken into account in decision-making | CAMPFIRE program in Zimbabwe (Hill 1996; Logan & Moseley 2002) |
| Participation through a share in benefits | People participate by gaining a share in the benefits created by a political process, and may contribute some resources to the process | Controlled Hunting Areas in Botswana (Phuthego & Chanda 2004) |
| Functional participation | People may form groups in accordance with procedures that are externally designed and such groups work toward meeting the predefined objectives of a political process | Joint Forest Management in India (Kumar 2002) |
| Interactive participation | People affected by decisions are involved in joint efforts with external actors to plan, direct, and shape the processes that are going to influence them | Locally Managed Marine Areas in the Pacific (Ferse et al. 2010); Creation and management of the Kaa-Iya del Gran Chaco National Park and Integrated Management Area through collaboration between a Guaraní peoples' group and the Wildlife Conservation Society (Arambiza & Painter 2006) |
| Self-initiated participation | People mobilize to form decision-making groups and assume the power to make decisions and such groups articulate with external actors on their own terms | Conservation via sacred groves (Colding & Folke 2001; Sheridan & Nyamweru 2008) |

missing. Part of the problem is that so many different factors affect the propensity to participate in locally, highly variable settings. Another aspect is the easy manner in which many conservationists defend participation as either good in itself or an obstacle to effective conservation without sufficient evidence of whether participation is valued by those who participate or whether and what forms of participation lead to more effective conservation outcomes.

## 6.4 Future Directions

We argue that political science has a great deal to offer conservation. "Without acute political analyses that take incentives and actions of multiple actors at different scales into account, there is no effective policymaking or governance related to biodiversity and, consequently, no protecting biodiversity" (Agrawal & Ostrom 2006, p. 682). Given the importance of politics and governance to biodiversity conservation, one might think that the application of the tools and theories of political science to this topic was widespread. Unfortunately, examples of fruitful collaboration between political science and conservation remain limited.

### 6.4.1 The Paradox of Contemporary Political Science and Conservation

Mainstream political science has seldom viewed environmental issues with much interest, let alone the more specific concern of biodiversity conservation. Conservation is not seen as an obvious arena for examination of political processes. Electoral systems and practices, democracy, political institutions, international regimes, public opinion, state–society relations, conflict, war, violence, race and ethnicity, policy making, strategic behavior, and policy outcomes are viewed as the more proper provinces of the discipline (Agrawal & Ostrom 2006). Furthermore, political scientists tend to focus on formal political processes at the level of the nation-state or in the international arena. Understanding and effectively addressing conservation problems, by contrast, requires concerted effort at many levels and necessitates understanding not only of formal (de jure) politics, such as laws and policies, but also informal and actual (de facto) political relations and processes.

More generally, political scientists tend to focus on policy issues only to the extent that the analysis of such issues helps cast light on basic puzzles related to political relations and behavior (see Box 6.1). However, analyses of conservation policies and processes can inform some of the most important theoretical puzzles in political science in a different way than other policy issues. For example, effective biodiversity policies must address the question of how decision makers can make long-term credible political commitments when the costs are borne by current generations and benefits will accrue to future generations. The surface answer to this question may appear to be that policy makers cannot make such commitments credibly. But the fact that decision makers around the world have set aside a significant proportion of the planet's terrestrial and marine area (Protected Planet 2021) shows that the surface answer is wrong. At the same time, biodiversity loss continues on a scale and at a speed that has scarcely a parallel in the long evolution of life on Earth. Conservation's successes and failures present theoretical puzzles whose solutions are likely to be of major interest to political scientists. But this case has yet to be made effectively and persuasively.

If political scientists have attended to conservation only rarely, conservationists have also devoted insufficient attention to the politics of conservation. Although many conservation practitioners well realize the importance of politics, formal political analyses of conservation success or failure remain rare. This is at least in part because relatively few people in the conservation field, prominent examples notwithstanding, are trained as political scientists. Most conservation practitioners seem to grapple with politics and political relationships intuitively and mainly on the basis of field experience.

This brief examination of the constraints on a conversation between those in the conservation field and political scientists points to two important insights. First, there is a significant unmet demand for better political science research on local as well as global environmental problems. Second, there is no dearth of specific conservation-related policy and practical issues that can be addressed in a way that better informs political science concerns with theory development. Nonetheless, to do so conservation practitioners must also educate themselves about how political scientists can help them address the problems they are interested in solving (Steinberg 2005, 2015).

### 6.4.2 Enhancing Engagement between Political Science and Conservation

There is a major opportunity for research within and across conservation science and political science to move beyond the current "dialogue of the deaf" (Agrawal & Ostrom 2006) for the mutual benefit of each field—and, ultimately, society and the biosphere. Some indication of the potential for enhanced dialogue is indicated by the fact that the first woman to receive a Nobel Prize in Economic Sciences—Elinor Ostrom—was a political scientist, and she won the prize in 2009 for her work on the relationship between institutions, politics, and local-level environmental and conservation processes. The field of natural-resource governance and local institutional relations continues to be a fruitful domain of inquiry in which political scientists interested in conservation may engage with their more mainstream political science colleagues, and conservation scientists and practitioners may work with political scientists.

There are hopeful signs that environmental issues, particularly climate change but also biodiversity conservation, will continue to become more important on the agenda of national governments and international policy-making institutions. As these issues become more prominent politically, they also are likely to become more important analytically for political scientists. Indeed, some of the recent meetings of major political science associations have included a growing number of environment-related sessions and have explicitly sought to reach across disciplinary boundaries. At the same time, the emergence of new fields of environmental inquiry within political science and social-scientific attention among conservationists means that there are at least incipient efforts under way to make the politics of conservation a more important field of study (Fisher et al. 2009; O'Neill 2009; Steinberg & VanDeveer 2012).

Several broad areas stand out as being particularly ripe for increased engagement across these two domains of inquiry. First, there is a considerable need to understand better how political dynamics, including policy choice, implementation, and performance, shape conservation outcomes at multiple scales. Even if practitioners are primarily interested in the local outcomes of their conservation projects, these outcomes are contingent upon a large variety of macro contextual and policy factors that they often overlook. A second need is research at the interface of science and politics. The relationships among science, politics, and policy can be vexing for conservation practitioners and political scientists alike. Research on environmental policy networks (e.g. Schneider et al. 2003; Lubell et al. 2017) is promising, but there remains little generalizable knowledge about how science influences political decision-making processes. A third important area of investigation is governance. Although significant work on this topic has already been carried out, enhanced analytical clarity for the term and improved understanding of how governance systems at multiple scales affect biodiversity outcomes continue to be frontiers for future research (Agrawal & Ostrom 2006). The decentralization of government authority, including in the natural resource sector, and variegated processes of democratization seen in myriad contexts around the globe represent a fourth area where further research blending political science with conservation may be especially fruitful. Finally, there is a need for better understanding of how norms of behavior, ideas, and institutions shape the politics of conservation and development efforts in different places over time.

The conversation between political science and conservation is clearly important for biodiversity. But there are also ways in which a deeper understanding of conservation processes can contribute to methodological and conceptual advancement relevant to political science. For example, most conservation processes—like political processes— yield multiple outcomes. But relatively little contemporary political science research analyzes relationships between different outcomes and the distinctive causal factors that are relevant to different outcomes simultaneously. As scholars of conservation attend more closely to the causes of and relationships among different outcomes of conservation processes, their methods and approaches will likely contribute to new ways of thinking about political processes and outcomes.

## For Further Reading

1 Enchantment and disenchantment: the role of community in natural resource conservation (Agrawal & Gibson 1999, *World Development* 27: 629–649).
2 Can information outreach increase participation in community-driven development? A field experiment near Bwindi National Park, Uganda (Buntaine, Daniels & Devlin 2018, *World Development* 106: 407–421).
3 *The Politics of the Environment: Ideas, Activism, Policy.* (Carter 2018, Cambridge University Press).
4 *Activists Beyond Borders: Advocacy Networks in International Politics* (Keck & Sikkink 1998, Cornell University Press).
5 Environmental Governance (Lemos & Agrawal 2006, *Annual Review of Environment and Resources*, 31, 297–325).
6 *Governing the Commons: The Evolution of Institutions for Collective Action* (Ostrom 1990, Cambridge University Press).
7 Institutional legacies explain the comparative efficacy of protected areas: Evidence from the Calakmul and Maya Biosphere Reserves of Mexico and Guatemala (Solorzano & Fleischman 2018, *Global Environmental Change* 50: 278–288).
8 *Weapons of the Weak: Everyday Forms of Peasant Resistance* (Scott 1985, Yale University Press).
9 *Who Rules the Earth? How Social Rules Shape Our Planet and Our Lives* (Steinberg 2015, Oxford University Press).
10 *Bureaucratic Landscapes: Interagency Cooperation and the Preservation of Biodiversity* (Thomas 2003, MIT Press).

## Acknowledgments

We gratefully acknowledge the following people for providing comments that allowed us to strengthen this chapter: Bill Adams, Liz Gerber, Debra Javeline, Michael Mascia, Spencer Piston, Diane Russell, and Ivan Scales as well as participants in the Interdisciplinary Workshop on American Politics at the University of Michigan and the Political Science Student–Faculty Seminar at the University of Illinois, Urbana-Champaign.

## References

Agarwal, B. (2001) Participatory exclusions, community forestry, and gender: an analysis for South Asia and a conceptual framework. *World Development* 29: 1623–1648.

Agrawal, A. (2005) *Environmentality: Technologies of Government and the Making of Subjects.* Durham, NC: Duke University Press.

Agrawal, A., Chhatre, A. & Gerber, E.R. (2015) Motivational crowding in sustainable development interventions. *American Political Science Review* 109 (3): 470–487.

Agrawal, A. & Goyal, S. (2001) Group size and collective action: third-party monitoring in common-pool resources. *Comparative Political Studies* 34: 63–93.

Agrawal, A. & Gupta, K. (2005) Decentralization and participation: the governance of common pool resources in Nepal's Terai. *World Development* 33: 1101–1114.

Agrawal, A. & Ostrom, E. (2001) Collective action, property rights, and decentralization in resource use in India and Nepal. *Politics and Society* 29: 485–514.

Agrawal, A. & Ostrom, E. (2006) Political science and conservation biology: a dialog of the deaf. *Conservation Biology* 20: 681–682.

Almond, G. & Verba, S. (1963) *The Civic Culture.* Princeton, NJ: Princeton University Press.

Anderson, B. (1991) *Imagined Communities: Reflections on the Origin and Spread of Nationalism.* London: Verso.

Andersson, K.P. & Gibson, C.C. (2007) Decentralized governance and environmental change: local institutional moderation of deforestation in Bolivia. *Journal of Policy Analysis and Management* 26 (1): 99–123.

Ansell, C. (2006) Network institutionalism. In: *The Oxford Handbook of Political Institutions* (ed. R.A.W. Rhodes, S.A. Binder & B.A. Rockman), 75–89. Oxford: Oxford University Press.

Arambiza, E. & Painter, M. (2006) Biodiversity conservation and the quality of life of indigenous peoples in the Bolivian Chaco. *Human Organization* 65: 20–34.

Aristotle (1992) *The Politics* (trans. T.A. Sinclair, rev. T.J. Saunders). London: Penguin Books.

Authelet, M., Subervie, J., Meyfroidt, P., Asquith, N. & Ezzine-de-Blas, D. (2021) Economic, pro-social and pro-environmental factors influencing participation in an incentive-based conservation program in Bolivia. *World Development* 145: 105487. https://doi.org/10.1016/j.worlddev.2021.105487.

Axelrod, R. (1986) An evolutionary approach to norms. *American Political Science Review* 80: 1095–1111.

Bachrach, P. & Baratz, M.S. (1962) Two faces of power. *American Journal of Political Science* 56: 947–952.

Barrett, C.B., Brandon, K., Gibson, C. & Gjertsen, H. (2001) Conserving tropical biodiversity amid weak institutions. *BioScience* 51: 497–502.

Barrett, C.B., Gibson, C.C., Hoffman, B. & Mccubbins, M.D. (2006) The complex links between governance and biodiversity. *Conservation Biology* 20: 1358–1366.

Bartels, L.M. (2008) *Unequal Democracy: The Political Economy of the New Gilded Age.* Princeton, NJ: Princeton University Press.

Baulenas, E., Baiges, T., Cervera, T. & Pahl-Wostl, C. (2021) How do structural and agent-based factors influence the effectiveness of incentive policies? A spatially explicit agent-based model to optimize woodland-for-water PES policy design at the local level. *Ecology and Society* 26 (2): 10. https://doi.org/10.5751/ES-12325-260210.

Berry, B.J.L., Kiel, L.D. & Elliott, E. (2002) Adaptive agents, intelligence, and emergent human organization: capturing complexity through agent-based modeling. *Proceedings of the National Academy of Sciences of the United States of America* 99: 7187–7188.

Botchway, K. (2001) Paradox of empowerment: reflections on a case study from northern Ghana. *World Development* 29: 135–153.

Bourdieu, P. (1985) The forms of capital. In: *Handbook of Theory and Research for the Sociology of Education* (ed. J.G. Richardson), 241–258. New York: Greenwood.

Burke, M., Hsiang, S.M. & Miguel, E. (2015) Climate and conflict. *Annual Review of Economics* 7 (1): 577–617.

Busby, J. (2018) Environmental security. In: *Handbook of International Security* (ed. A. Gheciu & W.C. Wohlforth), 471–486. New York: Oxford University Press.

Campbell, J.L. (2004) *Institutional Change and Globalization.* Princeton, NJ: Princeton University Press.

Chasek, P.S., Downie, D.L. & Brown, J.W. (2006) *Global Environmental Politics.* Boulder, CO: Westview Press.

Colding, J. & Folke, C. (2001) Social taboos: "invisible" systems of local resource management and biological conservation. *Ecological Applications* 11: 584–600.

Cooke, B. & Kothari, U. ed. (2001) *Participation: The New Tyranny?* London: Zed Books.

Crick, B. (1964) *In Defense of Politics.* London: Penguin Press.

Dahl, R.A. (1961) *Who Governs? Democracy and Power in an American City.* New Haven, CT: Yale University Press.

Dahl, R.A. (1971) *Polyarchy: Participation and Opposition.* New Haven, CT: Yale University Press.

Dahl, R.A., Shapiro, I. & Cheibub, J.A. eds. (2003) *The Democracy Sourcebook.* Cambridge, MA: The MIT Press.

Doyle, T., Mceachern, D. & MacGregor, S. (2015) *Environment and Politics*, 4th ed. London: Routledge.

Druckman, J.N., Green, D.P., Kuklinski, J.H. et al. eds. (2011) *Cambridge Handbook of Experimental Political Science.* Cambridge, UK: Cambridge University Press.

Easton, D. (1953) *The Political System: An Inquiry into the State of Political Science.* New York: Alfred A. Knopf.

Edwards, M. (2004) *Civil Society.* Cambridge, UK: Polity Press.

Farr, J. (1995) Remembering the revolution: behavioralism in American political science. In: *Political Science in History: Research Programs and Political Traditions* (ed. J. Farr, J.S. Dryzek & S.T. Leonard), 198–224. Cambridge, UK: Cambridge University Press.

Farr, J. & Seidelman, R. eds. (1993) *Discipline and History: Political Science in the United States.* Ann Arbor: University of Michigan Press.

Ferse, S.C.A., Costa, M.M., Máñez, K.S., Adhuri, D.S. & Glaser, M. (2010) Allies, not aliens: increasing the role of local communities in marine protected area implementation. *Environmental Conservation* 37: 23–34.

Fisher, B., Balmford, A., Green, R.E. & Trevelyan, R. (2009) Conservation science training: the need for an extra dimension. *Oryx* 43: 361–363.

Folke, C., Hahn, T., Olsson, P. & Norberg, J. (2005) Adaptive governance of social–ecological systems. *Annual Review of Environment and Resources* 30: 441–473.

Foucault, M. (1979) *Discipline and Punish: The Birth of the Prison.* New York: Vintage.

Foucault, M. (1991 [1978]) Governmentality. In: *The Foucault Effect: Studies in Governmentality* (ed. G. Burchell, C. Gordon & P. Miller), 87–104. Chicago: The University of Chicago Press.

Galik, C.S. & Jagger, P. (2015) Bundles, duties, and rights: a revised framework for analysis of natural resource property rights regimes. *Land Economics* 91 (1): 76–90.

Gibson, C.C. (1999) *Politicians and Poachers: The Political Economy of Wildlife Policy in Africa.* Cambridge, UK: Cambridge University Press.

Gibson, C.C., Andersson, K., Ostrom, E. et al. (2005) *The Samaritan's Dilemma: The Political Economy of Development Aid.* Oxford, UK: Oxford University Press.

Gibson, C.C., Ostrom, E. & Ahn, T.K. (2000) The concept of scale and the human dimensions of global change: a survey. *Ecological Economics* 32: 217–239.

Goldman, M. (2003) Partitioned nature, privileged knowledge: community-based conservation in Tanzania. *Development and Change* 34: 833–862.

Goodin, R.E. ed. (2011) *The Oxford Handbook of Political Science.* Oxford, UK: Oxford University Press.

Goodin, R.E., Rein, M. & Moran, M. (2006) The public and its policies. In: *The Oxford Handbook of Public Policy* (ed. M. Moran, M. Rein & R.E. Goodin), 3–36. Oxford, UK: Oxford University Press.

Granovetter, M. (1985) Economic action and social structure: the problem of embeddedness. *The American Journal of Sociology* 91: 481–510.

Green, J.F. (2013) *Rethinking Private Authority.* Princeton, NJ: Princeton University Press.

Haas, P.M. (1989) Do regimes matter? Epistemic communities and Mediterranean pollution control. *International Organization* 43: 377–403.

Hall, P. & Taylor, R. (1996) Political science and the three new institutionalisms. *Political Studies* XLIV: 936–957.

Hansen, J.M. (1985) The political economy of group membership. *The American Political Science Review* 79: 79–96.

Hardin, G. (1968) The tragedy of the commons. *Science* 162: 1243–1248.

Helmke, G. & Levitsky, S. (2004) Informal institutions and comparative politics: a research agenda. *Perspectives on Politics* 2: 725–740.

Hill, K. (1996) Zimbabwe's wildlife utilization programs: grassroots democracy or an extension of state power? *African Studies Review* 39: 103–123.

Hobbes, T. (1958 [1651]) *Leviathan.* New York: Macmillan.

Hoel, M. & Schneider, K. (1997) Incentives to participate in an international environmental agreement. *Environmental and Resource Economics* 9: 153–170.

Holden, M., Jr. (2000) The competence of political science: "progress in political research revisited". *American Political Science Review* 94: 1–20.

Inglehart, R. & Welzel, C. (2005) *Modernization, Cultural Change, and Democracy: The Human Development Sequence*. Cambridge, UK: Cambridge University Press.

Jacobs, J. (1961) *The Death and Life of Great American Cities*. New York: Random House.

Janssen, M.A., Smith-Heisters, S., Aggarwal, R. & Schoon, M.L. (2019) 'Tragedy of the commons' as conventional wisdom in sustainability education. *Environmental Education Research* 25 (11): 1587–1604.

John, P. (2017) *Field Experiments in Political Science and Public Policy: Practical Lessons in Design and Delivery*. New York: Routledge.

Jones, B. (2010) The evolution of Namibia's communal conservancies. In: *Community Rights, Conservation, and Contested Land: The Politics of Natural Resource Governance in Africa* (ed. F. Nelson), 106–120. Abingdon, UK: Earthscan.

Junn, J. (2009) Dynamic categories and the context of power. In: *The Future of Political Science: 100 Perspectives* (ed. G. King, K.L. Schlozman & N.H. Nie), 25–27. New York: Routledge.

Keck, M. & Sikkink, K. (1998) *Activists beyond Borders: Advocacy Networks in International Politics*. Ithaca, NY: Cornell University Press.

Keohane, R.O. (2008) Big questions in the study of world politics. In: *Oxford Handbook of International Relations* (ed. C. Reus-Smit & D. Snidal), 708–715. Oxford, UK: Oxford University Press.

Keohane, R.O. & Levy, M.A. (1996) *Institutions for Environmental Aid: Pitfalls and Promise*. Cambridge, MA: The MIT Press.

Khagram, S. & Ali, S. (2006) Environment and security. *Annual Review of Environment and Resources* 31: 395–411.

Kitschelt, H. & Wilkinson, S. (2007) *Patrons, Clients, and Policies: Patterns of Democratic Accountability and Political Competition*. Cambridge, UK: Cambridge University Press.

Koubi, V. (2019) Climate change and conflict. *Annual Review of Political Science* 22: 343–360.

Kumar, S. (2002) Does "participation" in common pool resource management help the poor? A social cost-benefit analysis of joint forest management in Jharkhand, India. *World Development* 30: 763–782.

Lackey, R.T. (2007) Science, scientists, and policy advocacy. *Conservation Biology* 21: 12–17.

Laitin, D.D. (2004) The political science discipline. In: *The Evolution of Political Knowledge: Theory and Inquiry in American Politics* (ed. E.D. Mansfield & R. Sisson), 11–40. Columbus: The Ohio State University Press.

Larson, A.M. & Soto, F. (2008) Decentralization of natural resource governance regimes. *Annual Review of Environment and Resources* 33: 213–239.

Lasswell, H.D. (1936) *Politics: Who Gets What, When and How*. New York: McGraw-Hill.

Lemos, M.C. & Agrawal, A. (2006) Environmental governance. *Annual Review of Environment and Resources* 31: 297–325.

Levi, M. (2002) The state of the study of the state. In: *Political Science: The State of the Discipline* (ed. I. Katznelson & H.V. Milner). New York: W.W. Norton.

Locke, J. (1980 [1689]) *Second Treatise of Government*. Indianapolis, IN: Hackett.

Logan, B.I. & Moseley, W.G. (2002) The political ecology of poverty alleviation in Zimbabwe's Communal Areas Management Programme for Indigenous Resources (CAMPFIRE). *Geoforum* 33: 1–14.

Lubell, M., Jasny, L. & Hastings, A. (2017) Network governance for invasive species management. *Conservation Letters* 10 (6): 699–707.

Lukes, S. (1974) *Power: A Radical View.* London: Macmillan.

Machiavelli, N. (1950 [1532]) *The Prince and the Discourses.* New York: The Modern Library.

Mansbridge, J. (2009) On the free rider problem. In: *The Future of Political Science: 100 Perspectives* (ed. G. King, K.L. Schlozman & N.H. Nie), 216–218. New York: Routledge.

McAdam, D., Tarrow, S. & Tilly, C. (2009) Comparative perspectives on contentious politics. In: *Comparative Politics: Rationality, Culture, and Structure: Advancing Theory in Comparative Politics*, 2nd ed. (ed. M. Lichbach & A. Zuckerman), 260–290. Cambridge, UK: Cambridge University Press.

McKean, M. (2000) Common property: what is it, what is it good for, and what makes it work? In: *People and Forests: Communities, Institutions, and Governance* (ed. C. Gibson, M. McKean & E. Ostrom), 27–55. Cambridge, MA: The MIT Press.

Meine, C. & Knight, R.L. eds. (1999) *The Essential Aldo Leopold: Quotations and Commentaries.* Madison: University of Wisconsin Press.

Mildenberger, M. (2019) The tragedy of the tragedy of the commons. Scientific American. https://blogs.scientificamerican.com/voices/the-tragedy-of-the-tragedy-of-the-commons.

Miller, D.C., Minn, M. & Sinsin, B. (2015) The importance of national political context to the impacts of international conservation aid: evidence from the W National Parks of Benin and Niger. *Environmental Research Letters* 10 (11): 115001. https://doi.org/10.1088/1748-9326/10/11/115001.

Moe, T.M. (1980) *The Organization of Interests: Incentives and the Internal Dynamics of Political Interest Groups.* Chicago: University of Chicago Press.

Moe, T.M. (1990) Political institutions: the neglected side of the story. *Journal of Law, Economics, & Organization* 6: 213–253.

Monroe, K.R. ed. (2005) *Perestroika! The Raucous Rebellion in Political Science.* New Haven, CT: Yale University Press.

Montesquieu, C.D.S. (1989 [1748]) *The Spirit of the Laws.* Cambridge, UK: Cambridge University Press.

Nie, N.H., Powell, G.B., Jr. & Prewitt, K. (1969) Social structure and political participation: developmental relationships, II. *The American Political Science Review* 63: 808–832.

North, D.C. (1990) *Institutions, Institutional Change, and Economic Performance.* New York: Cambridge University Press.

Nye, J.S. (2004) *Soft Power: The Means to Success in World Politics.* New York: Public Affairs.

Nzau, J.M., Gosling, E., Rieckmann, M., Shauri, H. & Habel, J.C. (2020) The illusion of participatory forest management success in nature conservation. *Biodiversity and Conservation* 29 (6): 1923–1936.

Olson, M. (1971) *The Logic of Collective Action: Public Goods and the Theory of Groups.* Cambridge, MA: Harvard University Press.

O'Neill, K. (2009) *The Environment and International Relations.* Cambridge, UK: Cambridge University Press.

Ophuls, W. (1977) *Ecology and the Politics of Scarcity.* San Francisco, CA: Freeman.

Ostrom, E. (1990) *Governing the Commons: The Evolution of Institutions for Collective Action.* New York: Cambridge University Press.

Ostrom, E. (1999) Institutional rational choice: an assessment of the IAD framework. In: *Theories of the Policy Process* (ed. P. Sabatier), 21–63. Boulder, CO: Westview Press.

Ostrom, E. (2005) *Understanding Institutional Diversity*. Princeton, NJ: Princeton University Press.

Ostrom, E. (2007) A diagnostic approach for going beyond panaceas. *Proceedings of the National Academy of Sciences of the United States of America* 104: 15181–15187.

Ostrom, E., Dietz, T., Dolšak, N. et al. eds. (2002) *The Drama of the Commons*. Washington, DC: National Academy Press.

Page, S. (2006) Path dependence. *Quarterly Journal of Political Science* 1: 87–115.

Peluso, N.L. (1993) Coercing conservation? The politics of state resource control. *Global Environmental Change* 3: 199–217.

Peters, B.G. (1999) *Institutional Theory in Political Science: The "New" Institutionalism*. London: Pinter.

Phuthego, T.C. & Chanda, R. (2004) Traditional ecological knowledge and community-based natural resource management: lessons from a Botswana wildlife management area. *Applied Geography* 24: 57–76.

Pierson, P. (2000) Increasing returns, path dependence, and the study of politics. *American Political Science Review* 94: 251–268.

Pierson, P. & Skocpol, T. (2002) Historical institutionalism in contemporary political science. In: *Political Science: The State of the Discipline* (ed. I. Katznelson & H.V.Milner), 693–721. New York: W.W. Norton.

Princen, T. (1994) The ivory trade ban: NGOs and international conservation. In: *Environmental NGOs in World Politics: Linking the Local and the Global* (ed. T. Princen & M. Finger), 121–159. London: Routledge.

Protected Planet (2021) World database on protected areas (WDPA) and world database on other effective area-based conservation measures (OECMs). https://www.protectedplanet.net/en (accessed July 17, 2021).

Putnam, R.D. (1995) Bowling alone: America's declining social capital. *Journal of Democracy* 6: 65–78.

Reed, M.S., Vella, S., Challies, E. et al. (2018) A theory of participation: what makes stakeholder and public engagement in environmental management work? *Restoration Ecology* 26: S7–S17.

Rhodes, R.A.W. (2006) Policy network analysis. In: *The Oxford Handbook of Public Policy* (ed. M. Moran, M. Rein & R.E. Goodin), 425–447. Oxford, UK: Oxford University Press.

Riker, W. (1980) Implications from the disequilibrium of majority rule for the study of institutions. *American Political Science Review* 80: 432–447.

Rode, J., Gómez-Baggethun, E. & Krause, T. (2015) Motivation crowding by economic incentives in conservation policy: a review of the empirical evidence. *Ecological Economics* 117: 270–282.

Roniger, L. (2004) Political clientelism, democracy, and market economy. *Comparative Politics* 36: 353–375.

Rosenau, J.N. & Czempiel, E.-O. eds. (1992) *Governance without Government: Order and Change in World Politics*. London: Cambridge University Press.

Rousseau, J.-J. (1987 [1762]) On the social contract. In: *The Basic Political Writings* (ed. & trans. D.A. Cress), 139–227. Indianapolis, IN: Hackett Publishing Company.

Sabatier, P.A. (1988) An advocacy coalition framework of policy change and the role of policy-oriented learning therein. *Policy Sciences* 21: 129–168.

Sanford, L. (2021) Democratization, elections, and public goods: the evidence from deforestation. *American Journal of Political Science.* https://doi.org/10.1111/ajps.12662.

Schlager, E. & Blomquist, W. (1996) A comparison of three emerging theories of the policy process. *Political Research Quarterly* 49: 651–672.

Schlager, E. & Ostrom, E. (1992) Property-rights regimes and natural resources: a conceptual analysis. *Land Economics* 68 (3): 249–262.

Schmidt, V.A. (2008) Discursive institutionalism: the explanatory power of ideas and discourse. *Annual Review of Political Science* 11: 303–326.

Schneider, M., Scholz, J., Lubell, M., Mindruta, D. & Edwardsen, M. (2003) Building consensual institutions: networks and the National Estuary Program. *American Journal of Political Science* 47: 143–158.

Scott, J.C. (1972) Patron-client politics and political change in Southeast Asia. *The American Political Science Review* 66: 91–113.

Scott, J.C. (1985) *Weapons of the Weak: Everyday Forms of Peasant Resistance.* New Haven, CT: Yale University Press.

Scott, J.C. (1998) *Seeing like a State: How Certain Schemes to Improve the Human Condition Have Failed.* New Haven, CT: Yale University Press.

Shepsle, K.A. (2010) *Analyzing Politics: Rationality, Behavior and Institutions,* 2nd ed. New York: W.W. Norton.

Sheridan, M.J. & Nyamweru, C. eds. (2008) *African Sacred Groves: Ecological Dynamics & Social Change.* Oxford, UK: James Currey.

Sikor, T., He, J.U.N. & Lestrelin, G. (2017) Property rights regimes and natural resources: a conceptual analysis revisited. *World Development* 93: 337–349.

Smith, R.J., Muir, R.D.J., Walpole, M.J. et al. (2003) Governance and the loss of biodiversity. *Nature* 426: 67–70.

Smith, R.J. & Walpole, M.J. (2005) Should conservationists pay more attention to corruption? *Oryx* 39: 251–256.

Somanathan, E., Prabhakar, R. & Mehta, B.S. (2009) Decentralization for cost-effective conservation. *Proceedings of the National Academy of Sciences of the United States of America* 106: 4143–4147.

Songorwa, A. (1999) Community-based wildlife (CBW) management in Tanzania: are the communities interested? *World Development* 27: 2061–2079.

Steinberg, P.F. (2005) Bringing political science to bear on tropical conservation. *International Environmental Agreements: Politics, Law and Economics* 5: 395–404.

Steinberg, P.F. (2015) *Who Rules the Earth? How Social Rules Shape Our Planet and Our Lives.* Oxford: Oxford University Press.

Steinberg, P.F. & VanDeveer, S.D. eds. (2012) *Comparative Environmental Politics: Theory, Practice, and Prospects.* Cambridge, MA: The MIT Press.

Steinhart, E.I. (2006) *Black Poachers, White Hunters: A Social History of Hunting in Colonial Kenya.* Oxford, UK: James Currey.

Stokes, L.C. (2020) *Short Circuiting Policy: Interest Groups and the Battle over Clean Energy and Climate Policy in the American States.* Oxford, UK: Oxford University Press.

Struthers, C.L., Scott, T.A., Fleischman, F. et al. (2021) The forest ranger (and the legislator): how local congressional politics shape policy implementation in agency field offices. *Journal of Public Administration Research and Theory* muab037. https://doi.org/10.1093/jopart/muab037.

Suiseeya, K.R.M. & Caplow, S. (2013) In pursuit of procedural justice: lessons from an analysis of 56 forest carbon project designs. *Global Environmental Change* 23 (5): 968–979.

Taylor, D.E. (2009) *The Environment and the People in American Cities, 1600s–1900s: Disorder, Inequality, and Social Change*. Durham, NC: Duke University Press.

Thomas, C.W. (2003) *Bureaucratic Landscapes: Interagency Cooperation and the Preservation of Biodiversity*. Cambridge, MA: The MIT Press.

Thoreau, H.D. (1993 [1849]) *Civil Disobedience and Other Essays*. New York: Dover.

Tilly, C. & Tarrow, S. (2006) *Contentious Politics*. Boulder, CO: Paradigm.

Truman, D.B. (1951) *The Governmental Process: Political Interests and Public Opinion*. New York: Knopf.

Tsebelis, G. (2002) *Veto Players: How Political Institutions Work*. New York: Russell Sage Foundation/Princeton University Press.

Ullman, R.H. (1983) Redefining security. *International Security* 8: 129–153.

Underdal, A. (2017) Climate Change and international relations (after Kyoto). *Annual Review of Political Science* 20 (1): 169–188.

Verba, S., Schlozman, K.L. & Brady, H.E. (1995) *Voice and Equality: Civic Voluntarism in American Politics*. Cambridge, MA: Harvard University Press.

Waldron, J. (2002) Justice. In: *Political Science: The State of the Discipline* (ed. I. Katznelson & H.V. Milner). New York: W.W. Norton.

Walker, K.L. (2009) Protected-area monitoring dilemmas: a new tool to assess success. *Conservation Biology* 23: 1294–1303.

Weingast, B.R. (2002) Rational-choice institutionalism. In: *Political Science: The State of the Discipline* (ed. I. Katznelson & H.V. Milner), 660–692. New York: W.W. Norton & Company.

WGI (Worldwide Governance Indicators) (2021) https://info.worldbank.org/governance/wgi.

Whyte, K.P. (2011) The recognition dimensions of environmental justice in Indian country. *Environmental Justice* 4 (4): 199–205.

Young, O. (2002) *The Institutional Dimensions of Environmental Change*. Cambridge, MA: The MIT Press.

# 7

# Psychology and Conservation

*Olin Eugene Myers Jr.*

## 7.1 Defining Psychology

Psychology is the scientific study of human thought, motivation, emotion, and behavior, and the application of the resulting knowledge not only to mental health but to every facet of human behavior. Psychology develops and tests theories about how individuals receive and process information, about what drives behavior, and about affective states (i.e. those relating to emotions, moods, and feelings) and mental illnesses. Psychological data include everything from the observation of behavior to verbal responses, to brain scans, to sampling the activity of the nervous system, endocrine, and other physiological systems. Psychology also is concerned with how all these phenomena, apparently "contained" in the individual, actually arise from and in turn affect the broader social context of which individuals are a part. Psychology helps explain human experience and behavior by winnowing commonplace ideas, generating novel hypotheses, inventing new ways to observe the psyche, and systematically reconstructing processes that may operate out of our conscious awareness. Psychology is a multiscale discipline that spans biological to cultural systems.

### 7.1.1 Major Themes and Questions in Psychology

Psychology has many subdisciplines that investigate the full range of human phenomena (Table 7.1).

Recent decades have seen an explosion of psychological research, but a set of key enduring questions continue to animate the discipline:

1) What is the relationship between matter and mind, or brain–body physiology and mental experiences, from sensation and perception to the apprehension of the most abstract concepts?
2) How do people perceive and process information about the world around them, and decide and act? What is the relationship between language and thought, and emotion and cognition? How should we understand top-down causality (e.g. a belief or stress causing a physiological illness)?

*Conservation Social Science: Understanding People, Conserving Biodiversity*, First Edition.
Edited by Daniel C. Miller, Ivan R. Scales, and Michael B. Mascia.
© 2023 John Wiley & Sons Ltd. Published 2023 by John Wiley & Sons Ltd.

**Table 7.1** Subfields in psychology.

| Subfield | Topics |
| --- | --- |
| Biological | Biological basis of behavior in brain and nervous system, and relations to other physiological systems |
| Clinical and counseling | Mental health and illness, with the aim of improving well-being through psychotherapy and medical interventions |
| Cognitive | Mental processes of attention, perception, memory, learning, language, emotion, and thought |
| Comparative | Behavior and mental life of non-human animals |
| Cultural and Cross-cultural | How cultural and psychological factors work together; understanding cultural variation; search for universals |
| Developmental | Change and continuity in mental process across the life span, from infancy through aging and death |
| Educational | Learning, teaching, and educational settings and methods |
| Environmental and Conservation | Effects of physical environments on behavior; determinants of conservation/environmentally relevant behavior |
| Industrial and organizational | Optimizing performance; how the work environment affects subjective aspects of work |
| Personality | Patterns of thought, behavior, feelings, or traits that are consistent across settings or time |
| Positive | Positive human traits, functioning, optimum performance, and fulfillment |
| Social | Effects of social context and influences on behavior; group dynamics and leadership; processing of social information |

3) How does the individual relate to social context? How do we select and shape our contexts? What motivates different forms of political participation?

4) How can an evolutionary perspective help explain human behavior, particularly preferences and tendencies in social interactions and groups? As behavior evolved, to what environment(s) has it adapted, and does this help explain behavior in newer settings? How are innate ("nature") and experiential influences ("nurture") deeply integrated across lifespan development?

## 7.1.2   Psychology and Conservation

Psychology is important to **conservation social science** because it describes the individual person level and how individuals work within **culture** (and create, modify, or challenge social institutions). Areas of theory and research related to conservation can be grouped into three areas: (i) cognition; (ii) values and attitudes; and (iii) social context and influence.

1) Cognition. How do people perceive and think about natural environments, species, ecological concepts, and conservation actions? Do the ways humans conceptualize conservation illustrate more general patterns in how people think about the world of living things? How do complex ideas and decisions reveal or depend on mental shortcuts (e.g. judging a taxonomic group such as insects unworthy of protection due to emotional disgust). It is important to note that cognition and other psychology topics vary greatly by cultural context. To date, psychology has overwhelmingly been developed in higher-income countries with more formal education. Existing theories serve as hypotheses to be explored in other cultures, but such studies need open-ended and "native"-led elements too. Thus, psychology's contributions to addressing conservation challenges may relate more to issues in richer countries, and to rapidly globalizing political and economic forces driving psychological motives for goods that require massive **ecosystem** degradation or species exploitation at "home" and elsewhere.

2) Values and attitudes. What evaluative beliefs—values and attitudes—do people hold toward nature and ways people treat it? What broad or specific human motivations initiate and sustain conservation behavior or undercut it? Some studies suggest that time in nature has physical and psychological benefits (and costs). Do these affect attitudes? Emotions are inherently evaluative—how do they affect people's behavior relating to conservation?

3) Social context and influence. How does social context influence behavior? Factors such as **norms** and **narratives**, perceived peer endorsement, modeling or copying others' behaviors, conformity with social expectations, trust, and credibility may explain behavior. What social constraints and facilitators characterize personal, organizational, citizenship, or activism behaviors? Identity is deeply tied to other people; in what ways is it related to conservation and the non-human world?

## 7.2 A Brief History of Psychology

Psychology is a relatively young but dynamic science. It has already made contributions to understanding and solving environmental problems but has much more to contribute toward conservation. Western psychology grew out of philosophies of the ancient Greeks, the Enlightenment, and other traditions. Psychology as a formal scientific pursuit emerged only in the late 1800s, diverging from an older matrix of philosophical inquiries into knowledge, ethics, and society (Table 7.2).

Psychology demonstrates a recurring tension between positivistic approaches modeled on the natural sciences (see also Chapter 2 for more on **positivism** and the scientific method) and hermeneutic (or interpretive) ones modeled on the humanities and more reflexive social sciences such as cultural anthropology. A minority of psychologists follow the latter hermeneutic or "interpretive turn" in the social sciences, often making critically insightful and context-sensitive contributions (Table 7.3). With either approach, evaluating any psychological theory or finding requires intricately unpacking the ideas and the details of the empirical methods used (e.g. how a theory was turned into particular questions, tests, observations; how it was administered to specific populations; and how ambiguities or observed extreme data points were dealt with in analysis), and then acknowledging these conditions as limitations on generalizability of the findings.

**Table 7.2** Milestones in psychology.

| Time period | Movements and early leaders in psychology fields |
| --- | --- |
| 4000 BCE forward | South Asian religions and traditions of mindfulness explore psychology and self. They influence Buddhist practices (fifth to fourth centuries BCE and later) and ideas of mind. These traditions do not explicitly interface with modern Western psychology until the mid-1900s works of Swiss psychologist Karl Jung. |
| fifth–fourth centuries BCE | Plato, echoing his teacher Socrates' influence, thinks the psyche (mind) is prepared to know the eternal nonphysical ideas holding ultimate truth. Such ideas reflect origins in the Upanishads and the South Asian philosophers residing in Athens then and before. Aristotle, Plato's student, distinguishes living things according to their organization and capacities; the mental is inseparable from the living body, except for the rational intellect. |
| seventeenth–eighteenth centuries | René Descartes explains most bodily & mental operations as material and mechanical, minimizing the role of transcendent soul and elevating reason. |
| | Enlightenment philosophers and the roots of knowledge. Rationalists (Baruch Spinoza, Gottfried Leibniz) see reason (deduction) as the primary root. Empiricists (John Locke, George Berkeley, David Hume, J.S. Mill) emphasize experience and establish the experimental method. These schools are later reflected in different schools of psychology. |
| Latter nineteenth century | Experimental psychophysics investigations such as Gustav Fechner's work on the relation between physical stimuli and sensation; Hermann von Helmholtz and Wilhelm Wundt study perception. |
| | First psychological laboratories set up by Wundt at University of Leipzig in 1879, and by William James at Harvard University. In 1890, James publishes *Principles of Psychology*. |
| 1900s–1930s | Behaviorism (Ivan Pavlov, Edward Thorndike, John Watson, B.F. Skinner) shows laws of conditioning; behaviorism attains dominance in American psychology. |
| | Gestaltists (Max Wertheimer, Wolfgang Köhler) show importance of pattern perception and insight learning. |
| | Psychoanalysis (Sigmund Freud) emphasizes dynamic relations between basic drives, unconscious thought, and the role of the ego in adjusting the person to social reality. Analytical psychology (Karl Jung) considers individuation of self from personal and group elements as life-long human developmental pathway. Both schools advance psychotherapy practice, which further differentiates with subsequent theories and systematic research. |
| | Social context (George H. Mead, James Mark Baldwin, John Dewey) recognized as important shaper of psychology. |

**Table 7.2** (Continued)

| Time period | Movements and early leaders in psychology fields |
| --- | --- |
| 1930s–1950s | Developmental psychology—earlier explored by G. Stanley Hall, Mark Baldwin, and John Dewey—is extended by Jean Piaget who models his "genetic epistemology" on the work of philosopher Immanuel Kant. In the Soviet Union, Lev Vygotsky shows developmental trends grow from social group. |
| | Field of life-span human development in biological and cultural contexts (Bernice Neugarten, Robert Havighurst, Robert Levine). |
| 1950s–1970s | Environmental and social psychology (Kurt Lewin, Roger Barker, James and Eleanor Gibson) focus on contextual influences on behavior, naturalistic methods and application of psychology to a built environment, and to social issues. |
| | Humanistic and existential psychology (Abraham Maslow, Rollo May, Erich Fromm, Viktor Frankl) emphasize the person at the center of subjective experience and human potential. |
| | Cognitivism (Noam Chomsky, George Miller, Ulrich Neisser, Albert Bandura, Jerry Fodor, Jerome Bruner, Howard Gardner) undercuts behaviorism and becomes dominant approach to mind and behavior in American psychology. |
| 1980s–2000s | Qualitative psychology rejects testing and lab experiments in favor of description and contextual validity. Phenomenological psychology (Ludwig Binswanger) focuses on qualities of lived experience, person-world co-constitution, and problems of intersubjectivity in methods. |
| | Cross-cultural (Ype Poortinga, Harry Triandis); cultural (Richard Shweder, Hazel Markus, Shinobu Kitayama); indigenous (Uichol Kim, Kwang-Kuo Hwang); and multicultural (Frederick Leong, Christine C. Iijima Hall) psychology schools of thought. |
| | In social/cognitive neuropsychology, noninvasive technologies and research methods developed for studying patterns of brain activation. |
| | Positive psychology (Martin Seligman, Mihaly Csikszentmihalyi) coalesces to study human strengths, well-being, positive emotions, engagement, self-actualization, positive psychotherapies. |

**Table 7.3** Comparison of positivistic and hermeneutic approaches to psychology.

|  | Positivistic psychology | Hermeneutic psychology |
|---|---|---|
| Ideal | Natural science theory testing | Humanistic understanding |
| Goal | Explanation | Interpretation |
| Scope of study | Narrowly defined problem | Broad or as yet undefined problem |
| Approach | Reductionism | Holism, synthesis |
| Methodological concern | Causal and external validity (generalize to theory and to larger populations) | Ecological validity, or ideography (generalize to similar settings, or no generalization) |
| Methodology | Experiment, controlled variables | Naturalistic, descriptive, contextual |
| Person | Individual variation as error; aggregate data patterns | Human lives irreducibly unique, active, agentic |
| Context | Behavior abstracted from context | Behavior in cultural–historical context |
| Values | Attempt to remove values from inquiry | Values unavoidably part of own preconceptions; reflexivity required |

Psychology has arisen as a largely Western discipline, although interest in the relations of culture and psyche—and influences from other cultures—go back to its earliest origins. Recent work has shown significant differences across cultures in cognition, emotion, morality, and social psychology. Other research delineates potential universals, although this may depend on the level of abstraction and method of comparison used. How well the majority of psychological findings generated in modern Western societies generalize across cultures remains unknown, and therefore they should be applied with caution.

## 7.2.1 Psychology and Conservation

In the latter half of the twentieth century, relations between human behavior and the environment became a clear focus in psychology. Lewin (1951) argued that the environment is a major determinant of behavior, which became a theme of environmental psychology in the early 1970s. Along with a focus on the built environment, and research on resource- and energy-use behaviors, some theories broke down the person–environment dualism. One approach is to identify how facets of the person and the environment are transactionally and holistically created through events that have meanings for participants from multiple perspectives. For example, people's front porches were found to provide more values to different users than encouraging neighborly interaction (Werner et al. 2002).

The vast majority of psychological research assumes that other humans are the most significant environments that shape a person's behavior. However, in the 1990s, new strands of psychology developed relating to nature. Ecopsychology critiques the psychology of Western domination of nature (Kahn & Hasbach 2012). Conservation psychology was inspired by conservation biology and has two main aims that partially overlap the broader field of environmental psychology described above: (i) to more thoroughly assess the psychological significance of natural phenomena; and (ii) to marshal subdisciplines and tools of psychology to aid **biodiversity conservation** and other sustainability challenges (Clayton 2003; Saunders 2003; Clayton & Myers 2015).

Psychologists have used theories concerning cognition, affect, identity, and social interaction to address environmental, sustainability, natural resource, and conservation challenges. Dialog between psychology, conservation biology, and the other social sciences, however, has been modest. Further, relatively few Western psychologists have adapted and applied concepts from the field to conservation challenges outside high-income, industrialized countries where substantial literatures have been built (with a recent emphasis on climate change). As a result, there is more psychological research relevant to higher-consumption societies' distant impacts to on-the-ground **biodiversity** in places where it is most threatened, as well as more participation in conservation-supportive institutions such as zoos and related organizations. Behaviors directly impacting biodiversity, such as habitat conversion or degradation, individual or collective private land management practices, or legal or illegal species harvesting, however, may have very different drivers in rich versus poor countries or communities.

Conservation has much to do with cultural systems and institutions such as markets and governance regimes, as studied directly by other disciplines in this volume. Psychology studies individual-level capacities that enable and constrain people's embeddedness and actions in these systems. It is individuals who help give these systems both stasis and dynamism. Individual behaviors are important: consumption decisions, child-socialization practices, playing a part in legal or illegal supply chains—the list of the ways individuals can affect conservation outcomes is long. For example, Dietz and colleagues (2009) estimated that, for US households, 17 actions could reduce greenhouse gas emissions by 7.4%, which would be a "behavioral wedge" similar in magnitude to reduction "wedges" in other sectors adding up to long-term, stabilized emissions.

Where potential environmental impacts are high and changes can be made feasible for large numbers of people, individual changes are key. For example, long-term carbon reductions are possible if many people purchase energy-efficient appliances (with subsidies for low-income households). Individuals can also influence broader conservation-relevant groups and organizations through citizenship participation, or becoming an activist or professional (or an opponent of conservation!). Psychology clearly has much to contribute to conservation social science, but there are significant barriers to overcome: the narrow training of psychologists, which obscures links to sociological variables; favored topics (such as neuroscience) with rare connections to conservation; career ladders that discourage psychologists to work in field settings; funding that flows largely to traditional topics; and unarticulated cultural assumptions. Nonetheless the range of possible behavioral science tools for applied conservation interventions is enormous (Balmford et al. 2021).

## 7.3 Concepts, Approaches, and Areas of Inquiry in Psychology

Three major conceptual areas in psychology and their associated analytic frameworks (Table 7.4) provide particularly valuable insight into understanding and promoting conservation-relevant actors and actions: cognition; need, motivation, and affect; and social context.

### 7.3.1 Cognition

How do people understand and think about nature, biodiversity, the environment? How do they process information and make decisions that affect conservation? Cognition

**Table 7.4** Three key psychological concepts and associated approaches.

| Concept | Approach | Description | Key references |
|---|---|---|---|
| Cognition | Information processing | How information is taken in, organized, thought about, and remembered via general and specialized aspects of cognition | Neisser (1967) |
| | Cognitive biases | Rather than thinking logically, humans use unconscious shortcuts in thinking, except when a task requires more attention. | Kahneman (2011) |
| | Culture and context | Psychology seeks ways to compare thought across cultures, but also has uncovered how deeply culture shapes cognition. | Atran and Medin (2008) |
| Need, motivation, and affect | Drives, motivations, and needs | People have bodily and psychological drives and needs that motivate them, and enable well-being when met and frustration if not. | Deci and Ryan (1985) |
| | Values and attitudes | People also have goals and enduring values that guide choices, and do not change readily. | Stern (2000) |
| | Affect and emotion | Emotions are evaluative and guide behavior, determining likes and dislikes, including about nature. | Lazarus (1991) |
| | Coping and efficacy | There are patterns to how people experience threats, risks, helplessness and loss, and ways they can be empowered in conservation work | Gardner and Stern (2002) |
| Social context | Social cognition and influence | Our social responsiveness is built into the brain and how we think. Others affect us through our ideas about them, and modeling, reference groups, and norms. | Bandura (1986) |
| | Identity | Identity is multifaceted and articulates the person to the society and nature. Environmental identity predicts some behaviors, and is a social identity. | Clayton and Opotow (2003) |
| | Conflict and cooperation | Group values and identities, intergroup contact, and moral inclusion vs. exclusion all affect conflict and cooperation. | Haidt (2012) |

concerns both the contents and underlying processes of thought. Three groups of concepts are particularly relevant to biodiversity conservation: (i) information processing that people use to make sense of experience and to deliberately and creatively solve problems; (ii) biases that affect decisions in conditions of uncertainty and risk; and (iii) how larger contexts such as culture shape cognition.

### 7.3.1.1 Information Processing and the Environment

Modern cognitive science sprang from the convergence of the computer with the philosophy of the mind: the brain is the hardware, and the mind is the software. The mind here refers to mental "representations" of experience, such as images, beliefs, concepts, and memories. These are "encoded" in neural circuitry and activation patterns, roughly like data stored on a computer's hard drive. The brain acts on these mental contents when we think (consciously or not). There is debate on how to conceptualize the mind–brain relationship, especially in light of new brain science, but to stick with the computing metaphor for now, information is received (perceived), stored (remembered), retrieved (recalled), operated upon (thought about, acted upon), and transmitted (communicated to others).

One kind of cognitive "operation" is assessment of the truth of mental contents. If someone accepts information as true, their psychological state about that information is one of belief. If we know what someone believes about nature, conservation, and the social world, we will understand their actions better.

One approach psychologists use to study information processing is cognitive constructivism, which holds that our knowledge is neither based primarily on some sort of innate knowledge, nor entirely acquired by experience. Rather, the person actively makes sense of experience, organizing information into patterns that provide something like mental models or theories about how the world works. A mental model then helps filter new information, and novel experiences may test it. The very act of thinking about a new piece of information "elaborates" it, weaving it into existing beliefs. Over time, stable mental models develop that function in the person's environments, with some parts relatively accessible to consciousness and other aspects not. Usually new information is selected that reinforces or is "assimilated" to these models, but in some circumstances new information can destabilize and cause a mental model to change or "accommodate" it. Cycles of these adaptive mental processes characterize cognitive development in early life but can occur at any life stage.

For conservation issues (like all others), mental models can mislead. For example, perception, the first step in cognition, poses challenges. It can be hard to grasp landscape-level changes such as forest degradation from the ground or through word-of-mouth; aerial images can bring the extent of deforestation into experience more directly (but with limitations as discussed in Box 5.5, Chapter 5). Likewise, conservation can involve small, slow, or hard-to-observe entities or processes inaccessible without technology, scientific representations, or historical data (Box 7.1). Not only experience, but ideas inform mental models. In ordinary experience, just as in the history of science, beliefs are founded on concepts that often turn out to be mistaken, incomplete, or overgeneralized. For example, ecological theories of succession, which assume that ecosystems develop toward steady-state "climax" biological communities, held sway through much of the twentieth century, and still tend to characterize the public's understanding of ecosystems. In ecology, however, their applicability is now more precisely delimited (see also Box 3.3, Chapter 3 for more on equilibrium and non-equilibrium models).

Until a person's existing ideas are fully shared and tested, they are not likely to change. Extensive direct experience, and instruction, are critical to constructing complex ecological concepts. Approximating a better understanding of the world is an uneven, surprise-filled, and never fully certain process on both individual and social (scientific knowledge) levels.

---

**Box 7.1    Crossing boundaries: Environmental generational amnesia**

One troubling runaway feedback cycle between individuals' ability to notice change in their natural environments, and those same actual environmental conditions (as measured by ecological methods) has been called "environmental generational amnesia" (Kahn 2002) or "shifting baseline syndrome" (Papworth et al. 2009). The composition of ecosystems (soils, plants, animals, etc.) and their functionality (water and nutrient cycles etc.) has changed over time in response to human pressures. Ecologists now realize that conditions that their young discipline saw as "baseline" and "natural" conditions in the early decades of the twentieth century were in fact already degraded. This problem of mistaken ecological health is in a sense universal, because each individual takes for granted the physical, social, and symbolic worlds of their childhood. Each generation thus cognitively constructs a "baseline" of nature that is successively more impoverished. For example, fewer bird species and numbers per species than in the past is "normal" for later generations. This syndrome thus may pose a challenge for maintaining long-term societal conservation goals. Its causes lie in reduced opportunities for interaction with nature, and reduced orientation to seek it out, driven by environmental change and more indoor behavior patterns, across time (Soga & Gaston 2018).

Measuring environmental generational amnesia requires data for perceptions of environmental change by observers of different ages, as well as independent biological data across time (Papworth et al. 2009). If data show differences across ages in perceived ecological change, but no change in biological data (or if a different actual change occurred from what was perceived) the older observers are demonstrating *memory illusion*. If biological data show change but all ages report recalling present conditions, then *personal* amnesia is indicated. If respondents all report currently perceiving past conditions, this would be a case of *change blindness*. Specifically, then, generational amnesia is suggested if older respondents report change consistent with changes in biological data, but younger ones don't perceive or understand any change to have taken place. This can help explain emerging non-response to worsening environmental conditions, and change in what is seen as worth protecting. Each generation may be taking the natural settings they experience as the norm, and tolerating worsening ecological conditions as mild, without having a full picture of the baseline.

| **Box 7.1** | **(Continued)** |
|---|---|

There are many questions regarding generational amnesia and environmental memory. How prevalent is this syndrome? A recent study suggests something like personal amnesia about extreme temperature anomalies, where the "remarkability" (though not the negativity) of such events declined rapidly for US residents, based on Twitter use 2014–15 (Moore et al. 2019). How strongly do personal or community baselines affect engagement in conservation? What is the role of direct experience in nature in understanding and protecting biodiversity? Can media such as scientifically sound documentary or personal narrative accounts compensate for deficits in experiencing change personally? Over what time periods? What are the implications for education? (Soga & Gaston 2018)

### 7.3.1.1.1 *Cognitive Specializations: Folk Biological System*

Domain specificity theory points to dedicated brain areas specialized for certain kinds of information, including the physical behavior of objects, biological categorization, understanding others' minds, and language learning. Typically, by the time children are five or six years old they hold concepts characteristic of several such domains. These concepts tend to be robust (reliably acquired and hard to dislodge), may govern thought throughout life, and underlie rough similarities across cultures. Of interest to conservation, humans appear wired to categorize nature in broadly comparable ways, an ability we transfer to many other kinds of things. As in other organisms, this human brain adaption may have evolved to help organize and remember certain foods and other elements of nature.

Individual and cultural patterns of biological and ecological knowledge constitute the folk biological system (FBS) (Atran & Medin 2008) (Table 7.5). FBS characteristics help explain beliefs related to ecology. Essentialism, or the belief that a species has an internal unchanging core "nature," for example, is compatible with creationism and may help to explain why evolution (which emphasizes change) is counterintuitive to many. People use similarity of body form or ecological role to infer the relatedness of different organisms, explaining how dolphins are often seen as fish rather than mammals.

While FBS springs from robust developmental neurology, it is responsive to both cultural elaboration and individual experience. Cultural elaboration implies shared knowledge, for example when a group socially or economically values an animal. But some individuals accrue further experiences, making them more knowledgeable. Superseding culturally transmitted understandings, they may become recognized as experts. An extended research program between anthropologist Scott Atran and psychologist Doug Medin demonstrated how features of FBS help account for different forest-clearing practices in three different cultural groups (Ixta', Q'eqchi', and Ladino) living near each other in the Guatemalan Department of Petén (Atran and Medin, 2008). The Ixta', actually indigenous to the area, were associated with the lowest deforestation rate. They also knew the most about the local ecology, for example having identified symmetrical helping relationships between pairs of plants and animals. Interestingly, all three cultural groups believed in forest spirits, but only the Ixta' believed they had to restrain their forest practices in reciprocity. Q'eqchi' also

**Table 7.5** Characteristics of folk biological systems.

| Characteristic | Description |
|---|---|
| Hierarchy | Categories of life forms range from broad and abstract (tree, mammal) to basic, readily learned, and perceptually similar (genus or species). |
| Living systems | Living things are understood to reproduce, grow, get ill, and die. |
| Essentialism | Biological categories are determined by inner, unchanging essences. Essentialism may be transferred to socially constructed social categories too. |
| Inferences | Behavioral and ecological similarities between groups of organisms are inferred from taxonomic relatedness or appearance. |

held such beliefs, but only applied them to the highland forests from which they had migrated. Although Western scientific conceptions are not compatible with spiritual beliefs about living things, it is only fair to point out that Linnaeus's original scientific taxonomy, *Systema Naturae* (1735–68, 1st to 12th editions), demonstrates FBS properties. It too was very much a product of human perceptual, cognitive, and (specific) cultural biases, plus observation.

### 7.3.1.1.2 General Cognitive Abilities

In addition to cognitive specializations, human general cognitive abilities work across domains and allow people to consider and integrate different facets or responses to a given stimulus or situation. For example, general abilities such as learning and the ability to inhibit impulses and desires are critical cognitive foundations for culture, since culture is transmitted across generations, and requires that we meet many needs by following complicated rules and conventions. The ability to envision future scenarios and deliberately analyze, plan, strategize, and evaluate depend on the "executive" integration of many areas of the brain.

Creativity is another phenomenon dependent on integrative use of brain abilities for pattern recognition and modification, analysis, mind–body–feeling integration, and participation in a community of practice. Some cognitive theories emphasize the capacity for deliberate, reasoned decision-making. Formal Western institutions and education emphasize this form of thinking with an emphasis on logic and evidence, or humanistic modes of thought like narrative and genre analyses in literature. But every culture's ways of deliberate transmission (whether called education or not) enhances the value of general intelligences, offering foundations for conservation problem solving. Education requires high cultural investment, but may promise high returns.

Other cognitive theories note the more complex, non-linear, or socially contextualized ways that adults think. In addition, humans have capacities for noticing mental activity and for thinking about thinking, which psychologists call "metacognition." This may begin with a recognition of conflicting authoritative claims, and the dependence of any knowledge on assumptions that may be made explicit and then examined. Metacognitive capacities are expressed differently in cultural or spiritual traditions of disciplined introspection or mindfulness, which offer deeper (although never absolute) self-awareness and a degree of internal freedom in how one responds to experience. Metacognition can be used manipulatively and harmfully (for example in extremist indoctrination). The social sciences offer various ways to understand and discipline the capacity for metacognition.

Finally, people strive for a sense of meaning or coherence among their ideas and between their views of how the world is and how it should be. People's concepts of biodiversity, for example, typically combine beliefs about balance, naturalness, and wildness with strong emotional and value dimensions, which have important implications for management (Fischer & Young 2007). Broadly speaking, all the general mental capacities discussed here underlie choice as a causal force in human existence (Baumeister et al. 2011). Choice, akin to **agency**, is something constituted and constrained by culture and not experienced equally. It is also a wellspring of constant cultural change, negotiation, and resistance to oppression. But it is not immune to various biases that we will discuss next.

### 7.3.1.2  Automatic versus Deliberate Cognition

Psychologists have described two different systems of human information processing: one is fast, easy, and automatic; the other is slow, effortful, and reflective. A key difference is that the first relies on mental associations and models that are strongly learned by experience, repetition, emotional meaning, or social modeling. The second relies on the careful search for relevant information, deliberate analysis, and self-checking. The "fast" system, underlaid by years of learning and habit, is adaptive in most situations the person inhabits, and results in choices that may be based on a quick read of superficial or compelling features of a situation and feel "intuitive." How a person deals with trash—toss it or recycle/compost it—probably is guided by this kind of thinking. Crucially, if a person lacks personal experience (for example if the problem is remote or hard to observe firsthand, such as the extent of land-cover change), their intuitive responses will not be informed or accurate.

The "slow" system takes effort and is "lazy," so it is not engaged unless something interrupts the smooth functioning of the intuitive system—perhaps a novel event, or a new goal. In our example, if a new item of trash presents a recycling puzzle, or if recycling is a new goal, a person may be pushed to consider and weigh alternatives, choose and track the consequences (or revert to habit because it is too challenging). But it is also this kind of slow learning that will put the new behavior into the intuitive system, just as so many items of culture lay there from long years of figuring out the ways and values of one's social group.

The slow system is more careful and thus also more logical. Because we must expend time and mental effort to solve a novel problem, we are aware that we are doing so. Revealingly, classical economic theories viewed humans as essentially rational actors who deliberately gather evidence and assess its reliability, logically weigh alternatives, reserve emotional reaction and judgment, and choose options that optimize one's interest (see Chapter 4 for more on humans as rational actors). This list includes many traits of the second, slow, system, but it is not predominant—merely self-conscious! Most of the time we go with the fast or easy route. That rational picture of ourselves is itself an overlearned, coherent story and causal explanation, and thus we are not prone to doubt it, or ourselves (Kahneman 2011). The research touched on here, however, has contributed key insights to behavioral economics (see also Chapter 4) with its arguably more psychologically realistic model of decision-making.

Psychologists have studied many mental shortcuts, also termed heuristics. They often involve substituting a psychologically easier or more salient question for the real question at hand. For example, it is easier and simpler to manage land for self-serving ends than to determine what would benefit entire existing ecological communities and integrate that into one's plans. Since these mental shortcuts are automatic and unconscious, humans have blind spots for their own cognitive biases, despite good intentions. These are general information-processing biases, not more specifically social biases, such as implicit bias toward certain groups, which depend on general biases but their contents originate in the social environment (Table 7.6).

### 7.3.1.2.1 Cognitive Biases and Psychology of Risk

Heuristics can influence risk assessment. People tend to underestimate some risks while overestimating others. Citizens in the USA, for example, are more concerned with low-probability but high-consequence events such as oil spills, or attacks by bears or mountain lions (often issues that are framed dramatically in the media). In contrast, risk professionals stress high-probability but incremental-impact events, such as global warming, invasive species, loss of wetlands, and decreased biodiversity.

**Table 7.6** Illustrative cognitive shortcuts identified in psychology.

| Cognitive shortcut and definition | Example |
| --- | --- |
| Anchoring and focusing:<br>Overreliance on one piece of information or consideration in decision-making | People may overfocus on a habituated level of consumption as a route to satisfaction, overlooking the impact of everyday routines and relationships. |
| Availability:<br>Overemphasis on what is salient, present, recent, vivid, unusual, or was prompted | People readily think of mammals when asked about endangered species because photographs of them are more common than of other taxa. |
| Affect:<br>Overinfluence of a feeling or emotion on perception of risks and benefits, judgments, decisions | People may endorse lethal control of large predators because of the fear they evoke. Or, overestimating the benefits and understating the costs of a liked alternative, and vice versa for a disliked one. |
| Loss aversion:<br>Overfavoring of options framed or perceived as a gain rather than a loss (even though the options are identical) | People may estimate the value of a species higher if asked how much they would have to be paid to give up a species compared to how much they would be willing to pay for its preservation. |
| Self-serving bias:<br>Favoring of self, due to the need to see self and one's efforts and beliefs in a good light | People are inattentive to evidence of climate change or biodiversity loss if it conflicts with their perceived self-interest and worldview. |
| Confirmation bias:<br>Favoring and seeking information that fits existing beliefs, and discounting that which challenges them | Conservationists overlooking negative effects of a program on local people, and emphasizing benefits aligned with goals of the program |
| Framing:<br>Selective choice of dimensions, perspectives, causes, effects, costs/benefits, emotions of an issue, constraining likely conclusions or solutions | Conservation statues analogous to an "ambulance service to save endangered species" direct resources away from proactive landscape-level ecosystem management. |

The under- or overestimation of risk seems to presume a correct estimation, but risk perception is **socially constructed** (see Sections 7.3.1.3 and 7.3.3; see also Chapter 5 for more on the social dimensions of risks and hazards). The public, as opposed to experts, may frame risks in relation to broader issues: concerns such as the equity and voluntariness of how risks are imposed, the inclusivity of decision-making, trust in government authorities, and beliefs about science may incline people to take a precautionary approach that asks that those proposing an action to first bear the burden of proof that it is safe. Experts also show biases. For example, authors of refereed works on grizzly bear conservation were asked whether the Yellowstone population should be listed under the Endangered Species Act, a decision which should depend on strictly biological criteria and the best available science. However, their responses were also influenced by certain social heuristics (mental shortcuts) related to their expectations of their peers' assessments of the situation (Heeren et al. 2017).

In addition to the cognitive shortcuts identified in Table 7.6, several specific biases may affect responses to conservation risks negatively. One is the "governance trap," where the public overwhelmingly believes that powerful institutional actors are responsible for climate change responses. This may remove personal responsibility for the drastic changes in lifestyle that effective climate mitigation may require (Pidgeon 2012). Many studies have found that people see climate change risks as distant in time, and likely to hit far away. Illustrating the latter, a study of 18 countries found a "spatial optimism" bias in the majority of countries, where subjects believed that local ecological conditions were better than national or global ones. Self-protective optimism, psychological distancing, and identification with one's nation may explain this. The exceptions to spatial optimism (India, Russia, and Romania) may reflect accurate assessments of local conditions and weak national identity effects. But across all but one of the 18 countries, respondents were pessimistic about changes at all scales (local, national, global) over the 25 succeeding years (Gifford et al. 2009). Both patterns would predict low motivation to act above the local level. Examples of "anchoring," or relying on one bit of information when making a decision (see Table 7.6), may include the practice of discounting the future in economic valuation (putting less value on benefits or losses in the future), and the "status quo bias" or risk aversion that favors present social behavior patterns despite long-term benefits of conservation (Weber 2017). Under-response to risks is predictable because the negative consequences of present activities and social arrangements are remote in both time and space.

### 7.3.1.2.2 Deeper Processing and Lasting Change

Socio-ecological systems present confounding "ill-structured" problems characterized by interconnection, complexity, pervasive uncertainty, social inequities, historical contingency, institutional constraints, and controversy. Addressing such challenges, however, presents conditions that can engage our slower ways of information processing, aided by education, intercultural understanding, and collective adaptive management.

One theory of behavior change, the Elaboration Likelihood Model (ELM, Petty & Cacioppo 1981), delineates communication routes that engage both slow and fast thinking. Temporary change may be attained through the fast route using factors like authority, emotion, social norms, rewards or incentives, and consistency (see discussion in Section 7.3.3.1 and Box 7.3). Longer and more generalizable changes can come from sparking mental "elaboration" or deliberation in the slower system, which can re-evaluate older mental models.

An example of this sort of "elaboration" can be seen in the deliberative democratic management of Marine Conservation Zones in the Sussex coast of England. Deep participation by those with direct experience was achieved by the creation and sharing of an interpretive film featuring divergent stakeholders talking about their relationships to the area, including discussing its transcendental values. This helped to identify common ground in workshops. The film method also allowed facilitators to avoid the voices of powerful community members from dominating. The workshops provided plenty of time and accessible and accurate information, management options for specific sites and how they are used, and decision criteria addressing probable impacts on community stakeholder interests (Ranger et al. 2016). Elaborated slow processing is also enhanced by raising taken-for-granted beliefs for examination, presenting multiple scientifically based longer-term scenarios, and addressing institutional supports (Petty et al. 1995). In short, education and thorough deliberative social processes offer ways to engage slow thought and forge more adequate intuitions.

### 7.3.1.3 Context and Culture

Social context, language, and culture always play some role in cognition. Debates over just how much, and how to study it, have a long history in psychology. Many psychologists have sought to discover consistent or universal concepts that apply to people everywhere. But since a large portion of psychological study samples are drawn from Western, educated, industrialized, rich, and democratic (WEIRD) societies, they may describe cultural norms rather than universals (Henrich et al. 2010). Even in cross-national studies, the subjects are usually from similar demographic groups (and often college students).

Cross-cultural psychology (conducted with either individual- or national-level data) has sought broad dimensions in which cultures can be compared. Comparative studies have established fairly well that environmental engagement is correlated with being female and having postmaterialist values, higher income and education levels, environmental knowledge, and concern about environmental impacts. But, as Milfont (2012) notes in reviewing this body of research, there are multiple possible problems with such comparative work. Some have to do with whether questions asked of respondents are adequately translated and contextually equivalent. For example, will a question about ecological concepts be valid among less Western-educated populations? A more basic problem is whether the concepts being explored are even relevant in a culture's own terms. To determine if a foreign-generated concept is valid, very different culturally specific studies of local knowledge and interpretations must be done. For example, Zinn and Shen (2007) asked Chinese participants many open-ended questions about emotions evoked by wildlife to test whether the wildlife values orientation concept (see Section 7.3.3.2) is valid there, concluding they could not rule out the existence of other possible dimensions, such as ones based in culturally specific symbolic meanings of species.

When such studies are done across a number of cultures, a new, more general scheme adequate to that specific set of cultures may be developed. Following a thorough development cycle that has been repeated extensively, very broad tendencies may be identified, for example between hierarchical vs. egalitarian cultures, or collectivist vs. individualist. This is the case for self-determination theory (Section 7.3.3.1), or Schwartz's values typology (Section 7.3.3.2) which are discussed later in the chapter. In these cases, stronger claims of broad applicability can be made.

At this time, the main examples of cross-cultural work in psychology are those producing broad cross-national comparisons of environmental attitudes, which may be of limited usefulness in more local conservation contexts where more specific customs and relationships may predominate. At such local intervention levels, the more important work may come from **qualitative** research with a robust base in local culture, language, and relationships, perhaps exploring how individual differences in values, sense of place, sense of competence, or other relevant ideas play into actors' choices.

Because individuals' mental models are constructed through experience, and because experience unfolds through particularities of time, place, social influences, and specialized activity, cognition (and all other parts of psychology) is context bound. What you believe, and thus think about, influences how you think. The close relationship between a culture's worldview and its most valued **epistemologies** reflects this at a broad level. Religious worldviews in particular may play a role in motivating conservation (Figure 7.1).

Several schools of psychological thought have developed that take culture seriously and study it up close, while also recognizing its dynamic, nuanced, changing, unique, and contested sides (Table 7.7). These approaches thus are not only less likely to reify culture or another culture's concepts, they may also be more likely to reveal psychology's unquestioned cultural foundations.

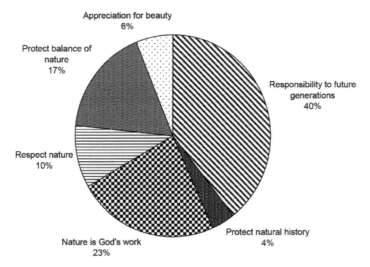

**Figure 7.1** American adults' responses to "Which is the most important reason for you personally to care about protecting the environment?" (Data from Biodiversity Project 2002, p.17).

**Table 7.7** Schools of thought in culture and psychology.

| School of thought | Core insight(s) | Key citation |
|---|---|---|
| Narrative psychology | "Story" holistically imbues human activity with meaning. | Bruner (1986) |
| Cultural psychology | Culture and language affect concepts and thought, and more deeply, viscerally felt or subtle emotions, health, and communication. | Shweder (1991) |
| Situated cognition | Knowledge is not transmitted so much as co-constructed in "communities of practice," and embodied in symbols, tools, and social relations. | Lave and Wenger (1991) |
| Multicultural psychology | Gives voice to lived realities and builds theory to authentically empower. Roots in ethnic psychologies. | David et al. (2014) |

As psychology moves to embrace multicultural psychology's insights about empowerment, previously unquestioned aspects of the relationship between researchers and those they study (including in conservation and other settings) are raised. A good example of how psychology's modes of inquiry can shift to include considerations of **positionality** and power while addressing culture and cognition holistically is the collaboration of Douglas Medin, Megan Bang, Menominee on-reservation tribe members, and users of the American Indian Center in Chicago (Medin & Bang 2014). Building on long-standing relationships, they broke down the separations between researchers and the community being researched, becoming deeply involved and reflecting overtly on their limitations and insights in their writing. Focusing on the underrepresentation in, and alienation around, Indigenous communities and Western science education they listened at many community meetings, learning to hear how, for example, forestry managers excluded some members from understanding by using technical terms and not fully explaining forest cutting decisions.

Tribal members with differing involvement in forestry had different perspectives, for example on watershed or timber harvest plans, but shared a common framing of tribal sovereignty and reinforced a core relational worldview and epistemology. Bang and Medin related this to the literature on the development of folk biological knowledge (including the work of Atran and Medin on FBS in Guatemala discussed in Section 7.3.1.1.1) and extended it through studies of ecological thinking among both on-reservation and urban American Indian participants and rural Euro-Americans. They found the Euro-Americans showed more "psychological distance" from nature in their outdoor activities (with nature often seen as a backdrop for sports), the way they talked about nature, and in illustrations in children's books (with animals and plants often depicted in separation from each other without visible connections). Direct evidence for cultural differences in ecological reasoning came from a developmental study with five- to seven-year-olds, where the American Indian children showed more sensitivity to ecological relations. They also showed a greater tendency to take multiple perspectives of animals and plants via spontaneous gesture and mimicry. Menominee adults (compared to their Euro-American counterparts) also demonstrated greater psychological closeness to other biological kinds (e.g. viewing other species as kin), and more relational or systems reasoning about the ecological world.

The researchers did not let these important findings stand only as abstract contributions to academic literature. As is clear throughout, the research was strongly motivated by an attempt to re-cast the culture of science in terms friendly to, or stemming from, Indigenous epistemology. They discovered that teachers at the American Indian Center were already embedding relationality in learning when designing curricula for urban American Indian students. For example, they used a strong focus on sensory processes and science inquiry (sensory observation, perspective taking), rather than a more alienating treatment of science as primarily an already produced body of knowledge. Finally, in recounting their research program, the authors use a less formal and more transparent and self-reflexive voice than one might encounter in a typical academic psychology monograph, again bridging different cultural epistemologies, and changing the science of culture (in this case psychology) to be more interdisciplinary and responsive to redress historical social inequities.

## 7.3.2 Motivation and Affect

Cognition, culture, and language interact deeply with motivation and affect. Motivation refers to the underlying "why" of a person's behavior: what led to its initiation; what determines its goal or direction, and its intensity; and what accounts for its persistence, or ceasing. Affect includes emotion and other feeling states that contain an evaluative (i.e. "I like it"/"I don't like it") dimension. These and related concepts like drives, goals, needs, values, and responses to stress help explain behavior, reactions to environments, and the people-to-people dynamics in conservation work.

### 7.3.2.1 Drives, Needs, and Motivation

What makes people start and continue, or cease, an activity is a fundamental question addressed in psychology by the concept of motivation. Drive theory proposes that some things reinforce behaviors without being learned. Thirst, hunger, excretion, and sleep are all examples of regulatory drives that affect behavior and aren't learned. Other drives are non-regulatory, not related to physiological survival, and often learned, such as social approval. Once a need is met, it no longer serves as a current motivation.

Un-met drives are felt strongly. In a conservation example, 69% of bushmeat hunters near Haut-Niger National Park in Guinea said they would stop hunting if they had enough money and food (Pailler et al. 2009). Drives also may be mentally associated with other stimuli that then become learned motivators, as elaborated in the classical and "operant" learning theories of Ivan Pavlov, John B. Watson, and B. F. Skinner. In operant conditioning, satisfaction of a drive "reinforces" the behavior that led to it, and it is "learned." For example, some drivers of hybrid cars alter their driving to keep the miles per gallon feedback readout high, which translates to saved money and carbon, all of which reinforce the driving behavior. In theory, cost-benefit approaches to conservation (such as payment for **ecosystem services** [PES] systems, which monetarily compensate communities or landowners who practice conservation) can function as reinforcement. These might work if conservation motives become learned as the reason for the behavior, but might not if they don't. It is always important to consider the rewarding or punishing contingencies of behavior in understanding their causes.

#### 7.3.2.1.1 *Maslow's Hierarchy of Needs*
Maslow (1943) proposed a hierarchy of needs in which satisfaction of a lower level is prerequisite for pursuit of the next level (Figure 7.2). Maslow's hierarchy of needs helped lead psychology toward an expanded concept of human well-being. The lower levels of human need (e.g. food, water, clothing, and shelter) are considered basic to personal and societal welfare and essential. The upper triangle includes various ways the human need for "self-actualization" (i.e. achieving one's full potential) may be met.

Where does biodiversity conservation fit in this hierarchy? In contrast to development theories (see Section 5.3.1.2, Chapter 5) that assume people will not be motivated to protect the environment until a certain level of affluence is met, Dunlap and Mertig (1995) found environmental concerns across 24 countries with widely varying gross domestic products. Thus, environmental welfare may be a basic need. Indeed, other theories disagree that "higher" needs (such as problem solving, creativity, and ethics) come later. It is not hard to find examples of people who sacrifice "lower" needs like security for the sake of doing what is right or meaningful. Thus, it makes sense to think of all Maslow's needs overlapping each other.

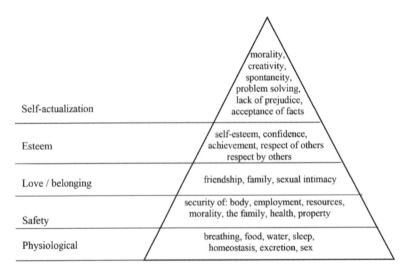

**Figure 7.2** Maslow's hierarchy of needs. Adapted from Psychology Wiki 2021, Psychology Wiki 2021 / Fandom, Inc / CC BY SA 3.0

#### *7.3.2.1.2 Self-determination Theory and Intrinsic vs. Extrinsic Motivations*

Self-determination theory (SDT; Deci & Ryan 1985; Ryan & Deci 2017) research across cultures has found that three universal, basic psychological needs must be met for personal adjustment, integrity, and growth. They are: (i) autonomy, or the experience of volition and willingness, which allows a sense of authenticity and integrity (or if frustrated, a sense of pressure or conflict); (ii) competence, or experiencing effectiveness and mastery from using skills; and (iii) relatedness, or a feeling of belonging, care, and warmth from connecting to significant others. It should be noted that autonomy in the sense used here is not the same as Western notions of individualism, and has been shown to hold in more collectivist-minded and less egalitarian-minded cultures. Further, universalism is not the same as uniformity, and individual differences are always found.

SDT distinguishes between different types of motivation based on whether an activity meets these needs. An activity that is autonomously chosen and increases competence is *intrinsically* motivated, meaning that the person consciously endorses it and wants to do it, often for its own sake. An intermediate form is *internalized* motivation, where the person assimilates a value expressed by those around them and comes to take ownership of it or identify with it. The opposite of intrinsic and internalized motivation is *controlled* motivation, where behavior is externally monitored and rewarded, threatened, or punished. In an intermediate from, an activity might be regulated by approval, avoidance of shame, or the need for self-esteem.

The basic intrinsic vs. extrinsic motivation contrast makes a huge difference. Compared to external motivation, autonomous motivation is associated with greater persistence, higher performance, conceptual understanding, positive affect, healthier lifestyles, and psychological well-being. The differences are more pronounced when competence and relatedness needs are simultaneously also fulfilled or frustrated in the activity. So, for example, we might expect that conservation practices that derive from and are controlled by a person's and their community's own values and deliberations, are felt to belong to them, and facilitate competences will be embraced persistently, enjoyed, and contribute to the generation of individual life aspirations. On the other hand, schemes that are imposed and policed, undercut by large external rewards (e.g. financial incentives), or require action that might threaten group bonds will not take hold or persist, and may be harmful to people's overall well-being.

Stemming from SDT is the notion of autonomy-supportive environments (ASEs) that support basic needs. An ASE avoids controlling interactions and external motivators, instead creating interpersonal or institutional conditions conducive to autonomy, competence, and belonging. ASEs for conservation develop psychological empowerment by fostering intrinsic and internalized motivations aligned with a person's core interests and identity (Table 7.8). These forms of motivation are especially important when regulatory or economic incentive approaches fail to promote conservation. ASE's provide individual and group self-determination but are not necessarily at odds with conservation organizations or agencies having a central role. Rather, they provide foundations for collaboration and for the integration of different groups' agendas. For example, protected area conservation projects that furthered at least one intrinsic motivation (often community-oriented) were found to be three times more likely to meet social and ecological goals (Cetas & Yasué, 2017).

**Table 7.8** Autonomy-supportive environments (ASEs).

| Characteristic of an ASE | Effects |
| --- | --- |
| Empathic understanding | Conveys respect; ensures all parties' interests are represented |
| Provision of choice | Individuals feel free to solve problems for themselves. |
| Transparent administration and access to decision makers | Clear rationales, necessity, and value of rules |
| Non-controlling communication, voluntary participation | Conveys approachability; provides ideals that are worth buying into |

*Source:* Adapted from DeCaro and Stokes (2008).

The field of positive psychology, initially a corrective response to a bent in psychology to focus on pathology and human weakness and now offering fuller understanding of many areas of psychology, has investigated many intrinsic sources of motivation. Some intrinsic motivations, as well as enabling well-being and not depending on material consumption, also happen to be relevant to conservation (Table 7.9). Interestingly, after "immersing" themselves in images of nature, research subjects placed more value on intrinsic life aspirations and less on extrinsic ones such as money, image, and fame (Weinstein et al. 2009). Motives need not be altruistic to intrinsically motivate conservation: thoughtful frugality, community participation, and competence are examples of "selfish" intrinsic motivations (De Young 2000).

**Table 7.9** Intrinsic motivations and their relevance to biodiversity conservation.

| | Conservation relevance | | | |
| --- | --- | --- | --- | --- |
| Intrinsic motivation | Lowers material consumption | Promotes the common good | Endures in the face of difficulty | Allows identification with nature |
| Friendships, intimacy | X | | X | |
| Leisure pursuits | X | | | |
| Affiliation with nature | X | | | X |
| Feeling needed | | X | X | |
| Higher social goals | | X | X | |
| Self-acceptance | X | | X | |
| Mindfulness | X | | X | X |
| Sense of meaning | X | X | X | X |

### 7.3.2.2 Values and Attitudes

People hold values and goals for how they want to live. Some goals are deep, abiding, and general, and others more superficial. We learn (and sometimes examine) these values and goals and express them in actions, choices, judgments, or statements. Psychologically speaking, goals that include values, attitudes, and preferences are *evaluative beliefs*: networks of ideas and feelings that guide our judgments about what is good or bad, admirable or not, and may unite us in social groups. Values are enduring and apply to behavior across many areas, whereas attitudes are more changeable and more specific to a given target. Preferences are even more specific and can be satisfied in various ways.

Values are enduring and are transmitted across generations during socialization through participation in culture. Youths in a culture that values social hierarchy, for example, are not only instructed in this value but their lives are immersed in it. They may see others express appropriate deference to certain individuals and groups, notice prominent images of authority figures at school, inhabit spaces arranged to emphasize social rank, follow certain dining or marriage customs, and in a myriad of daily experiences have this value reinforced.

#### 7.3.2.2.1  *Wildlife Orientations and Economic Modernization Theory*

Manfredo, Bruckotter, and colleagues (2017) suggest that value systems change slowly only as new factors make other existing or modified values more adaptive in society's socio-ecological system (SES). Most notably, Inglehart (1995) argues that as industrialization and economic modernization (increased education, wealth, and urbanization) have spread to emerging economies around the globe since World War II, people's needs have changed from basic security and subsistence to more self-expressive ones. These SES changes coincided with a shift from materialist values to postmaterialist values and from "domination" values of wildlife to "mutualism" values that see wildlife as deserving of human care and being capable of more complex relations with humans (Manfredo 2008; Manfredo, Bruckotter et al. 2017). Value systems can resist change, revert after a stress, or may be captured in broader traditionalist populist cultural backlash (Manfredo, Teel et al. 2017). Thus, the task for conservation is not so much to *change* values as to understand and work with or around them (Manfredo, Bruckotter et al. 2017). Steg and colleagues (2014) recommend reducing the conflict between environmental values and personal-gain values by either making environmentally beneficial activities pleasurable and money-saving (e.g. providing convenient, clean, low-cost public transport) or making environmentally damaging activities less attractive (e.g. speed bumps to slow down drivers and road tolls).

#### 7.3.2.2.2  *Norm Activation and Value-belief-norm Theory*

One influential model of values underlying Manfredo and colleagues' work is Schwartz's typology of nine major values, which has been refined through extensive cross-cultural studies (Schwartz et al. 2012) (Figure 7.3). The values are displayed in a circle because it has been found that values placed opposite each other tend to be mutually exclusive psychologically. For example, it is unusual to express high levels of both benevolence values (such as altruism), and hedonistic values (such as self-aggrandizement). Adjacent values on the circle are also related and grouped into two pairs of generally opposite values: openness to change vs. conservation (referring to stable social order); and self-enhancement vs. self-transcendence.

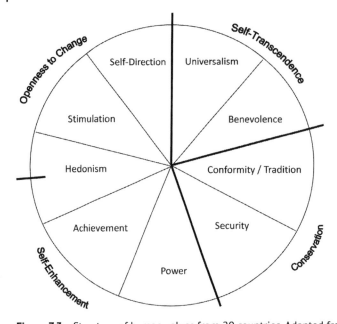

**Figure 7.3** Structure of human values from 20 countries. Adapted from Schwartz 1992.

Stern and colleagues adapted Schwartz's work in the value-belief-norm (VBN) model of environmental values and action (Stern 2000). The model holds that whether to carry out a particular conservation behavior is rooted in internalized values that line up with the self-transcendent and/or self-enhancing quadrants of Schwartz's typology. If people's socially or ecologically altruistic (self-transcendence) values are activated by a particular situation, and if they believe that because of ecological interconnectedness their choices will have consequences for the values concerned (and that they can do something effective), then they may feel a personal sense of obligation to act and may then do so (Figure 7.4). For example, the amount of participation in restoration actions among landowners correlated with the degree of awareness of consequences and sense of personal responsibility. Notably, participants did not differ in underlying values structure, but rather activation of those values depended also on awareness of the need for restoration, a sense of ability to do the work, and a felt personal responsibility to put their values into action (Johansson et al. 2013).

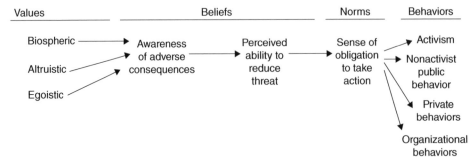

**Figure 7.4** Value-belief-norm model (Stern 2000, reprinted with permission of John Wiley & Sons).

### 7.3.2.2.3  *Theory of Planned Behavior*

The theory of planned behavior holds that target-specific beliefs, attitudes, and intentions predict behavior best (Figure 7.5). Ajzen (1991) considers three types of underlying beliefs: (i) those about a specific behavior's consequences; (ii) those about other people's norms regarding the behavior; and (iii) those about whether one can actually carry out the behavior. Attitudes derived from these beliefs should predict the person's intention to carry out (or not carry out) the behavior, which is the proximate step before actual behavior.

For example, in deciding how to manage a forest plot, beliefs about options would be elaborated in terms of effects on values—income, wildlife, aesthetics, or other specifics—leading to attitudes favoring or disfavoring various options. One's general beliefs about what one "should" do (**normative** beliefs) might become more specific in terms of what other forest owners are doing. One's general sense of control or efficacy would help shape one's attitude of capability for each option. The net effect would be an intention to carry out the action that best lines up with these specific attitudes.

Strong attitudes are predictive, but there are many ways they may not be acted upon, including external factors. Attitudes may be more changeable than values, but Heberlein (2012) warns that people may resist efforts to change them. As with values, rather than change attitudes, sometimes advocates should accept them and find ways around them. For example, they should first determine a specific audience's values attitudes, and then frame a message congruently. In the USA, messages built around the public's strong self-enhancement and conservatism values (such as success, material wealth, accomplishment, family, and security) might frame land conservation around quality of life, benefiting one's children, health, property values, and opportunity (Schultz & Zelezny 2003).

### 7.3.2.2.4  *Cognitive Dissonance Theory*

Value and attitude theories assume that beliefs and values normally precede behavior—individuals have certain beliefs and values and then act according to them. In contrast, cognitive dissonance theory explains how behavior may sometimes actually precede attitude and belief and that, counterintuitively, if people change behavior they will change what they think and feel too. The theory assumes people are motivated to have consistency across their psyche. People's beliefs, attitudes, and behaviors are often inconsistent. When attention is brought to this inconsistency, an unpleasant feeling of dissonance may be experienced (Festinger 1957). For example, two beliefs can conflict: riparian set-asides are good ecologically but they reduce crop or livestock income. Behavior can be at odds with an

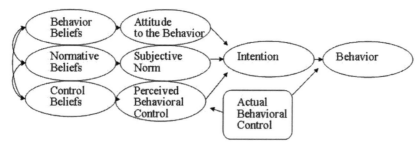

**Figure 7.5**  Theory of planned behavior. Adapted from Ajzen 2006.

attitude: a farmer may value water quality in a stream but may not set aside riparian areas from their fields. Mental consonance can be re-achieved in several ways: (i) by avoiding discrepant information and bolstering belief or attitude with new information to make the inconsistency less serious; (ii) by changing the attitude to fit the existing behavior; or (iii) changing the behavior to fit the belief or attitude. The greater the dissonance, the harder one needs to work to reduce it.

In understanding cognitive dissonance, an additional factor is to whom or what one attributes such a change in one's behavior. If one sees it coming from one's own changed thoughts, feelings, or reasons, with little external influence, the change is more likely to last. This is especially true if people tell others about their choices in some kind of public commitment (perhaps they display a decal or sign, or say something to their neighbors). Conversely, behaviors may disappear if strong external incentives were used to kick start it. For example, if villagers attribute their willingness to not cut nearby forests to the money offered for the health clinic and staff provided in a PES scheme, the curtailment of cutting is likely to cease when the compensation ceases. The villagers may have their own intrinsic motivations to preserve the forest, but in this case those risk being "crowded out" by strong extrinsic motives from the exchange.

Rode and colleagues (2015) provide a review of empirical evidence of motivational "crowding out" vs. "crowding in," key concepts that have gained currency in other social science disciplines. An important footnote on cognitive dissonance theory, however, is that people vary in their tolerance of inner inconsistency. In less individualistic cultures that tend to place more emphasis on harmonious relationships and collective well-being, inner consistency may not matter as much (Kim 2002). This variation underscores the importance of culture in attempts to change conservation behaviors.

### 7.3.2.3 Affect and Emotion

Psychologists use the term *affect* to mean the realm of feelings including emotions and moods. Emotions involve specific, comparatively brief feelings (designated by names such as sadness, happiness, etc.) plus some pattern of physiological arousal (adrenaline release, heart rate changes, sweating) and a tendency to express the emotion and thus communicating it (possibly through words, specific facial expressions, or a conventional action). Theoretical traditions in psychology looking at emotion are several and complex, but each tells part of the story. Although previously regarded as irrational and distracting, emotions are now generally seen as functioning to serve a person's core needs and relationships, and having adaptive evolutionary roots related to sociality and well-being. Emotion-related information is processed faster by the brain and may dominate a person's response compared to other information. If the person is highly aroused physiologically, this primes mid-brain centers to encode the information strongly and long term so that later similar events quickly trigger responses.

Emotion depends on two stages of information processing (cognition), which are often unconscious but may be re-examined. The first stage quickly determines if the event is good or bad for one's main goals or relationships. The second step involves deciding what response to make, including action, expression, and communication (Lazarus 1991).

Different emotions shape behavior in different ways. Negatively toned emotions (e.g. anger and fear) involve narrowed thought patterns and action tendencies, such as fleeing a danger or confronting it. Positive emotions occur in non-threatening contexts and work very differently, adding to adaptation by broadening perception and building resources such as physical wellness and high-quality relationships with others (Frederickson 2005).

Several emotions have been considered basic and universal: happiness, sadness, anger, fear, disgust, and surprise (Ekman & Friesen 1969). But cultures vary in how they are experienced, labeled, displayed, shared—and by whom in what contexts (Kitayama & Markus 1994). There are many emotions, and some are part of some cultures' repertoires but not others. In addition, there are individual variations in emotion and affect, which also include moods (longer lasting), sentiments (dispositions to feel certain ways, like shyness), and affective disorders.

Cultural worldviews shape emotion because they determine what is real, of highest value, and of importance to the self. Emotions are a predominant way that we interact with the world (social and natural) because they link our goals and motivations to what happens around us.

Some of the ways affect and emotion intersect with conservation are in motivating conservation, learned benefits of natural settings, and emotional connection to natural places. We will look briefly at theories explaining benefits from natural settings and at emotion as an element in people's attraction to or dislike of different species and thus their support for or opposition to conservation.

### 7.3.2.3.1 Emotions Motivating Conservation

Emotional connections to natural places may motivate conservation. Extensive early experience of the outdoors (often facilitated by a mentor with knowledge and practical know-how) has been identified upon retrospection by adult conservation educators and activists as a "significant life event" leading to the careers of conservationists and environmental educators (Chawla 1998). Survey data also suggest early experience relates to environmental concern across the general US public (Wells & Lekies 2006). Positive experiences, and anger or sadness about the loss of a natural place, play important roles in protective behavior. In an example of how negative emotions can constrain human responses, Markowitz and colleagues (2013) found that the capacity for compassion of people responding to hypothetical environmental scenarios drops as the number of needy individuals increases, the victims become less easy to identify with, or as the percentage of people that can be helped decreases.

### 7.3.2.3.2 Health Benefits and Connection to Nature

There is a burgeoning literature on psychological and physical health benefits of nature. Emotional connections may be learned by associating various benefits with the spaces where they occur. Benefits of green spaces include improved physical health, more social contact, and better psychological health (e.g. higher subjective well-being and increased task performance). However, the benefits of natural areas are not equitably distributed (Shanahan et al. 2015), and accessibility is affected by real and perceived safety. Research on this topic has also been geographically biased toward high-latitude and wealthier societies.

*Place* is a multifaceted concept, but always laden with meaning and emotion (see Section 5.1.1.3, Chapter 5 for more on the concept of place). Intersecting with social constructions of place, individual place attachment may be based on emotional memories, feeling related to a place, and concrete use of resources. Psychology might explore pathologies related to place: alienation, placelessness, and loss of significant places. Dislocation due to development, conservation, climate change, or other causes has social-psychological consequences such as loss of social connections and confidence, vulnerability, a sense of injustice, victimization, and stigma.

### 7.3.2.3.3 Theories Explaining Psychological Benefits of Nature

Two broad theories have been suggested for people's attraction to and benefits from natural settings. Kaplan and Kaplan (1989) suggested people are attracted to and benefit from settings that have *coherence* (i.e. are understandable and not confusing or too simple) and *complexity* (i.e. which invite exploration, are accessible, and contain possible resources). A dense jungle or a salt flat might represent extremes on these variables that would make them less preferable to most people.

Attention restoration theory (Kaplan 1995) further proposes a cognitive explanation for the psychological benefits of nature: our capacity for effortful directed attention (as opposed to "involuntary" attention) is limited and readily used up by everyday demands. But it can be restored by things that capture and hold our attention involuntarily. The complexity and coherence of natural environments are restorative if they capture our attention. Thus, an environment is more likely to be restorative if it feels psychologically "away" from routine, stressful settings, is fascinating, seems to extend in time and space, and is compatible with the person's current goals.

An evolutionary perspective, however, points to an alternate explanation for people's attraction to certain places. In research studies, landscapes that people prefer tend to have certain features, such as water, views, shelter, and tree cover. Prospect and refuge theory suggests that attraction to these features was adaptive to our ancestors, and as a result, still retain stress-reducing effects for people (Heerwagen & Orians 1993).

### 7.3.2.3.4 Emotions and Responses to Animals

Emotion may dominate people's responses to animals, spanning the range from highly averse to extremely positive, with some species much more emotionally salient than others (Castillo-Huitrón et al. 2020). Negative reactions tend to occur when animals conflict with humans for a resource or when they threaten human welfare (Opotow 1994). Greater understanding or liking of a species only partially mitigates these effects. Positive reactions to animals tend to correlate with perceived physical or mental similarity to humans. Species with "cute" (often juvenile) facial features elicit caring from humans. Is this "love?" Probably not, if love includes intimacy and commitment. Feelings of love, as well as a sense of connection and amusement are, however, strongly correlated with self-reported motivation to preserve a species, but people are "highly selective" about the targets of these emotions (Figure 7.6). On the other hand, wonder and respect are "equal opportunity" emotions—people are less selective about what can evoke them. But despite that, these feelings are also related to a desire to conserve the target species. Despite inspiring words that "we will save only what we love and love only what we understand," these findings suggest love is a high-investment emotion. The step from understanding to love is also not automatic (Box 7.2).

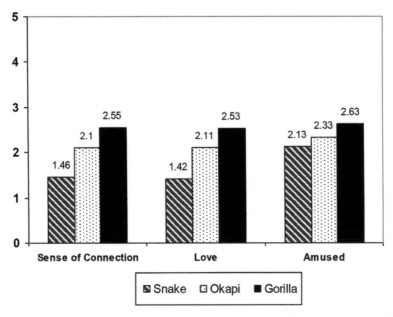

**Figure 7.6** Significant differences separated self-rated level of love and sense of connection across three animal species (Reproduced with permission from Myers et al., 2004 / John Wiley & Sons).

---

**Box 7.2 Debates: Evolutionary psychology and biophilia**

The biophilia hypothesis, developed by sociobiologist E. O. Wilson (1984), proposes that evolutionary pressures led to biophilia—genetically based predispositions to take a strong interest in and affiliate with nature (including plants, animals, and landscape features). This hypothesis illustrates evolutionary psychology: the theory that psychological traits and brain specializations are responses to adaptive challenges that human ancestors faced. The concept of biophilia remains slippery. So far, no strong and specific biophilic tendencies or "learning rules" that Wilson proposed have been demonstrated conclusively. Human attraction to nature is supported by landscape feature preferences and cross-species comparison but human groups' occupation of virtually every biome suggests landscape preferences were too weak to constrain cultural adaptation. By contrast, biophobia—such as quickly learned and hard to unlearn aversions to spiders and snakes—lends itself to, and is supported by, rigorous behavioral research (Ulrich 1993).

Kahn (1999) argued that biophilia (and its opposite, biophobia) are due to neither culture and socialization nor native disposition (i.e. "nature") alone. Rather, they need to be examined through a developmental lens that reveals how both influences are integrated over an individual's life span and through an individuals' active building of understanding from experience. Kahn (1999) showed that some people come to integrate biophilic and biophobic feelings and thoughts in a more mature understanding. Such an integration might be based on fundamental similarities of humans to other organisms, and acknowledge that people are equal participants in ecological wholes rather than standing outside them.

(Continued)

---

**Box 7.2   (Continued)**

   More fundamental criticisms have been levelled at evolutionary psychology. A common assumption in adaptationist explanations of human behavior is that the conditions among hunter-gatherers recorded **ethnographically** by Western anthropologists in the present day resemble the Pleistocene environments humans evolved in. But they may not. Archaeological reconstructions of the evolutionary social environment provide only speculative explanations. Further, psychology does not have only evolutionary theory to draw on to generate its hypotheses. It has drawn on insights just as diverse as the problems it tackles. The first task of psychology is a descriptive one—to describe the structure and function of the mind—and it is not finished. Logically speaking, questions of the origin of the mind are secondary to this task (Hardcastle 2002). Nonetheless, an evolutionary angle is often contemplated in contemporary psychology.

---

### 7.3.2.4   Coping and Efficacy

Motivation and emotion always interact with cognition in leading to behavior. This broad psychological principle is central to people's responses to threats and risks. A threat to well-being, oneself, or one's core goals or values evokes feelings and motivations like fear and escape or uncertainty and desire to remove the threat. But one's response to a threat is complicated by how one cognitively assesses the situation and one's abilities to cope, or deal with the threat. Enhancing people's ability for practical, effective coping responses may be desirable, but when threats are seen as too great, emotionally intelligent work may be a prerequisite to work though fear, self-protective coping, pessimism, disempowerment, or grief.

#### 7.3.2.4.1   *Protection–motivation Theory*

People may perceive some conservation efforts as threats. A study of villagers living around Tarangire National Park in Tanzania found that people living closer to the park perceived greater threats from conservation (e.g. that policies would limit their land use, or that the park might expand) and wildlife (e.g. that wildlife would destroy crops or livestock) (Baird et al. 2009). But across villages, responses varied, with some villages expanding agriculture to reduce the perceived threat of park expansion, and others feeling there was nothing they could do to prevent it.

   Such varied reactions may be elucidated by protection–motivation theory (PMT; Gardner & Stern 2002). PMT applies emotion theory (discussed earlier) to situations of stress or threat. When something one values is threatened, the emotion system first reacts by quickly assessing the degree and likelihood of the threat, probably resulting in fear or anxiety. This is followed by an assessment of coping ability: Is there an effective action to take? Can I take it? What are the resources and barriers? Depending on these appraisals, a person will respond with a coping strategy. This will be either a problem-focused or emotion-focused response (usually chosen very quickly, but a slower deliberate response is possible). Emotion-focused strategies mainly just help the person feel "okay" and include avoidance or denial of the threat, wishful thinking, religious faith, or fatalism. These strategies allay anxiety but don't deal with the threat. Problem-focused coping strategies include analysis and efforts (feasible or not) to stop or mitigate the harmful situation. For example, Tarangire area villagers who pre-emptively cultivated more land to prevent park expansion followed such a strategy.

An important aspect of PMT is that it predicts people will underreact if they think the risk is low, or if the perceived risk is high but they lack resources to respond practically. Thus, underreaction might result from either under- or overestimation of the risk. PMT explains the consistent failure of campaigns that use vivid images or other "scare tactics" to convey the severity of a threat. Evoking too much fear will produce emotion-focused coping, unless the message is perceived as legitimate, strong arguments substantiate the threat, and concrete, effective, doable actions are given (Hoog et al. 2005). But fear may be effective at motivating future prevention after an event has happened, and when collective responses are highly visible. If efforts to tie biodiversity loss to threats to personal well-being are not convincing, it is unlikely that threat appraisal will even be engaged.

### 7.3.2.4.2 *Explanatory-style Theory of Optimism and Pessimism*

"Learned helplessness" develops when efforts to mitigate a threat are repeatedly thwarted, or the threat is too large. Motivation might be present, but belief that one's practical abilities are low preclude acting on it. In one prominent theory of optimism, Abramson and colleagues (1978), suggested that persistence in the face of adversity is determined by the "explanatory style" of how one explains a threat or bad event to oneself. Two basic explanatory styles—optimistic or pessimistic—are distinguished by whether they see the causes of events as: (i) internal, essential, or inherent about something in the situation versus brought on by external factors; (ii) stable and unchanging versus unstable and changeable; or (iii) global (or pertaining across a variety of situations) versus specific to the specific instance. When confronted with a negative experience, an optimistic explanation would focus on external factors that are perceived as specific and changeable ("I was misinformed"). In comparison, a pessimistic explanation would emphasize factors that are internal, stable, and global ("human nature is stupid and selfish"). Interestingly, the opposite configuration occurs when explaining a positive outcome (Gillham et al. 2001).

As an illustration of how the two styles might explain negative versus positive events differently, consider the following scenario and two events, and the different explanatory-style interpretations shown in Table 7.10. You are a wildlife officer working to prevent poaching in a biodiverse area. You are familiar with a range of conditions influencing whether poachers get away with it or not, but it is always uncertain exactly what the causes are, so what you focus on is partly subjective. Negative event: you find evidence some poaching has occurred. Positive event: you apprehend some poachers.

Obviously, all these causes may be important variables in the real world. Regardless of which is true, explanatory-style theory predicts that people may consistently focus on different types of cause, leading to a greater or lesser sense of efficacy about their ability to deal with threats or achieve goals. While the example is about an agent with a key role, the theory applies well to assessments of members of the broader public about socio-ecological threats and opportunities.

**Table 7.10** Optimistic and pessimistic explanatory style interpretations of good and bad events by law officer.

| Explanatory style | Event | Possible causes officer thinks explain the event | | |
| --- | --- | --- | --- | --- |
| | | Internal or External | Stable or Unstable | Global or Specific |
| Optimistic style | Negative | External | Unstable | Specific |
| | | Poachers were offered quick cash, and prompted by a trader | Markets change; trader can be arrested; focus on conditions that can change | Market for this species is narrow; poachers not aware of penalties; enforcement low that day |
| | Positive | Internal | Stable | Global |
| | | Officer has traits, skills, knowledge, social networks to outsmart poachers | Officer's skills, knowledge, networks are built over time | Officer sees self as generally intelligent, persistent, brave; antipoaching system is robust |
| Pessimistic style | Negative | Internal | Stable | Global |
| | | Poaching is a livelihood and infiltrated by organized crime. | Drivers lie in human nature—greed and security. Policies have not stopped poaching. | Poaching is motivated by long-term poverty that leads to many illegal acts. |
| | Positive | External | Unstable | Specific |
| | | The poachers were careless, weather was helpful. | Officer was just lucky to be there and notice them. | Animal-made noise; poachers were trapped by terrain. |

#### 7.3.2.4.3 *Empowerment and Collective Hopefulness*

The opposite of helplessness is psychological empowerment. Its components include a motivation to have control, a general belief that events result from one's own actions as opposed to forces external to oneself (fate, chance, powerful others, or forces), and self-efficacy (one's belief that one can in fact carry out a given action). Self-efficacy may be undercut if people feel that others are trying to control them. Then they may display reactance, or doing the opposite, in an attempt to reassert control (Brehm 2000). Many top-down conservation policies may have fallen prey to this hazard, as may self-righteous attitudes or conservation-by-command (Figure 7.7).

There are important collective-level dynamics of hope and empowerment. Christens and colleagues (2013) suggest that a hopeful vs. hopeless emotional orientation interacts with a critical vs. uncritical cognitive view of problem causes. A critical view recognizes the deep and systemic nature of the causes. One may be uncritically hopeful, for example wishfully thinking an information campaign may reduce overfishing when its roots lie in poverty and social norms. Or one might be critical and unhopeful, seeing the broad drivers of collective resource management system failures but unable to

**Figure 7.7** A famous injunctive-as-proper noun made a good target for reactance (doing the opposite).

break them down to actionable steps, or falling into an easy, general cynicism. Or one might be neither critical nor hopeful, seeing no hope in small steps and perhaps disengaging. But one might be critical but still hopeful. Christens and colleagues (2013) argue that the "critical and hopeful" position may be maintained by collective action, and by feeling the self is united with a long lineage of others who have contributed to the cause. This also gives the effort larger meaning. The self is neither the first nor last to sacrifice for the better: a lesson well taught by movements among Indigenous, People of Color, and others who have suffered oppression. This suggests that education about the history of conservation and intergenerational collaboration may be important in conservation work.

### 7.3.2.4.4 Hope, Grief, and the Emotional Challenges of Conservation Work

Perhaps not surprisingly, pervasive pessimism may be chronic among conservation professionals (Swaisgood & Sheppard 2010). But optimism (an expectation of better outcomes) or hope (acting according to one's highest values) may be requisite for consistent action. If so, the conservation scientists and practitioners captured by pessimism may be less than fully effective, look unattractive to recruits, and convey a disempowering frame to key audiences. Chronic high levels of anxiety or grief about ecological degradation may be immobilizing and affect mental health (Doherty 2018). In a survey study of 182 conservation biologists, educators, and environmental advocates by Fraser and colleagues (2013), 88% reported the highest levels of concern about future environmental conditions on the planet (self-reporting 6 or 7 on a 1 to 7 scale, where 7 was "constant worry"). These are people with a high level of attachment to nature. A high level of emotional distress over an environmentally harmful event witnessed in the subjects' lives was found. As with any psychological stress, social support mattered. Respondents who perceived that family or co-workers did not share their environmental values felt higher levels of fear or panic.

A second study used interview data to further describe the emotionally taxing nature of the work of environmental interpreters, and found their resilience strategies helped only somewhat. Fraser and colleagues (2013) speculate that they observed a cognitively mediated traumatic distress reaction. That is, thorough and intimate knowledge of biodiversity crisis is not traumatic in the usual sense, but rather comes about exactly through that knowledge, and the values that led to the choice of career. One implication is that for these subjects, nature experiences may not be restorative but rather re-stimulating. Perhaps more troubling, as with other traumas, one way of coping may be to pass the trauma on to others. In this case, telling tales of biocatastrophe might at best give the sense others share the distress but might also disempower them.

More broadly, "eco-anxiety" is on the rise globally (Manning & Clayton 2018), but social norms tend to deny such negative emotions. There are alternatives. One is to work through the emotions of hopelessness and grief, a route to justified anger and action (Atkinson 2021; see also Eaton 2017). Another is to counteract the stories of hopelessness with many well-documented and current solutions-focused stories from around the globe, such as those assembled by marine conservation biologist Elin Kelsey and her collaborators (OceanOptimism 2017).

### 7.3.3   Social Context

One of the major lessons of both cognition and motivation research is that social context has a powerful and pervasive effect on the individual. Social contexts at every scale of social organization—from friendships, to families, to age or generational cohorts, to socially imposed categories such as race, to voluntary associations such as clubs or political parties, to institutional affiliations (e.g. schools, work places, nation-states), to subculture and broader culture—affect knowledge, emotion, motivation, and empower-ment and generate or affect identity and other social characteristics of people. Larger system contexts are addressed by other disciplines in this volume. Psychology has explored many of them too, and helps elucidate the reciprocal shaping of individuals and groups.

#### 7.3.3.1   Social Cognition and Influence

Social cognition concerns how people perceive, remember, and make sense of their social world. It includes the brain processes and conceptual schemas used to detect, understand, explain, and predict others' behaviors, and one's own. Social cognition demonstrates var-ious biases. For example, when asked to explain others' behaviors, people tend to refer to internal or enduring reasons like personality. But when asked to explain one's own behavior, people refer to situational factors like an event that just put them in a bad mood. In fact, psychological studies suggest that internal factors are important, but their effects depend greatly on context.

One theory for the asymmetry in people's explanations of behaviors is the difference between being the actor (who knows one's reasons, sees and feels the influence of sur-roundings, and knows that one's own behaviors vary across situations) vs. being an observer (who is cognitively and visually outside the actor, and has sampled the actor's behaviors in only a few situations). Another explanation is that people from cultures that believe people are in charge of their own destinies will tend to overattribute behavior to personality, adding yet another twist of context to the theory itself! The general lesson is to consider the person's context first and interpret the effects of individual beliefs and motivations within that.

##### 7.3.3.1.1   *Mirroring, Empathy, and Social Modeling*
The social brain hypothesis proposes that in the evolution of animals with high degrees of interdependence in meeting basic survival needs, close social groups become an adaptive

strategy. These groups in turn generate new selective pressures such as positions that influence access to food or mates (Dunbar & Shultz 2007). There is then a premium on processing social information.

It is likely that strong interdependence pushed early *Homo* lineages to build on how other primates use the self as a model to understand and anticipate others' behaviors. This "inner modeling" is underlain by the "mirror neuron system," which enables internal mimicking of others' experience (Decety 2005). Empathy gives us the feeling of being the other. Mirroring may be automatic but is more accurate if one deliberately learns the perspective of the other. In such "cognitive empathy" the observer may better understand the actual as opposed to the apparent situation of the observed. Active brain imaging studies have revealed that when a person sees someone make a motion, express an emotion, or even communicate an idea, some of the same brain areas are activated as if that person did, experienced, or imagined it themselves. Much of this may occur out of awareness and explains social influence.

Clearly linked to mirroring, observational learning occurs when people learn by watching others' modeled behavior and vicariously grasping the consequences of that behavior. The prevalence of social learning frees humans somewhat from learning only by trial and error, but it also reveals why relatively independent creative or critical thinking benefits from deliberate cultivation. Social learning is fundamental to socialization, the life-span developmental process of internalizing the beliefs, symbolic systems, norms, social positions, and behaviors of one's groups, from family to global culture.

### 7.3.3.1.2   *Reference Groups*

Humans have a strong desire to be accepted by their groups, even in ephemeral or artificial groupings, generating strong tendencies to conformity (Sunstein 2019). But other people are particularly influential to an observational learner who regards them as a reference group—the group he or she aspires to belong to or resemble. Social comparison tells people how they stand within their group, with strong effects on identity. Many conundrums of conservation are underlain by motivations for acceptance, approval, esteem, status, or achievement. For example, for some West African urban dwellers, bushmeat—which threatens the populations of engendered mammals—maintains its value as a status symbol and connection to traditional village life (Redmond et al. 2006). Similarly, in high-income countries, a major reason for high material consumption is social comparison: if one's friends show a certain status-based standard of eating, for example, eating "lower" may feel like risking rejection. This applies whether the valued food is vegan or bushmeat.

Greater physical proximity and social similarity to others predict mirroring and greater influence. The transmission of social norms can be accounted for by perceived peer endorsement. For example, in experiments on the attitudes of groups of high school girls to toxic substances in household products, informative lectures alone had no effect on their attitudes toward the environment. Adding discussion, however, did. The change was explained by how much group members expressed preferences for non-toxic products, asked questions, shared knowledge, and gave toxic products little praise. Praising toxic products, however, could quickly undercut the shift (Werner 2003).

Getting those adopting conservation attitudes to observably express their new commitments among their associates thus makes the information salient and readily available. Much social influence is more subtle than "pressure," and based on conformity out of a need to belong. For example, in a study of domestic energy use, when households' utility statements showed household energy consumption in comparison to the average in their neighborhood, households shifted toward the mean, whether that meant decreasing or increasing use—energy conservers conformed by using more (Schultz et al. 2007).

### 7.3.3.1.3 Social Norms

The subjects in Schultz and colleagues' (2007) study of energy use were responding to a social norm—a group's implicit rule about what is appropriate or inappropriate to think, value, or do in a given type of situation, as was communicated in their power bills. According to Cialdini and Goldstein's (2004) theory of normative conduct, such "descriptive" norm information can be quite powerful. Messages may inadvertently convey anti-conservation norms if they imply social behaviors that lead to biodiversity loss are common, or may increase conservation if conservation behaviors are described as the norm. For example, interpretive signage intended to discourage feeding wildlife that mentioned that doing so is prevalent (and harmful) may in fact increase such behavior by normalizing it. "Prescriptive" norm information, on the other hand, conveys an obligation directly, such as a message to recycle (Figure 7.8). If the observer sees somebody else sanction (correct or disapprove of) behavior that does not comply with a norm, this is called an "injunctive" norm communication. In littering experiments, seeing someone pick up litter was more powerful than simply being urged not to litter.

**Figure 7.8** Signage that combines a prompt with an implicit prescriptive norm. Photo by Gene Myers

### 7.3.3.2 Social Marketing

Many of the foregoing ideas from social psychology have been incorporated into social marketing, which is now common in conservation campaigns. Applications of these ideas to conservation include improving knowledge and attitudes toward local laws among fishers and local leaders in Madagascar, leading to decreases in destructive fishing methods and better enforcement (Andriamalala et al. 2013); and reducing forest reduction for fuel wood and increasing efficient stove use in the golden snub-nose monkey range in China (DeWan et al. 2013). Social marketing is a useful but not wholly sufficient tool; in these and other examples, governance institutions play key roles.

Social marketing borrows a "consumer"-centric view from commercial marketing, except that what is being "sold" is not a product but an idea, practice, or behavior. Notably these targets are more specific than broad "value" concepts (Andreasen 1995). The term *customer* is used advisedly, and primarily implies that people change voluntarily because they see it as a net benefit. Conservation programs deal with "high-involvement" behaviors, or ones "about which individuals care a great deal, where they see significant risks, where they think a lot before acting, and where they frequently seek the advice of others" (Andreasen 1995, p. 38). Thus, it is critical to deeply understand the behavior setting and social world of the actor in creating a marketing program.

Program teams must first do research to determine what factors operate for and against a conservation goal in the specific setting before a social marketing program can be designed. Then they may apply whichever of a variety of social psychological tools are indicated to nudge behaviors (Table 7.11). McKenzie-Mohr and Smith (1999) emphasized such

**Table 7.11**  Social marketing concepts and tools.

| Tool or concept | Application recommendations |
| --- | --- |
| Incentives | Incentives should be just enough to generate the behavior. If they are too large, they can inhibit "owning" the decision to change behaviors. |
| Norms | Communicate norms in personal and positive terms, using both descriptive, prescriptive, and injunctive norm messages; clearly state that "it's the right thing to do." |
| Commitments | Ask people to commit to an easier action before a harder (target) action ("foot in the door" technique). Commitments must be voluntary and have strongest effect if made in the person's community, publicly, or to a leader. |
| Framing | Messages should be framed carefully and be vivid, personal, and concrete. Use cognitive shortcuts like the desire to avoid losses. |
| Information | Determine which information is actually a limiting factor, and deliver it effectively. This often may be "how-to" information. |
| Information sources | When possible, trusted, high-credibility reference group members should endorse or model the behavior. |
| Prompts | Place a catchy reminder where it will be noticed at the right time and place for the behavior. |
| Feedback | Share results of efforts at both individual and community levels. |

*Source:* Adapted from McKenzie-Mohr and Smith (1999).

background research as community-based social marketing. But the questions of who initiates a project and maintains it, and of how engagement is built are complexly tied to values and context. Sterling and colleagues (2017) thematically analyzed nearly 300 studies of stakeholder engagement in biodiversity conservation and recommend that participation be seen as complex and continually under negotiation. Underscoring the variability in benefits and costs of participation in protected-area governance in three Madagascar villages, Ward and colleagues (2018) found that knowledge about who could participate and how to do so was unevenly distributed within and between communities. Also, perceived costs and benefits varied and reduced participation among men, poorer households, and those living in remote villages. Communities are far from homogeneous in knowledge and access to institutions, posing challenges for equitable participation (see Section 2.5.8, Chapter 2 for more on participatory methods and Section 3.3.2.1 in Chapter 3 for more on the complexity of communities and participation).

To change behavior, the central insight is to understand the barriers and benefits (often indirectly related to the new behavior) which affect adoption of the behavior. Infrastructure gaps or barriers may exist, such as lack of convenient alternate transportation services. These may be controlled at a different system level such as by a regional jurisdiction, calling for "upstream" marketing to audiences at that level (Box 7.3.).

---

**Box 7.3   Application: Social marketing for conservation in South Africa**

Author: Angelika Wilhelm-Rechmann, Nelson Mandela Metropolitan University

In landscape conservation, key behavioral choices are often made by mid-level managers in agencies. Social marketing in this context means identifying the high-impact management behaviors and understanding, both analytically and empathically, the realities of those who undertake them. This requires an investment in front-end research. The audience's behaviors may be driven by cultural, legal, economic, institutional, and technological factors. Analysis of these contexts helps displace marketers' naïve theories, reveal segments of the audience, and pinpoint the behaviors that lead to environmental degradation as well as the maximally impactful alternative behaviors (Wilhelm-Rechmann & Cowling 2008).

The Eastern Cape Province of South Africa harbors a large variety of species and relatively untouched environments. The Cape is under considerable development pressure, especially along the coastline, as wealthy individuals look for unspoiled locations near the sea and poor communities desperately seek socioeconomic development. To advance conservation in the Cape, biologists have developed maps highlighting the relative importance of various lands to biodiversity conservation. Social marketing to mainstream the use of these biodiversity maps in local land-use planning processes represents a logical way to advance conservation planning and action.

Front-end research eventually identified local government officials as the most suitable target audience. The desired social intervention to be adopted by them was the behavior of consulting and using the maps when any land-use change was proposed in the municipality. Contrary to expectations, interviews with land-use planners revealed they did not object to using biodiversity maps. Rather, the key obstacle was that

**Box 7.3 (Continued)**

locally elected councilors would not support the use of maps because of their own ignorance of land-use planning procedures and environmental considerations.

Thus, while still working with the planners, the marketing program expanded the target group and investigated how locally elected councilors see the role of the environment in land-use planning. Technically speaking, this is an "upstream" social marketing approach, meaning it included an additional target group because they have a considerable direct or indirect influence (such as through approval, policy, or regulations) on the behavior-to-be-changed of the original target group.

Interviews revealed that most councilors do value their environment personally as well as for its tourism potential. However, the term "biodiversity" holds almost no meaning and "conservation" is perceived negatively for various reasons. For example, conservation is often considered a "white" issue and a legacy of the politics and policies of apartheid. Furthermore, communication by conservationists is often perceived by councilors as disrespectful because it disregards other arguments. Thus, conservation outreach is often counterproductive. Moreover, conservation registers very low relative to other political issues facing local municipalities. One of the key priorities for the councilors, however, is "service delivery," including housing, water, and electricity (but not ecosystem services).

This research concludes that subsequent development of computerized map-based planning tools should emphasize "options" rather than a "veto" approach to situating development away from biodiversity priority areas, and that maps should identify relevant local ecosystem services. This case highlights how social marketing can be adapted to the complexities of local land-use planning. Customer research is absolutely essential for identifying the optimal target group, for discerning the explicit behavior to be addressed, and for unearthing the key psychological factors that generate the behavior-to-be-changed (Wilhelm-Rechmann et al. 2013).

### 7.3.3.3 Identity

Identity is a potent product of social interaction. It includes ideas consciously held about the self. Identity encompasses personal characteristics, diverse social relations, roles, and social categories. Identities are both internal (how the self sees itself) and external (how others see the self), although these may differ. One's identities may be as multiple as one's social contexts. Identity has been extensively explored in psychology in relation to natural and especially social environments. Three perspectives on the significance of identity relevant to conservation are (i) mainstream Western social identities and nature, (ii) understanding identification with nature, and (iii) examining the ways collective identity affects behavior and group relations.

#### 7.3.3.3.1 *Mainstream Western Social Identities and Nature*

Psychologist Erik Erikson saw identity as a psychosocial developmental process. Cultures are typified by the identities they encourage. Individuals engage with the **discourses** and

**ideologies** of their culture, and construct identities—personal narratives, revised over the life span—that reflect or react to those larger discourses (Hammack 2008). This can be commonly seen when conflicts polarize the identities of the members of their respective cultures. It is possible too that cultural identities might be polarized *against* nature, as it and humanity are constructed by that culture (e.g. ideas of wilderness as frightening, full of danger, and therefore something to be tamed).

Some psychologists critique the deeply embedded domination-over-nature themes in anthropocentric, often Western, discourses and identities that place nature as an "out-group," similar to social divisions as described by social identity theory (Fielding & Hornsey 2016). Derogatory stereotypes of animals as dirty, dangerous, or non-sentient, attitudes of dominance over nature as merely a resource for humans, and the replacement of native vegetation with more "beautiful" exotics may be characteristic of anthropocentric identities that underlie unsustainable behavior. Even more directly, social identities built around, expressed through, or justifying materialism lead to economic and ecological over-consumption (Crompton & Kasser 2009). Crompton and Kasser (2009) hold that conservation or sustainability strategies using selfish motivations and economic incentives (e.g. green **capitalism**) only reinforce the utilitarian premises of such identities. Alternatively, they propose using identity processes in environmental campaigns to activate empathy and perspective, reinforce humans as part of nature, decrease psychological defenses that maintain human supremacy, and highlight non-material goals as centers of human identity.

### 7.3.3.3.2  Identification with Nature

People whose identities incorporate the natural world behave with greater consideration toward nature. Such an identity would come about through developmental processes that occur via the person's particular interactions with the context concerned—in this case with their environments, specifically the non-human aspects. Chawla (1992) traced the expanding range of place attachments children might construct in three expanding spheres from infancy and early childhood, to middle childhood, and then adolescence. Each sphere is defined by developmental changes that create inward and outward pulls, as well as context.

Researchers use several ways of conceptualizing and assessing such identities in adults and have shown clear variations in identification with nature, and their possible consequences (Box 7.4). Measures such as inclusion of nature in self (Schultz 2001), connectedness to nature (Mayer & Frantz 2004), and relatedness to nature (Nisbet et al. 2009) show correlations with pro-environmental worldviews, behavior, and time spent in natural settings. A more general measure of environmental identity that included questions about interactions with nature, feeling one is a part of nature, positive emotions, and collective identity predicted college students' positions on an environmental conflict and support for animal rights (Clayton 2003). Adding identity to Ajzen's theory of planned behavior increases its ability to explain environmental behavior.

While identity is a partly self-conscious phenomenon, regular self-report scale measures may be inaccurate because research subjects may avoid responding in ways they think people might disapprove—a **methodological** pitfall called "social acceptability bias."

---

**Box 7.4   Methods: Measuring identification with nature**

Diverse data can reveal identity. Qualitative methods include ethnographic (see Section 2.5.3, Chapter 2) and interview techniques such as spending time with a group and recording the practices and language used to define and talk about their identity in relation to values and other groups. This can yield detailed portraits of the meanings of places that are relevant to conservation and highly valid in their context.

**Quantitative** methods for measuring identity include questionnaires, scales, or lists of statements with which a person can agree or disagree, place themselves on a continuum, or provide answers via other methods. Scales for identity include multiple questions probing issues such as the importance of the identity to the person, feelings associated with that identity, **ideological** commitments such as nature as benign, frequency of interaction with nature, places, or species, and behavior consistent with the identity. Before the scale is used in research, its validity (Does it measure what it intends to?) and reliability (Does it measure it consistently?) are assessed in preliminary studies. Subscales (sets of questions within the instrument) that are embedded within a scale draw out different dimensions that may have been theorized or discovered by statistical factor analysis.

Although various measures have been developed with the intention of highlighting specific dimensions, Tam (2013) compared the performance of nine such instruments and found that they were strongly intercorrelated with each other and with criterion variables such as self-reported behavior. On the other hand, some dimensions of general models of identity were not tapped by any measure. Like other psychometric instruments (data-gathering tools intended to measure a psychological variable), these scales have external validity only for populations with whom they were developed, and results thus should not be generalized to other groups.

---

Implicit association tests (IATs) require subjects to respond as quickly as possible to questions presented on a computer screen and that do not reveal what is being measured (by also requiring subjects to respond to "distractor" topics). Stronger underlying mental associations between two entities allow the subject to respond more quickly than for weaker ones, revealing unintentional and unconscious associations. These have been used widely on topics such as gender or racial bias. Schultz and colleagues (2004) developed an IAT for association of self with nature. It is administered via computer, with precise response times indicating how strongly a person's unconscious mental models associate the self with nature vs. the built environment.

### 7.3.3.3.3   Collective Identity, Behavior, and Group Relations

Collective identity develops in the context of groups, including social movements. By presenting the self in a way consistent with group identity, members fulfil needs for belonging. Collective identity fosters strong bonding within the group, development of a coherent set of values, a sense of individuality within group identity, and motivation for action, for instance among zoo volunteers (Fraser 2009). Identities are influenced by larger-level social symbolic systems, norms, and values (see Chapter 3). Collective identities may be harmonious with mainstream values, or oppositional to them, encouraging activism, radicalism, or retreat. Low perceived public esteem of the group, or external threats, motivate supportive actions and stronger within-group bonding, or sometimes within-group conflict. Identities based around socially patterned inequalities associated with race, class, gender, ethnicity, and other social categories may interact with conservation priorities.

Just as some such identities are imposed externally, people hold views of the prototypical environmentalist. In an IATs study, Ratliff and colleagues (2017) discovered that people's implicit picture of environmentalists is significantly less attractive, cool, fun, or intelligent than their self-reported explicit environmentalist prototype, although equally judgmental. Those with positive prototypes were more likely to report engaging in pro-environmental behavior.

The polarization besetting conservation can be explained by "protective cognition," which is the tendency of people to endorse positions that "reinforce their connection to others with whom they share important commitments" such as hierarchical vs. egalitarian social values (Kahan 2010, p. 296). Thus, positions on otherwise unconnected issues (such as climate change and abortion) line up in nearly identically polarized fashion. For science communication to escape this fate, its communication must be disassociated from cues people use to identify information aligned with their social worldviews, or its messages need to be carried by in-group members.

#### 7.3.3.4 Psychology of Conflict and Cooperation

Conflict and its close relation, cooperation, are ubiquitous in human affairs. Conflict occurs when the interests of two or more parties are in competition, and at least one party is perceived by another to promote its interests to the detriment of the other. Sherif (1966) termed this "negative goal interdependence," where parties' goals are mutually exclusive. Conflict may exist covertly or openly in any degree of escalation. Conflicts often entail power differences, political-economic institutions and other social factors, manipulated perceptions, as well as "objective" conditions on the ground. For example, environmental degradation and resource scarcity (or abundant but unregulated resources) have been identified as contributory drivers as well as consequences of conflicts. However, some social scientists are wary of explanations of conflict that focus narrowly on environmental factors (see Section 5.3.1.4, Chapter 5 for more on so-called environmental determinism).

Biodiversity conservation is affected in numerous ways by conflict. Human interests can conflict directly with those of other species, but often the situation is a conflict between humans that concerns or affects natural systems. Some of the psychological dimensions that are most relevant to conservation are (i) the roles of beliefs and values, (ii) group effects on identity, (iii) conditions for cooperation, and (iv) the psychology of moral conflict.

#### 7.3.3.4.1 The Role of Beliefs and Values

Conservation conflict inherently entails social cognitions of others' values, beliefs, and intentions regarding the object of conflict (White et al. 2009), as illustrated by the way disagreements about wolf recovery in the US West revolve around different stereotypes

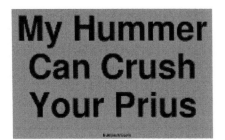

**Figure 7.9** Sensitivity to others' perceived self-righteous identities can fuel conflicts. Owners of both "Hummer" (large SUV) and "Prius" (small hybrid) cars have been shown to feel they hold the high moral ground (Adapted from Luedicke et al., 2009).

about the role of government: to stay out of individuals lives, or to protect the vulnerable. How much the groups in wolf recovery controversies disagree on deeper beliefs influences how strong the conflict is perceived to be. The accuracy of such perceptions is influenced in turn by actual differences as well as by communication between parties. Communication in turn is mediated by social institutions as well as psychological factors. Organizations that bring parties together are critical in intergroup relations because they provide opportunities to test agreement on beliefs and values, and to build trust.

#### 7.3.3.4.2 *Group Effects on Identity*

Psychological research has shown that being in a group increases competitive behavior. Whereas individuals may feel inhibited when alone, in a group (even if randomly assigned, with patently artificial group interests) competitive motives are shared and inhibition is decreased (Insko et al. 1987). When groups are in competition, in-group solidarity increases, negative stereotyping of the out-group increases, the situation is construed in win/lose terms, and hostile messages and actions are likely to increase (Figure 7.9). Such dynamics readily move competition due to resource constraints and/or value and belief differences toward conflict. As identity becomes moralistic and conflict becomes hostile, negative emotions are triggered and overlay events with readily elicited emotional memories.

Environmental identity, like all other sorts of identities, is salient in conflicts between groups. Opotow and Brook (2003) studied conflicts surrounding an endangered mouse species that polarized ranchers and environmentalists against each other. Both ranchers and environmentalists used psychological techniques to de-legitimize the others' concerns and validate their own, for example, by attributing false motives or painting the other group as powerful and privileged. This led to reactance and subterfuge, for example, in a "shoot, shovel, and shut up" attitude of some ranchers who suggested others clandestinely kill mice. The first overt manifestations of conflict may disguise deeper issues such as threatened identities. In human–wildlife conflict, for example, psychological complexities include embedded emotional memories of earlier conflicts, individual's symbolic meanings and interpretations of conflict, social trust, and interrelationships between conflict and identity, and other psychological needs. Addressing these interacting levels offers chances to transform and reconcile wildlife-related conflicts (Madden & McQuinn 2014).

#### 7.3.3.4.3 *Conditions for Cooperation*

Psychology offers insight into cooperation. Positive goal interdependence is characterized by a goal both parties highly value, but that cannot be attained without cooperation (Sherif 1966). Indeed, mutual environmental welfare goals can bring together parties involved in non-environmental conflicts, as in the establishment of transboundary protected areas to celebrate peaceful relations or serve as buffers and ecological refuges between states in conflict (Ali 2007). Examples include the Red Sea Marine Peace Park between Israel and Jordan, the Cordillera del Condor Peace Transborder Reserve between Peru and Ecuador, and a potential park on the Korean peninsula (UN CBD 2017). Cooperation requires that existing identities be affirmed, not threatened. Instead, they may be brought under a wider "umbrella identity" based on shared values and superordinate goals, as has been successfully done by various rangeland conservation coalitions (Opotow & Brook 2003).

Since groups in conflict interpret all interactions through the lens of conflict, positive goal interdependence should be made unambiguous. Nonetheless, progress from conflict toward cooperation is not linear. In approaching conflicted situations, psychology, as well as common sense, places a premium on overarching values such as democracy, the right to be heard, the obligation to listen, willingness to entertain self-doubt, and elevating coexistence despite disagreement above narrow consistency with ideology or single principle. Intergroup contact, especially living and working together for a shared goal, and the development of friendships across groups reduces stereotypes and dehumanization that fuel conflicts, and builds positive attitudes (Pettigrew & Troop 2006).

### 7.3.3.4.4 The Psychology of Moral Conflict

A full psychology of morality requires engagement with philosophy to determine what morality and ethics are. Lacking that, psychology's role in describing people's moral ideas and behavior must always carry the qualifier that what is empirically observed (what "is") cannot determine what is good or right (what "ought to be"). Morality and ethics (used roughly interchangeably here for simplicity) refer to people's ideas, feelings, and behavior around what is right or good, what is permissible or impermissible, and what is a mere matter of preference, obligatory, or beyond the call of duty. Ethics includes a fundamental individual component of choice and responsibility and cannot be reduced to culture or social norms. On the other hand, we cannot make sense of morality outside the need for social coordination, where culture has strong influences. Morality is so fundamental to human psychology that almost all people feel a strong need to perceive themselves as "good," and try to ensure others see them that way too. To do otherwise is to risk falling outside others' sphere of inclusion among those to whom moral consideration is owed. Of course, the latter is not ensured—at either the extreme of convicted criminals who lose some measure of consideration of their full rights, or the extreme of moral heroes who exceed what a "normal" society, or a corrupt regime, considers acceptable. While cultures vary substantially in their moral codes and how they are expressed, even strong cultural relativists concede that abstract principles of justice and the avoidance or harm (or more positively, welfare) are universal. We can only touch on a few elements of this complex relationship among culture, society, individual, and philosophy here.

Moral psychology draws on reasoning. Moral action is not possible if people cannot recognize when their self-interests or desires, or those of someone else, conflict with duties to others, including possibly nature. This is underlaid by "reversibility," or the ability to cognitively see the situation from the other's (or society's) point of view. Arguably this is a core part of justice. Thinking through one's actions, or providing a justification for them, is also cognitive, and can be used critically and correctively—or for self-deception. Because our feelings can be strongly self-centered, some thinkers hold that only actions that are impartial are truly moral. Moral action, that is, occurs when one acts against self-interest, or incurs a real cost to follow an impartial principle.

Emotion, nonetheless, plays a big role in moral psychology. It provides the motive to do more than think about information. Part of what would make one take the perspective of the other is empathy—feeling the other's distress, distress that might be due to one's own actions. In childhood socialization, caregivers who merely point to the distress of another whom a child hurt evokes empathy that helps the child understand what they did was wrong—and does so without blaming the child directly. This is a key dynamic in moral socialization.

Haidt (2012) has hypothesized five universal moral-emotional "foundations," each with evolutionary roots in threats that close-knit groups of humans faced. Each describes (but does not provide an ethical argument for!) a psychologically distinct "morality." Haidt expresses each with positive and negative poles: care/harm, fairness/cheating, loyalty/betrayal, authority/subversion, and sanctity/degradation. Recent research supported their applicability in 17 WEIRD and 13 non-WEIRD cultures (Doğruyol et al. 2019; WEIRD is explained in Section 7.3.1.3). But cultures vary in whether and how these moralities are elaborated and socialized in their moral codes, and thus whether the relevant emotions are felt powerfully and intuitively—sparking judgment and action. For example, liberals and conservatives in the USA differ in which foundations they include (Haidt 2012). Strong conservatives include all five moralities and weigh each about the same. Strong liberals, however, consider only care and fairness as really moral, and are skeptical of how in-group loyalty, obedience to authority, and moralization of sanctity and purity (e.g. around control of others' bodies) can lead to power and violations of justice and welfare. Haidt points out that liberals do also depend on social order (the thrust of loyalty and authority) and they do moralize some issues around purity (e.g. food beliefs and environmental protection). The net effect of moralities on the groups observing them is to bind the members into a team around highly protected values, often prompting them to out-group and polarize those who disagree. In Haidt's view, reasoning or conscious cognition is activated mostly after the fact, in the form, or rationales and justifications.

There is little doubt that moral/ethical differences make conflicts more intractable. Opotow and Weiss (2000) showed how conflicts are more destructive when fed by forms of denial that exclude others from moral consideration. Stakeholders can be outright excluded, obviously among humans but also non-humans (Opotow 1994). The seriousness of others' actions can be denied, as in denying that climate change cannot be mitigated or denying one's own responsibility or displacing the blame. All these actions influence the interparty dynamics of the conflict.

Another school of moral psychology emphasizes the social disequilibrium of moral transgressions rather than conflicted principles (cognitive) or emotions (affective). Central to these social-psychodynamics of moral conflict is understanding moral exclusion and dehumanization, and the kinds of moral conflict strategies that reinforce it or open ways out. Haan and colleagues (1985) observed how the former, or "defensive strategies," maintain enmity, leave options unexplored, displace responsibility, or refuse to engage. The latter, or "coping strategies," on the other hand, keep things more open, involve acknowledging the self's own involvement and moves, avoid blaming, and enact moral inclusion (Table 7.12).

**Table 7.12** Coping and defensive strategies in moral conflict.

| Coping strategies | Defensive strategies |
|---|---|
| Trying to be objective | Self-righteousness |
| Focusing on ideas or interests, not persons | Withdrawal and isolation |
| Logical analysis, analogical thinking | Intellectualization |
| Tolerance of ambiguity | Projection |
| Empathy and perspective taking | Self-effacement |
| Reviewing what happened | Giving up |
| Concentration | Denial |
| Integration of feelings and ideas | Displacement of feelings |
| Consolation and acceptance | Blaming, demonization |
| Forgiveness | |

## 7.4 Future Directions

Though psychology's focus on conservation has been relatively limited to date, it has been influential in resource-use conservation and climate change communication and policy in wealthier Western countries. Its theories and principles may provide foundations for future research and application to conservation in other settings. In particular, several areas stand out. How might a psychologically realistic understanding of human choice inform institutions for conservation? How does the interplay of individual agency and social context explain conservation behavior, values, and identity? How robust are psychological findings with Western populations (more precisely WEIRD ones) in other cultural settings with high-stakes conservation issues? What interindividual processes explain and can be applied in how conservationists interact with their constituents to improve outcomes? What are the consequences of the environmental generational amnesia of nature for human well-being and for the experiential bases of values important to conservation? How do the dynamics of moral feeling and thought facilitate and thwart valuing nature while also respecting human values? Finally, how can the field of positive psychology be recruited so conservation efforts contribute directly to people's well-being?

### 7.4.1 Conservation Institutions Informed by Realistic Views of Human Choice

Decisions related to conservation are ubiquitous and occur at every level of space, time, and social organization. How can psychology's description of the factors affecting choice be better translated to conservation? Humans have multiple pathways by which to make decisions. The classic "rational" decision maker of rational choice theory appears far from universal and quite context-limited. Most decision-making is more constrained, un-self-aware, and path-dependent. Many choices are fast, driven by biases and cognitive and emotional shortcuts. Because such choice is based largely on learned mental associations rather than universal rationality, these insights may need to inform conservation institutions on a culture- and site-specific basis.

### 7.4.2 Social Psychology, Individual Agency, Governance, and Culture

Individual variation in beliefs and values may affect the ways people interact with com-mon-property governance institutions (see Chapter 6), but little work in this area has been done. How information can best be used and communicated in different contexts—from United Nations offices, to corporate board rooms, to urban centers and rural villages—can be informed by studies of cognitive biases, social influences, social learning for effective priority setting, research use, problem solving, or adaptive management, and many other topics. Further, broad-scale institutions channel behavior but are ultimately constituted and influenced by individuals' choices.

Social influences should be researched in applied conservation contexts. Environmental identities are associated with deeply internalized values, but also with polarizing inter-group dynamics. What factors lead to conservation behaviors being part of a society-wide collective identity in the same way as patriotism or citizenship? How can interventions be best designed to work with emotion-guided, socially swayable, satisficing, over- (or under-) confident, but well-intentioned humans? And with those acting from an unmitigated self-serving perspective?

Conservation psychology faces challenges and opportunities in translating its theories and findings to cultures in conservation-priority areas. Psychology leaves the relation-ship between culture and nature open. Findings about the development of ecological concepts, natural and social identities, social influence, and moral and psychological val-uations of nature should be investigated cross-culturally to determine their contours and dynamics.

Accomplishing conservation is about human–human interactions. Face-to-face dynamics affect group and institutional functioning. For example, emotion regulation, trust/distrust, integrity, leadership, creativity, emotional intelligence, and consensus-building, affect social norms, organizations, and intergroup relations. Application to conservation is ripe.

### 7.4.3 Individuals' Base of Experience in Natural Systems

Collectively, humans' multiple information processes provide human adaptiveness (Weber & Johnson 2009). In particular, the human mind may consciously and uncon-sciously monitor, select, and control which processes are governing it. In conscious awareness, "mindfulness" occurs when we observe not just the outer world, but our own inner responses, allowing us to choose our response more reflectively. However, all human responses, including those to nature, are constrained by unconscious internal processes and limited experience. The human experience of nature is in precipitous decline due to ecosystem losses, urbanization, and new technologies. To the extent this decline results in lower appreciation of the value of biodiversity and nature, the effect may be maladaptive responses, however mindful. Thus, an urgent basic research task for psychology is to describe and categorize the human experience of nature, its psychological benefits, and its expressions in thought, motivation, and emotion (Kahn & Hasbach 2012).

### 7.4.4 Positive Psychology of Conservation

Psychology should help conservation use the best human motivations possible. Humans are not only motivated by fear, scarcity, power, and self-interest. We are also powerfully motivated by fairness, reciprocity, trust, self-determination, being understood, feeling connected, feeling competent and helpful, moral elevation, and many other aspects of "positive psychology." Well-being in life comes from positive emotions, deep engagement in activities, accomplishment, relationships, and contributing to something one values that is greater than oneself (Lopez 2020). These should be embedded in conservation applications and tested. While it has seemed that conservation is mostly a matter of "sacrifice" and "curtailment behaviors" it can also be conceptualized in terms of higher human aspirations.

## For Further Reading

1 *The Native Mind and the Cultural Construction of Nature* (Atran & Medin 2008, MIT Press, Cambridge, MA).

2 *Oxford Handbook of Environmental and Conservation Psychology* (Clayton, ed. 2012, Oxford University Press, New York).

3 *Conservation Psychology: Understanding and Promoting Human Care for Nature*, 2nd ed. (Clayton & Myers 2015, Wiley/Blackwell, New York).

4 *Identity and the Natural Environment* (Clayton & Opotow, eds. 2003, MIT Press, Cambridge, MA).

5 *Environmental Problems and Human Behavior* (Gardner & Stern 2002, Allyn & Bacon, Boston, MA).

6 *The Human Relationship with Nature* (Kahn 1999, MIT Press, Cambridge, MA).

7 *Children and Nature* (Kahn & Kellert, eds. 2002, MIT Press, Cambridge, MA).

8 *Social Marketing to Protect the Environment: What Works* (McKenzie-Mohr, Lee, Schultz, & Kotler 2012, Sage, Thousand Oaks, CA).

9 Special issue: Conservation Psychology (Saunders & Myers, eds. 2003, *Human Ecology Review* 10 (2)).

10 *Psychology for Sustainability*, 5th ed. (Scott, Amel, Koger, & Manning 2021, Routledge, New York).

## References

Abramson, L.Y., Seligman, M. & Teasdale, J.D. (1978) Learned helplessness in humans: critique and reformulation. *Journal of Abnormal Psychology* 87: 49–74.

Ajzen, I. (1991) The theory of planned behavior. *Organizational Behavior and Human Decision Processes* 50: 179–211.

Ajzen, I. (2006) The theory of planned behavior. http://www.people.umass.edu/aizen/tpb.diag.html (accessed March 30, 2010).

Ali, S.H. ed. (2007) *Peace Parks: Conservation and Conflict Resolution*. Cambridge, MA: MIT Press.

Andreasen, A. (1995) *Marketing for Social Change*. San Francisco, CA: Jossey-Bass.

Andriamalala, G., Peabody, S., Gardner, C.J. et al. (2013) Using social marketing to foster sustainable behaviour in traditional fishing communities of southwest Madagascar. *Conservation Evidence* 10: 37–41.

Atkinson, J.W. (2021) Facing it: a podcast about love, loss, and the natural world. https://www.drjenniferatkinson.com/facing-it (accessed May 16, 2021).

Atran, S. & Medin, D. (2008) *The Native Mind and the Cultural Construction of Nature*. Cambridge, MA: MIT Press.

Baird, T., Leslie, P. & McCabe, J. (2009) The effect of wildlife conservation on local perceptions of risk and behavioral response. *Human Ecology* 37: 463–474.

Balmford, A., Bradbury, R.B., Bauer, J.M. et al. (2021) Making more effective use of human behavioural science in conservation interventions. *Biological Conservation* 261: 109256.

Bandura, A. (1986) *Social Foundations of Thought and Action*. Englewood Cliffs, NJ: Prentice-Hall.

Baumeister, R.F., Crescioni, A.W. & Alquist, J.L. (2011) Free will as advanced action control for human social life and culture. *Neuroethics* 4: 1–11.

Biodiversity Project & Belden, Russonello & Stewart (2002) *Americans and Biodiversity: New Perspectives in 2002: A Cluster Analysis of Findings from a National Survey*. Washington, DC: Belden, Russonello & Stewart [Madison, WI].

Brehm, J. (2000) Reactance. In: *Encyclopedia of Psychology*, vol. 7 (ed. A.E. Kazdin), 10–12. London: Oxford University Press.

Bruner, J. (1986) *Actual Minds, Possible Worlds*. Cambridge, MA: Harvard University Press.

Castillo-Huitrón, N.M., Naranjo, E.J., Santo-Fita, D. et al. (2020) The importance of human emotions for wildlife conservation. *Frontiers in Psychology* 11: 1277. https://doi.org/10.3389/fpsyg.2020.01277.

Cetas, E.R. & Yasué, M. (2017) A systemic review of motivational values and conservation success in and around protected areas. *Conservation Biology* 31 (1): 203–212.

Chawla, L. (1992) Childhood place attachments. In: *Human Behavior and Environment: Advances in Theory and Research*, vol. 12 (ed. I. Altman & S. Low), 63–88. New York: Plenum Press.

Chawla, L. (1998) Significant life experiences revisited: a review of research on sources of environmental sensitivity. *Journal of Environmental Education* 29 (3): 11–21.

Christens, B.D., Collura, J.J. & Tahir, F. (2013) Critical hopefulness: a person-centered analysis of the intersection of cognitive and emotional empowerment. *American Journal of Community Psychology* 52: 170–184.

Cialdini, R.B. & Goldstein, N.J. (2004) Social influence: compliance and conformity. *Annual Review of Psychology* 55: 591–622.

Clayton, S. (2003). Environmental identity: a conceptual and an operational definition. In: *Identity and the Natural Environment* (ed. S. Clayton & S. Opotow), 45–65. Cambridge, MA: MIT Press.

Clayton, S. & Myers, O.E., Jr. (2015) *Conservation Psychology: Understanding and Promoting Human Care for Nature*, 2nd ed. New York: Wiley-Blackwell.

Clayton, S. & Opotow, S. eds. (2003) *Identity and the Natural Environment*. Cambridge, MA: MIT Press.

Crompton, T. & Kasser, T. (2009) *Meeting Environmental Challenges: The Role of Human Identity*. Surry, UK: World Wildlife Fund–United Kingdom.

David, E.J.R., Okazaki, S. & Giroux, D. (2014) A set of guiding principles to advance multicultural psychology and its major concepts. In: *The APA Handbook of Multicultural Psychology*, vol. 1 (ed. F.T.L. Leong), 85–104. Washington, DC: American Psychological Association.

DeCaro, D. & Stokes, M. (2008) Social psychological principles of community-based conservation and conservancy motivation: attaining goals within an autonomy-supportive environment. *Conservation Biology* 22 (6): 1443–1451.

Decety, J. (2005). Perspective taking as the royal avenue to empathy. In: *Other Minds: How We Bridge the Divide between Self and Others* (ed. B.F. Malle & S.D. Hodges), 143–157. New York: Guilford.

Deci, E. & Ryan, R. (1985) *Intrinsic Motivation and Self-determination in Human Behavior*. New York: Plenum.

DeWan, A., Green, K., Li, X. et al. (2013) Using social marketing tools to increase fuel-efficient stove adoption for conservation of the golden snub-nosed monkey, Gansu Province, China. *Conservation Evidence* 10: 32–36.

De Young, R. (2000) Expanding and evaluating motives for environmentally responsible behavior. *Journal of Social Issues* 56 (3): 509–526.

Dietz, T., Gardner, G.T., Gilligan, J. et al. (2009) Household actions can provide a behavioral wedge to rapidly reduce U.S. carbon emissions. *Proceedings of the National Academy of Sciences USA* 106 (44): 18452–18456.

Doğruyol, B., Alper, S. & Yilmaz, O. (2019) The five-factor model of the moral foundations theory is stable across WEIRD and non-WEIRD cultures. *Personality and Individual Differences* 151: 109547.

Doherty, T.J. (2018). Individual impacts and resilience. In: *Psychology and Climate Change* (ed. S. Clayton & C. Manning), 245–266. Cambridge, MA: Academic Press.

Dunbar, R.I.M. & Shultz, S. (2007) Evolution and the social brain. *Science* 317 (5843): 1344–1347.

Dunlap, R. & Mertig, A. (1995) Global concern for the environment: is affluence a prerequisite? *Journal of Social Issues* 51 (4): 121–137.

Eaton, M. (2017). Navigating anger, fear, grief and despair. In: *Contemplative Approaches to Sustainability in Higher Education: Theory and Practice* (ed. M. Eaton, H.J. Hughes & J. MacGregor), 40–54. New York: Routledge.

Ekman, P. & Friesen, W.V. (1969) The repertoire of nonverbal behavior: categories, origins, usage and coding. *Semiotica* 1 (1): 49–98.

Festinger, L. (1957) *A Theory of Cognitive Dissonance*. Stanford, CA: Stanford University Press.

Fielding, K.S. & Hornsey, M.J. (2016) A social identity analysis of climate change and environmental attitudes and behaviors: insights and opportunities. *Frontiers in Psychology* 7: 121. https://doi.org/10.3389/fpsyg.2016.00121.

Fischer, A. & Young, J. (2007) Understanding mental constructs of biodiversity: implications for biodiversity management and conservation. *Biological Conservation* 136: 271–282.

Fraser, J. (2009) An examination of environmental collective identity development across three life-stages: the contribution of social public experiences at zoos. Doctoral dissertation, Antioch University New England.

Fraser, J., Pantesco, V., Plemons, K. et al. (2013) Sustaining the conservationist. *Ecopsychology* 5 (2): 70–79.

Frederickson, B. (2005). Positive emotions. In: *Handbook of Positive Psychology* (ed. C. Snyder & J. Lopez), 120–134. New York: Oxford University Press.

Gardner, G. & Stern, P. (2002) *Environmental Problems and Human Behavior*, 2nd ed. Boston, MA: Allyn & Bacon.

Gifford, R., Scannell, L., Kormos, C. et al. (2009) Temporal pessimism and spatial optimism in environmental assessments: an 18-nation study. *Journal of Environmental Psychology* 29 (1): 1–12.

Gillham, J., Shatté, A., Reivich, K. et al. (2001) Optimism, pessimism and explanatory style. In: *Optimism and Pessimism: Implications for Theory, Research and Practice* (ed. E.C. Chang), 53–75. Washington, DC: American Psychological Association.

Haan, N., Aerts, E. & Cooper, B. (1985) *On Moral Grounds: The Search for a Practical Morality*. New York: New York University Press.

Haidt, J. (2012) *The Righteous Mind: Why Good People Are Divided by Politics and Religion*. New York: Vintage Books.

Hammack, P.L. (2008) Narrative and the cultural psychology of identity. *Personality and Social Psychology Review* 12 (3): 222–247.

Hardcastle, V.G. ed. (2002) *Where Biology Meets Psychology: Philosophical Essays*. Cambridge, MA: Bradford/MIT Press.

Heberlein, T.A. (2012) *Navigating Environmental Attitudes*. New York: Oxford University Press.

Heeren, A., Karns, G., Bruskotter, J. et al. (2017) Expert judgment and uncertainty regarding the protection of imperiled species. *Conservation Biology* 31 (3): 657–665.

Heerwagen, J. & Orians, G. (1993). Humans, habitats, and aesthetics. In: *The Biophilia Hypothesis* (ed. S. Kellert & E.O. Wilson), 138–172. Washington, DC: Island Press.

Henrich, J., Heine, S. & Norenzayan, A. (2010) The weirdest people in the world? *Behavioral and Brain Sciences* 33: 61–135.

Hoog, N., Stroebe, W. & de Wit, J.B.F. (2005) The impact of fear appeals on processing and acceptance of action recommendations. *Personality and Social Psychology Bulletin* 31: 24–33.

Inglehart, R. (1995) Public support for environmental protection: objective problems and subjective values in 43 societies. *PS: Political Science* 15: 57–71.

Insko, C., Pinkley, R., Hoyle, R. et. al. (1987) Individual versus group discontinuity: the role of intergroup contact. *Journal of Experimental Social Psychology* 23: 250–267.

Johansson, M., Rahm, J. & Gyllin, M. (2013) Landowners' participation in biodiversity conservation examined through the Value-Belief-Norm Theory. *Landscape Research* 38 (3): 295–311.

Kahan, D. (2010) Fixing the communications failure. *Nature* 463: 296–297.

Kahn, P.H., Jr. (1999) *The Human Relationship with Nature*. Cambridge, MA: MIT Press.

Kahn, P.H., Jr. (2002). Children's affiliations with nature: structure, development and the problem of environmental generational amnesia. In: *Children and Nature: Psychological, Sociocultural and Evolutionary Investigations* (ed. P.H. Kahn Jr. & S. Kellert), 93–116. Cambridge, MA: MIT Press.

Kahn, P.H., Jr. & Hasbach, P.H. (2012) *Ecopsychology: Science, Totems and the Technological Species*. Cambridge, MA: MIT Press.

Kahneman, D. (2011) *Thinking, Fast and Slow*. New York: Farrar, Straus & Giroux.

Kaplan, R. & Kaplan, S. (1989) *The Experience of Nature: A Psychological Perspective*. Cambridge, UK: Cambridge University Press.

Kaplan, S. (1995) The restorative benefits of nature: toward an integrative framework. *Journal of Environmental Psychology* 15: 169–182.

Kim, M. (2002) *Non-Western Perspectives on Human Communication*. Thousand Oaks, CA: Sage.

Kitayama, S. & Markus, H.R. eds. (1994) *Emotion and Culture: Empirical Studies of Mutual Influence*. Washington, DC: American Psychological Association.

Lave, J. & Wenger, E. (1991) *Situated Learning: Legitimate Peripheral Participation*. Cambridge, UK: Cambridge of University Press.

Lazarus, R.S. (1991) *Emotion and Adaptation*. New York: Oxford University Press.

Lewin, K. (1951) *Field Theory in Social Science*. New York: Harper.

Lopez, S.J. ed. (2020) *Oxford Handbook of Positive Psychology*, 3rd ed. New York: Oxford University Press.

Luedicke, M., Thompson, C. & Giesler, M. (2009) Consumer identity work as moral protagonism: how myth and ideology animate a brand-mediated moral conflict. *Journal of Consumer Research* 36: 1016–1032.

Madden, F. & McQuinn, B. (2014) Conservation's blind spot: the case for conflict transformation in wildlife conservation. *Biological Conservation* 178: 97–106.

Manfredo, M. (2008) *Who Cares About Wildlife?* New York: Springer.

Manfredo, M., Bruckotter, J., Teel, T. et al. (2017) Why social values cannot be changed for the sake of conservation. *Conservation Biology* 31 (4): 772–780.

Manfredo, M., Teel, T., Sullivan, L. et al. (2017) Values, trust, and cultural backlash in conservation governance: the case of wildlife management in the United States. *Biological Conservation* 214: 303–311.

Manning, C. & Clayton, S. (2018). Threats to mental health and wellbeing associated with climate change. In: *Psychology and Climate Change* (ed. S. Clayton & C. Manning), 217–244. Cambridge, MA: Academic Press.

Markowitz, E.M., Slovic, P., Västfjäll, D. et al. (2013) Compassion fade and the challenge of environmental conservation. *Judgment and Decision Making* 8 (4): 397–406.

Maslow, A. (1943) A theory of human motivation. *Psychological Review* 50 (4): 370–396.

Mayer, F.S. & Frantz, C.M. (2004) The connectedness to nature scale: a measure of individuals' feeling in community with nature. *Journal of Environmental Psychology* 24: 503–515.

McKenzie-Mohr, D. & Smith, W. (1999) *Fostering Sustainable Behavior: An Introduction to Community-based Social Marketing*. Gabriola Island, BC, Canada: New Society.

Medin, D.L. & Bang, M. (2014) *Who's Asking? Native Science, Western Science and Science Education*. Cambridge, MA: MIT Press.

Milfont, T. (2012). Cultural differences in environmental engagement. In: *Oxford Handbook of Environmental and Conservation Psychology* (ed. S. Clayton), 181–200. New York: Oxford University Press.

Moore, F.C., Obradovich, N., Lehner, F. et al. (2019) Rapidly declining remarkability of temperature anomalies may obscure public perception of climate change. *Proceedings of the National Academy of Sciences USA* 116 (11): 4905–4910.

Myers, O.E., Jr., Saunders, C. & Birjulin, A. (2004) Emotional dimensions of watching zoo animals: an experience sampling study building on insights from psychology. *Curator* 47 (3): 299–321.

Neisser, U. (1967) *Cognitive Psychology*. New York: Appleton-Century-Crofts.

Nisbet, E.K., Zelenski, J.M. & Murphy, S.A. (2009) The nature relatedness scale: linking individuals' connection with nature to environmental concern and behavior. *Environment and Behavior* 41 (5): 715–740.

OceanOptimism (2017) Stories. http://www.oceanoptimism.org/stories (accessed September 20, 2021).

Opotow, S. (1994) Predicting protection: scope of justice and the natural world. *Journal of Social Issues* 50: 49–64.

Opotow, S. & Brook, A. (2003). Identity and exclusion in rangeland conflict. In: *Identity and the Natural Environment: The Psychological Significance of Nature* (ed. S. Clayton & S. Opotow), 249–272. Cambridge, MA: MIT Press.

Opotow, S. & Weiss, L. (2000) Denial and the process of moral exclusion in environmental conflict. *Journal of Social Issues* 56 (3): 475–490.

Pailler, S., Wagner, J., McPeak, J. et al. (2009) Identifying conservation opportunities among Malinké bushmeat hunters of Guinea, West Africa. *Human Ecology* 37: 761–774.

Papworth, S.K., Rist, J., Coad, L. et al. (2009) Evidence for the shifting baseline syndrome in conservation. *Conservation Letters* 2 (2): 93–100.

Pettigrew, T.F. & Troop, L.R. (2006) A meta-analytic test of intergroup contact theory. *Journal of Personality and Social Psychology* 90 (5): 751–783.

Petty, R.E. & Cacioppo, J.T. (1981) *Attitudes and Persuasion: Classic and Contemporary Approaches*. Dubuque, IA: W.C. Brown Publishers.

Petty, R.E., Haugvedt, C.P. & Smith, S.M. (1995). Elaboration as a determinant of attitude strength: creating attitudes that are persistent, resistant, and predictive of behavior. In: *Attitude Strength: Antecedents and Consequences* (ed. R.E. Petty & J.A. Krosnick), 93–130. Mahwah, NJ: Lawrence Erlbaum.

Pidgeon, N. (2012) Public understanding of, and attitudes to, climate change: UK and international perspectives and policy. *Climate Policy* 12: S85–S106.

Ranger, S., Kenter, J.O., Bryce, R. et al. (2016) Forming shared values in conservation management: an interpretive-deliberative-democratic approach to including community voices. *Ecosystem Services* 21 (B): 344–357.

Ratliff, K.A., Howell, J.L. & Redford, L. (2017) Attitudes toward the prototypical environmentalist predict environmentally friendly behavior. *Journal of Environmental Psychology* 51: 132–140.

Redmond, I., Aldred, T., Jedamzik, K. et al. (2006) *Recipes for Survival: Controlling the Bushmeat Trade*. London, UK: Ape Alliance & World Society for the Protection of Animals.

Rode, J., Gomez-Baggethun, E. & Krause, T. (2015) Motivation crowding by economic incentives in conservation policy: a review of the empirical evidence. *Ecological Economics* 117 (C): 270–282.

Ryan, R.M. & Deci, E.L. (2017) *Self-Determination Theory: Basic Psychological Needs in Motivation, Development, and Wellness*. New York: Guilford Publishing.

Saunders, C. (2003) The emerging field of conservation psychology. *Human Ecology Review* 10: 137–149.

Schultz, P.W. (2001) Assessing the structure of environmental concern: concern for the self, other people, and the biosphere. *Journal of Environmental Psychology* 21: 327–339.

Schultz, P.W., Nolan, J.M., Cialdini, R.B. et al. (2007) The constructive, destructive and reconstructive power of social norms. *Psychological Science* 18: 429–434.

Schultz, P.W., Shriver, C., Tabanico, J.J. et al. (2004) Implicit connections with nature. *Journal of Environmental Psychology* 24: 31–42.

Schultz, P.W. & Zelezny, L. (2003) Reframing environmental messages to be congruent with American values. *Human Ecology Review* 10 (2): 126–136.

Schwartz, S.H. (1992) Universals in the content and structure of values: theory and empirical tests in 20 countries. In: *Advances in Experimental Social Psychology*, vol. 25 (ed. M. Zanna), 1–65. New York: Academic Press.

Schwartz, S.H., Cieciuch, J., Vecchione, M. et al. (2012) Refining the theory of basic individual values. *Journal of Personality and Social Psychology* 103 (4): 663–688.

Shanahan, D.F., Fuller, R.A., Bush, R. et al. (2015) The health benefits of urban nature: how much do we need? *Bioscience* 65 (5): 476–485.

Sherif, M. (1966) *In Common Predicament: Social Psychology of Inter-Group Conflict and Cooperation*. Boston, MA: Houghton-Mifflin.

Shweder, R.A. (1991) *Thinking through Cultures: Expeditions in Cultural Psychology*. Cambridge, MA: Harvard University Press.

Soga, M. & Gaston, K.J. (2018) Shifting baseline syndrome: causes, consequences and implications. *Frontiers in Ecology and the Environment* 16 (4): 222–230.

Steg, L., Willem Bolderdijk, J., Keizer, K. et al. (2014) An integrated framework for encouraging pro-environmental behaviour: the role of values, situational factors and goals. *Journal of Environmental Psychology* 38: 104–115.

Sterling, E.J., Betley, E., Sigouin, A. et al. (2017) Assessing the evidence for stakeholder engagement in biodiversity conservation. *Biological Conservation* 209: 159–171.

Stern, P. (2000) Toward a coherent theory of environmentally significant behavior. *Journal of Social Issues* 56: 407–424.

Sunstein, C. (2019) *Conformity: The Power of Social Influences*. New York: New York University Press.

Swaisgood, R.R. & Sheppard, J.K. (2010) The culture of conservation biologists: show me the hope! *Bioscience* 60 (8): 626–630.

Tam, K.-P. (2013) Concepts and measures related to connection to nature: similarities and differences. *Journal of Environmental Psychology* 34: 64–78.

Ulrich, R. (1993). Biophilia, biophobia, and natural landscapes. In: *The Biophilia Hypothesis* (ed. S. Kellert & E.O. Wilson), 73–137. Washington, DC: Island Press.

UN CBD (UN Convention on Biological Diversity) (2017) Peace parks. https://www.cbd.int/peace/about/peace-parks (accessed July 31, 2021).

Ward, C., Holmes, G. & Stringer, L. (2018) Perceived barriers to and drivers of community participation in protected-area governance. *Conservation Biology* 32 (2): 437–446.

Weber, E.U. (2017) Breaking cognitive barriers to a sustainable future. *Nature Human Behavior* 1: 0013.

Weber, E.U. & Johnson, E.J. (2009) Mindful judgment and decision making. *Annual Reviews of Psychology* 60: 53–85.

Weinstein, N., Przybylski, A. & Ryan, R. (2009) Can nature make us more caring? Effects of immersion in nature on intrinsic aspirations and generosity. *Personality and Social Psychology Bulletin* 35 (10): 1315–1329.

Wells, N. & Lekies, K.S. (2006) Nature and the life course: pathways from childhood nature experiences to adult environmentalism. *Children, Youth, and Environments* 16 (1): 1–24.

Werner, C.M. (2003) Changing homeowners' use of toxic household products: a transactional approach. *Journal of Environmental Psychology* 23: 33–45.

Werner, C.M., Brown, B.B. & Altman, I. (2002). Transactionally oriented research: examples and strategies. In: *Handbook of Environmental Psychology* (ed. R. Bechtel & A. Churchman), 203–221. New York: Wiley.

White, R., Fischer, A., Marshall, K. et al. (2009) Developing an integrated conceptual framework to understand biodiversity conflicts. *Land Use Policy* 26: 242–253.

Wilhelm-Rechmann, A. & Cowling, R. (2008). Social marketing as a tool for implementation in complex social-ecological systems. In: *Exploring Sustainability Science: A Southern African Perspective* (ed. M. Burns & A. Weaver), 179–204. Stellenbosch, South Africa: SUN Press.

Wilhelm-Rechmann, A., Cowling, R.M. & Difford, M. (2013) Using social marketing concepts to promote the integration of systematic conservation plans in land-use planning in South Africa. *Oryx* 48 (1): 71–79.

Wilson, E.O. (1984) *Biophilia*. Cambridge, MA: Harvard University Press.

Zinn, H.Z. & Shen, X.S. (2007) Wildlife value orientations in China. *Human Dimensions of Wildlife* 12 (5): 331–338.

# 8

# Sociology and Conservation

*Jennifer Swanson, Steven R. Brechin, and J. Timmons Roberts*

## 8.1 Defining Sociology

### 8.1.1 Major Themes and Questions in Sociology

In the late 1950s, the well-known sociologist C. Wright Mills (1959) advocated that sociologists use what he called a "sociological imagination"—the ability to understand how an individual's life and personal circumstances are influenced by larger social and historical forces. Mills argued that doing so requires asking three kinds of questions about society. First, it is important to understand the structure of human society, investigating its various components and how these components are organized. Second, society should be understood in relation to history to gain a sense of how social change occurs. Third, it is critical to examine who has power in society, how they acquired it, and the ways this power is contested or accepted.

Broadly speaking, these three major areas of inquiry still animate the field of sociology. Contemporary sociologists try to understand how various social formations emerge, evolve, and influence individual and collective experience and how social arrangements create inequalities or influence other social outcomes. Sociologists also study how members of groups assign collective meaning to objects, events, and situations; how groups interact and take collective action; and how social forces at a variety of scales influence people's lives. Finally, sociologists remain concerned about power and inequality and how social processes can either privilege or marginalize different groups.

Since the discipline's inception, sociologists have developed a robust arsenal of methods to study social life. Some sociologists study society using **quantitative** methods—statistical analyses to discern trends and differences in the attitudes, behavior, or composition of society. Quantitative methods can be used with both large and small datasets, and many possible approaches exist, including testing hypotheses through surveys and questionnaires, using computational techniques to identify trends or correlations in data or to see how particular variables change over time, and quantitative modeling. For example, sociologists might conduct a formal survey of a cross-section of a population to understand views about a targeted issue. Or they might use a large dataset, like a national census, to

*Conservation Social Science: Understanding People, Conserving Biodiversity*, First Edition.
Edited by Daniel C. Miller, Ivan R. Scales, and Michael B. Mascia.

analyze the relationships between particular variables, such as comparing levels of education and income in a given year or looking at changes in population density over time.

Sociologists also use **qualitative** methods to generate thick descriptions of social life that can lead to insights about how people perceive or act in the world. Qualitative researchers often study social phenomena directly, gathering empirical evidence by conducting interviews and **participant observation** (see Chapter 3 for more on participant observation). Using an **ethnographic** approach, for example, researchers might directly observe the behavior, language, and dynamics of a given population. Or, they might conduct in-depth individual interviews or conduct focus groups to identify patterns in the ways people assign meaning to the events, objects, and relationships in their lives and to understand how they act upon these ideas. Other approaches include analyzing contemporary texts, photos, or visual artifacts, collecting oral histories, exploring how individuals interact within the context of specific institutions, or carrying out archival research and historiographical analysis.

Sociologists have a long history of refining both quantitative and qualitative approaches and sometimes combine them to better explain complex phenomena. Whatever the particular approach, sociology can be described as the systematic study of human society, a practice that uses analytical methods to understand patterns and differences in social life.

### 8.1.2   Sociology and Conservation

Natural resource and environmental management issues emerged as important topics within sociology in the 1970s. Work within rural sociology focuses on the social, economic, and political consequences of natural resource management regimes on individuals and communities (Field & Burch 1991; Rudel 1998). The study of human ecology, which arose from pioneering sociologists at the University of Chicago, explores how humans adapt to changing social and environmental conditions. Scholars in this field originally focused on the forces shaping urban life, but the field has evolved to incorporate broader analyses of the interactions between humans, the natural environment, **biodiversity**, technology, and social organizations (Duncan 1959; Humphrey et al. 2002; Tàbara & Pahl-Wostl 2007).

The subfields of environmental sociology and organizational sociology have addressed **biodiversity conservation** quite directly (see Box 8.1). Environmental sociologists explore many aspects of human–environment interrelationships and analyze the rise of environmentalism and conservation efforts as a broad social phenomenon (Shabecoff 2003; Lewis 2016; Farrell 2020). Organizational sociologists investigate the ongoing practices of conservation organizations, looking how they function and how they can shape social and economic realities (Brechin et al. 2002, 2003; K. MacDonald 2010). So, albeit slowly, biodiversity conservation has emerged as a critical field of study within sociology over the last few decades (York et al. 2003; Hoffman 2004; Clausen & York 2008; Besek & McGee 2014; Dunlap & Brulle 2015; Besek & York 2018). That said, there are many subdisciplines in sociology that remain largely untapped and offer knowledge, insights, and approaches relevant to biodiversity conservation, including subfields focused on race, class, gender, development, inequality, migration as well as historical sociology, economic sociology, and political sociology, to name a few.

---

**Box 8.1   Methods: Using sociological methods to understand environmental attitudes**

---

Sociologists collect and analyze data about social phenomena using both quantitative and qualitative methods, sometimes mixing the two in order to benefit from the advantages of each. Survey research is one of the most common means of collecting and analyzing quantitative data. When conducting surveys, respondents are chosen deliberately using sampling techniques to ensure that survey subjects represent a larger population, such as a community or nation-state (Fowler 2013). After respondents provide answers to the survey questions, their answers are coded into numbers that can be analyzed using statistics, which can help identify patterns across large sets of data.

In the early 1990s, for example, sociologists analyzed cross-national survey data gathered by the Gallup polling organization to better understand people's attitudes about the environment in both rich and poor nations. At the time, researchers assumed that relatively rich individuals and countries would more likely prioritize concern for the environment, making it a luxury good that people worried about only after their basic needs were met. The data showed that concern for the environment was a global phenomenon, however, since respondents from both wealthy countries and poorer countries demonstrated strong concern for the environment (Dunlap et al. 1993; Dunlap 1994). In fact, respondents from poorer countries were, on average, statistically significantly more concerned about the environment than respondents from rich countries, something important to know when designing conservation strategies.

Unlike quantitative methods, where the goal is to gather a sample of respondents to enable conclusions that can be generalized to a larger population, qualitative methods aim to gain a deeper understanding of phenomena relevant to a particular population. Thus, qualitative methodologies involve closely observing and analyzing the attitudes and actions of individuals or groups within a given population. Interviews, participant observation, and content analysis are just a few of the main approaches that researchers use to gather qualitative data. Once such data have been collected, what individuals or groups have said, done, or written can be analyzed, using different techniques, including coding the data to discover emergent themes.

To illustrate, researchers seeking to understand the attitudes of impoverished rural communities living around national parks and other protected areas of Costa Rica and Honduras used semi-structured interviews (Schelhas & Pfeffer 2009). They found that rural farmers (Figure 8.1) were aware of global **discourses** about the importance of biological diversity and the need for nature protection efforts. These rural farmers connected the global messaging with values, meanings, and material needs relevant to themselves and their local communities. The interviews revealed that while there is a global transfer of ideas, people also mediate global discourses through their own experiences so such discourses come to have specific meanings for them. Being able to better understand the variation in people's views about conservation goals and projects in this way is critical for biodiversity conservation practitioners.

**Figure 8.1** Left: A resident of San Jose de Los Planes, a buffer zone community of Cerro Azul Meambar National Park, Honduras, stands at the entrance to the community in 2005. Right: A community park ranger (right) stands with a landowner and his work crew in the official buffer zone of the Cerro Azul Meambar National Park, Honduras, in 2005. Courtesy of John Schelhas.

## 8.2 A Brief History of Sociology

Sociology emerged as a field of study during the late nineteenth century in Europe and the USA. Before this time, sociology was not differentiated from other social sciences such as history, economics, and political science. From the mid-nineteenth century into the early twentieth century, however, scholars and intellectuals such as Auguste Comte, Karl Marx, Herbert Spencer, Thorstein Veblen, Max Weber, Emile Durkheim, Jane Addams, and W.E.B. DuBois began formally analyzing society to better understand the causes and effects of the rapid social changes afoot. This early "science of society" was often embedded in an underlying belief that careful analysis of society could prompt improvements in social and political life—a way of thinking that is still promoted by many within the discipline today (Burawoy 2005).

By the end of the nineteenth century, there were concerted efforts to legitimize the fledgling field of study into a formal and professional discipline. The first formal sociology department was established at the University of Chicago in 1892, and schools across Europe soon followed. Because sociology developed parallel in a variety of countries, including Germany, France, England, and the United States, the intellectual trajectories of sociology varied among countries. That said, a few key areas of study characterized the discipline as a whole. Scholars in this period attempted to analyze the distinctly *social* aspects of society, including the influences that social institutions such as the state, the economy, and religion had on individual actions. Early sociologists also tried to understand systems of power and the ways they shape and constrain individual choices, opportunities, and life chances.

From the early decades of the twentieth century until roughly the 1960s sociologists advanced a variety of theoretical and empirical approaches within the discipline. Some, heavily influenced by the ideas of Karl Marx, focused on how social institutions created systemic inequality between members of society, giving some more wealth, opportunities, or power than others. Other researchers conducted ethnographic fieldwork, focusing on the analysis of human behaviors and interactions within specific communities and subcultures. By mid-century, largely in reaction to the chaos and social strife caused by two world wars, scholars also began to theorize the causes of social conflict and harmony, as well as the nature of deviance (Munch 1994a). Advances in statistical analysis at this time also allowed scholars to study and characterize significant trends in society using quantitative data.

From the end of the 1950s into the 1970s, sociologists began to question both existing methodological approaches and dominant theoretical paradigms in sociology. Emerging revolutions, **decolonization**, and growing social movements and political change world-wide prompted new questions and new ways of thinking. In response, the discipline confronted a host of new questions, such as whether Western notions of progress and **development** adequately described varied national experiences; whether one group could legitimately speak on behalf of another; and whether existing sociological theory adequately explained inequality and power relations. As more diverse social groups fought for recognition and rights during this time, sociologists began to ask whether it was even possible to develop overarching sociological theories that hold true across all social contexts, and started shifting their focus to middle-range theories that reflected particular contexts (Munch 1994b). During this period, entirely new subfields arose in response to social change, including feminist approaches to sociology and environmental sociology.

From the 1980s, sociologists have continued to advance sociological analysis in new ways, exploring such issues as the nature of power and knowledge, the construction of nationalism, the boundaries between humans and machines, the intersectional nature of social location, the relationship of humans to the environment, and other topics relevant to a changing contemporary world. The number of subfields within the discipline has grown tremendously in recent decades, as scholars pursue new intellectual pursuits and objects of study (see Table 8.1). Sociology remains the systematic study of human society but is now characterized by a range of specialties and significant diversity in both methods and theoretical paradigms.

**Table 8.1** Subfields in sociology.

| Subfields | |
| --- | --- |
| Aging and the Life Course | International Migration |
| Alcohol, Drugs and Tobacco | Inequality, Poverty and Mobility |
| Altruism, Morality and Social Solidarity | Labor and Labor Movements |
| Animals and Society | Latino/a Sociology |
| Asia and Asian America | Sociology of Law |
| Sociology of Body and Embodiment | Marxist Sociology |
| Children and Youth | Mathematical Sociology |
| Collective Behavior and Social Movements | Medical Sociology |
| Communication and Information Technologies | Sociology of Mental Health |
| Community and Urban Sociology | Methodology |
| Comparative and Historical Sociology | Organizations, Occupations, and Work |

*(Continued)*

**Table 8.1** (Continued)

| Subfields | |
| --- | --- |
| Sociology of Consumers and Consumption | Peace, War, and Social Conflict |
| Crime, Law, and Deviance | Political Economy of the World-System |
| Sociology of Culture | Political Sociology |
| Sociology of Development | Sociology of Population |
| Disability and Society | Race, Gender, and Class |
| Economic Sociology | Racial and Ethnic Minorities |
| Education | Rationality and Society |
| Sociology of Emotions | Sociology of Religion |
| Environmental Sociology | Science, Knowledge, and Technology |
| Ethnomethodology and Conversation Analysis | Sociology of Sex and Gender |
| Evolution, Biology, and Society | Sociology of Sexualities |
| Family | Social Psychology |
| Global and Transnational Sociology | Sociological Practice and Public Sociology |
| History of Sociology | Teaching and Learning in Sociology |
| Human Rights | Theory |

*Source:* Data from https://www.asanet.org/communities-sections/sections/current-sections.

## 8.3 Concepts, Approaches, and Areas of Inquiry in Sociology

The practice of biodiversity conservation can be viewed from many different sociological vantage points. One way to consider conservation is to see it as a process that *engages a multitude of stakeholders*, with different priorities and values, as well as different cultural and symbolic understandings of places and resources. A second way to consider conservation is to think of it as *a process of social negotiation*, since it is a process that requires collective human action. Third, conservation can be viewed as *a process that occurs in relation to a variety of macroscale social forces*, which can impact conservation planning. With these different vantage points in mind, we highlight six major areas of study in sociology that offer key concepts, perspectives, and approaches relevant to the practice of biodiversity conservation: (i) social structures and patterned inequality; (ii) symbolic interaction; (iii) social connections; (iv) collective social action; (v) population; and (vi) global social forces.

### 8.3.1 Social Structures and Patterned Inequality

Societies are organized in distinct, culturally specific ways that can generate recurrent patterns of social interaction. Sociologists often refer to these patterns as social structures since they influence or "structure" people's behavior, experiences, and even life chances. In addition, social location—how someone is positioned in terms of attributes such as race, ethnicity, gender, age, sexuality, nationality, ability, or other social markers—can also affect or direct social relations in consistent ways over time. While these patterns are never entirely permanent and can be challenged and changed, they can be persistent and influential, generating inequality and impacting the degree of power that individuals and groups have in society. Understanding these power dynamics is critical when planning and implementing conservation projects.

### 8.3.1.1 Social Institutions

Social institutions are sets of social positions knitted together over time into recognizable and relatively stable assemblages. Each position within a social institution has roles, rights, and obligations associated with it, and social institutions as a whole have formal and informal **norms** and values that guide the behavior of members. Some examples of social institutions include families, government, the economy, the media, the military, and systems of education and religion. Each contains recognizable social positions, such as mother, husband, professor, religious leader, worker, soldier, or politician. Each position is associated with a set of expected roles, such as caregiving, teaching, or policy making. More broadly, particular social norms and values embedded within each institution guide the behavior of members, such as "freedom of the press" in the US media or "due process of law" within the American judicial system. Social institutions endure as organizing structures over time, even as particular individuals within them come and go. Despite their ubiquity throughout the world, however, social institutions are not generic. Social institutions are culturally specific, and the expected roles of members within any given social institution can differ significantly across different societies.

Sociologists have developed many methods for understanding how social institutions function in society (see Chapters 4 and 6 for other ways the term institution is used in the social sciences). Perhaps the most famous analysis of a social institution was Karl Marx's examination of the economy under **capitalism**. Marx noted that in a capitalist economy, owners (which he referred to as the *bourgeoisie*) and workers (which he referred to as the *proletariat*) were part of the same economic system but had different positions within it, influencing their relative ability to accumulate wealth. He further observed that these disparities create distinct social classes in society, leading to unequal degrees of social power and an economic system that perpetuates and reproduces class difference over time. Sociologists after Marx continued to analyze class stratification, investigating the ways that power, prestige, and practices differ between classes and mark class identification (Weber et al. [1948] 1998; Marx 1976; Bourdieu 1984). Since then, contemporary sociologists have developed a measure called socioeconomic status (SES) to understand social class. SES is a composite measure that integrates indicators about income, wealth, levels of education, and occupation. Similarly, sociologists use a variety of approaches to evaluate, analyze, and understand many other social institutions.

Many social institutions are relevant to biodiversity conservation, since negotiations about nature and the environment, work and labor, consumption and resources, and the use and management of natural resources take place within a variety of arenas, including households, communities, workplaces, and government. Transforming institutions so they support conservation goals is a critical step in moving conservation from an agenda to a common practice. For example, the economy is a key social institution relevant to conservation success. Not only do class stratification and socioeconomic status influence how people use and depend on resources, but biodiversity conservation efforts can also affect people's livelihoods. Furthermore, socioeconomic status can affect whether people have the access and ability to participate in the process of conservation planning.

Other social institutions are also critical to the success of biodiversity conservation. For example, in a landmark study, researchers showed how a variety of social institutions from law enforcement to the courts play a key role in effectively deterring illegal wildlife trade. Prior to this study, many conservation planners assumed that increased surveillance would prevent wildlife crimes. This study, however, likened the process of enforcement to links in a chain. Looking beyond detection efforts, they assessed the likelihood of detection, arrest, prosecution, conviction, and penalties using qualitative and quantitative data in three countries, revealing how weak spots throughout the chain of enforcement can cause enforcement efforts to fail. Since the location of the weak links differed from country to country, the study underscored how targeted, site-specific conservation investments in a variety of social institutions can potentially do more good than simply ramping up patrols and surveillance activities (Akella & Cannon 2013).

The media—a key vehicle for spreading new ideas—is another important social institution that can be harnessed for conservation efforts. Recognizing this, conservation practitioners often try to change people's understanding of biodiversity and the necessity of its conservation through education and communication campaigns (Veríssimo et al. 2018). Religion is also a social institution that can be an effective lever for conservation. Box 8.2 provides an example of the critical role Buddhist monks play in forest conservation in Cambodia. These examples demonstrate how conservation efforts on the ground inherently involve social institutions, since people's behaviors occur within the contexts of institutions and because institutions offer a platform to shape behavior change.

---

**Box 8.2  Applications: Religion and conservation in Cambodia and beyond**

In 2000, a well-respected Cambodian monk named Venerable Bun Saluth moved back to his home region in Oddar Meanchey as the country struggled to rebuild after decades of civil war, poverty, and government corruption. Ven. Saluth was shocked by the degree of forest destruction in the region, where encroachment, land concessions, and rampant logging of high-value timber were increasingly taking a toll. From 2002 to 2006, the province's annual deforestation rate was 2.1%, and the few conservation organizations working in the country were struggling to gain ground given the complex social context.

In the absence of any conservation efforts, Ven. Saluth took the initiative to protect the area. On February 7, 2002, he was granted formal permission to oversee an 18,261-hectare forest. In 2008, it was legally designated as a "community forest," referred to as the Monks Community Forest (MCF), or Sorng Rokavorn in the local language. This legal designation indicates a class of forest protection where forests are actively managed to prevent illegal activity but still provide communities with legal access for sustainable livelihood activities. Under the agreement, Ven. Saluth had the authority to establish the rules governing resource use and extraction within the MCF. Subsequently, logging and hunting were banned, but community members were still allowed to harvest non-timber forest products and fish using traditional methods. To enforce the new management rules, Ven. Saluth enlisted monks from the Samraong

**Box 8.2 (Continued)**

Pagoda to patrol the forest by foot and to engage in outreach and awareness efforts with local communities as well as provincial, district, and other government departments.

In Oddar Meanchey, the main threat to forests was illegal logging by the military and others backed by businessmen and government officials. Because many of the loggers were former Khmer Rouge soldiers, initial protection efforts were met with angry and sometimes armed confrontation. The monks persisted, however, peacefully confronting those committing forest crimes and working to raise awareness about the new protected area. Over time, they built a protection system based on Buddhist principles that successfully stopped forest crime and secured access rights for local villagers. In 2018, the Cambodian Ministry of Environment declared MCF a wildlife sanctuary and expanded the area of protection to over 30,000 hectares—a testament to their success.

Buddhist monks may be surprising protectors of forests (Figure 8.2), especially to those who advocate that fences, guns, and guards are the most effective means of conservation. However, the monks' efforts have been effective because they are widely respected in Cambodian society, and the management rules they established are consistent with and connected to deeply held Buddhist beliefs and values. They were able to create a conservation ethos within the villages surrounding MCF, and villagers joined in with the monks' patrols. This kind of engaged Buddhism is a relatively new phenomenon in Cambodia and its application to environmental protection has only surfaced in the last few decades (Elkin 2009).

Worldwide, faith groups are poised to play a critical role protecting habitats, wildlife, and biodiversity. Not only are faith groups key players in education and community issues, they own approximately 7%–8% of habitable land and 5% of commercial forests around the world, and many have significant financial assets and investments (Palmer & Finlay 2003). The growing recognition that religious institutions have a significant role to play in conservation efforts has spurred increased interest and investment in this sector.

*Source*: Adapted from Elkin 2009, with permission

**Figure 8.2** Left: Buddhist monks in Cambodia engage in a tree ordination ceremony, sanctifying the tree as if it were a monk, with the goal of protecting it and the forest from illegal logging. Right: Buddhist monks and local villagers patrol the protected forest to combat poaching and illegal logging in 2015. Their effective protection of the forest demonstrates the power of religion in conservation. *Source:* Monks Community Forest / Chantal Elkin.

Social institutions can be the source of challenges as well as opportunities for conservation, so understanding the specific dynamics of existing social institutions in a given setting should be a critical focus in conservation planning efforts. Even if a particular social institution at first appears to be separate from or unrelated to biodiversity conservation goals, it might be a key feature on the social landscape that could either block or advance conservation efforts. Sociology offers approaches for identifying, analyzing, and mapping social institutions, as does the development sector, which has a long history of working through existing social institutions to affect change (Newing 2011).

#### 8.3.1.2 Social Location

Differences between people are often celebrated as forms of diversity and can serve as important aspects of identity that individuals cherish. However, differences also can be used to confer privileges and benefits to some while denying them to others, creating inequality in society. Figure 8.3 depicts some of the social markers that can shape people's experiences, including class, gender, race, ethnicity, ability, and age, among others. Some aspects of diversity and social location are internal and unchangeable, while others are based on life choices and life chances, evolving as individuals live their lives within various social structures. This complex and intersecting matrix of factors can generate patterns of

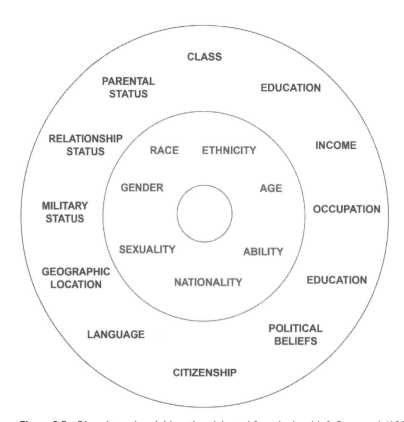

**Figure 8.3**   Diversity and social location. Adapted from Loden, M. & Rosener, J. (1991).

privilege and oppression, inclusion and exclusion, and visibility and invisibility that can be critical in conservation practice, since they can influence whose views are heard, whose interests are considered, and who has the legitimacy and authority to participate in conservation planning.

One example of social location is gender—the culturally specific roles, behaviors, and attributes that become associated with men and women. Gender is **socially constructed**, meaning that it is not predetermined by biology or nature but instead is created by society. People's ideas and assumptions about gender generate patterns of behavior and expectations that often structure social life as a whole.

Many feminist scholars analyze how ideas about gender stratify society. For example, gendered hiring practices and divisions of labor, as well as assumptions and expectations about gender roles can constrain and/or marginalize women in a variety of settings, from the household to the workplace and beyond. Also, the knowledge and lived experiences of women are sometimes discounted or marginalized because of the dominant position of males in society, despite the fact that women and men alike have only partial perspectives on social life. To acknowledge and address this unevenness, feminist scholars proposed a framework called "standpoint theory," which emphasizes the fact that women's understanding of the world should be recognized as a source of knowledge that is both relevant and different from what men might know or understand about the world (Harding 2004). While its origins are rooted in feminist scholarship, the approach also can be more generally deployed in recognition that all marginalized groups have particular, situated knowledge and experience that should not be overlooked.

Race and ethnicity are other examples of social location. Despite the common presumption that race is based on physical or biological characteristics, there is no scientific basis for distinguishing one race from another. Racial classifications might appear to be rooted in physiological characteristics, but in fact, societies designate racial categories, which in turn produce racial differences in culturally and historically specific ways. Racial categories have varied over time and change across locations, demonstrating how the meaning of race is not fixed and can be transformed and modified according to social conventions (Omi & Winant 1986). Ethnicity, a related concept, typically encompasses a shared suite of customs, beliefs, and practices that distinguish groups from one another through shared **cultures**, histories, and identities (Schelhas 2002).

In practice, race and ethnicity have real effects on people's lives. The rights and privileges granted to members of particular racial and ethnic categories vary across societies, and race and ethnicity influence power relations. Racial and ethnic discrimination and marginalization operate at both the individual level and the institutional level; they are perpetuated through everyday actions as well as through institutional policies and practices (Omi & Winant 1986). In the USA, for example, race relations and ethnic inequalities have led to different opportunities for whites, blacks, Native Americans, immigrants, and various ethnic groups throughout history. In ways that have shifted over time, these labels have determined who could hold property and accumulate wealth, who had access to favorable housing and lending, and who was granted the full rights of citizenship, including the right to vote (Adelman 2007). Similarly, race and ethnicity affect power and inequality in a variety of ways globally.

Social location can play a critical role in conservation planning and the effectiveness of conservation initiatives. To some degree, the importance of acknowledging social location is already underway. For example, efforts to recognize the voices and perspectives of women have been promoted for decades within conservation and environmental movements, with varying success. The role of women has been addressed in the tenets of many international treaties and agreements as well as in the organizational priorities of a variety of conservation and development organizations (Vencatesan 2008). Because of heightened recognition that women play key roles in natural resource management and have access to important knowledge about the environment, concerted efforts have been made to include women in conservation planning. Government agencies such as the US Agency for International Development (USAID) as well as nonprofits and nongovernment organizations (NGOs) have developed guidelines and policies to better account for women's experience.

Still, many conservation efforts overlook or ignore the issue of gender, sometimes even exacerbating women's marginal status within particular communities. In some cases, efforts to address gender in conservation planning are oversimplified, such as participatory planning efforts that involve women but do not effectively use the opportunity to gain a solid understanding of women's knowledge, their experience with natural systems, or the ways conservation efforts might impact their lives (Bandiaky 2008; Bandiaky-Badji 2011). Adequately addressing gender as a factor in conservation requires an acknowledgment of the importance of women's experience; the recognition of women's differing roles within local, national, and international institutions; attention to site-specific social dynamics; and the recognition of potentially uneven power relations as a result of gender in any given setting.

Similarly, race and ethnicity play a role in conservation-related policies and practices around the world and can affect conservation efforts. Conservation projects focused on issues such as land tenure, zoning, and resource management, for example, can correct or reinforce racial and ethnic marginalization, depending on whether projects include relevant groups in planning efforts and address how different groups use and value natural resources (Timko & Satterfield 2008). In a conservation project in the Vilcabamba region of Peru, for example, indigenous Machiguenga communities were interested in protecting indigenous land from oil extraction, but also saw participation in a conservation effort sponsored by a large international NGO as a way of confirming their right to participate in policy discussions (Roca 2004). Additionally, the development aspects of the project helped them advocate for health services they were legally entitled to, which also helped affirm their legitimacy as citizens. Effective participatory planning processes can help reveal these types of site-specific social dynamics (see Chapters 3 and 6 for more on participation).

Finally, the makeup of the conservation community itself can influence how projects are carried out. Recent research has demonstrated the lack of racial (as well as class and gender) diversity in all major US environmental and conservation organizations, offering strategies for overcoming these shortcomings (Taylor 2018). This is critical, since effective organizations, including conservation organizations, benefit from diverse members who can better reflect the real world and expand the pool of ideas.

## 8.3.2 Symbolic Interaction and the Construction of Meaning

While some sociologists focus on the structure of society in an effort to understand human experience, others emphasize human **agency** and the many subjective factors that influence how people live and act in the world. For example, a classic theoretical perspective in sociology called symbolic interactionism (SI) focuses on how the continuous process of social interaction affects human behavior (Blumer 1986). Scholars who work in the symbolic interactionist tradition study the ways people assign symbolic meaning to things, behaviors, and events and how they act based on those ascribed meanings.

One key aspect of this perspective is the recognition that the process of creating meaning is fundamentally social. People adopt and adapt the meanings of things as they encounter new people and situations, modifying their actions and their understanding of the world in response. A second important aspect of the SI framework centers on the idea that people actually understand themselves from a symbolic perspective as well, fashioning their identities in ways that shape behavior and understanding (Goffman 1956; Sandstrom et al. 2013).

### 8.3.2.1 Symbols, Language, and Interaction

The SI approach focuses on the processes through which people come to understand and act in the world. The perspective is rooted in three premises that distinguish it from other sociological approaches (Blumer 1986). First, the symbolic interactionist approach begins with the assumption that people's actions are based on the meaning they ascribe to the world. This includes the meanings attached to physical objects, locations, other people, other people's actions, institutions, values, as well as events and situations that arise throughout life. For example, someone might attach particular meaning to the location where he or she was born, the day he or she met a spouse, or a prized ambition such as being financially secure. These meanings influence people's resulting plans and actions. For SI scholars, understanding symbolic meaning is critical because it shapes resulting decisions, behaviors, and opinions.

A second working principle in SI is that meaning is not inherent in objects or things, nor is it generated solely from the individual's psyche or internal self. Instead, SI scholars assert that meaning is generated through social interaction. From this vantage point, people do not arrive to situations and simply act out their beliefs, as if social interaction is merely a venue for people to express personal ideas about things. Rather, the interaction between individuals and groups is an influential and productive site where people develop, align, and rethink their own ideas and behaviors in light of those around them. Finally, SI practitioners assert that people have agency and actively construct their world and actions using the filter of ongoing interactions and interpretations, rather than having their behavior entirely determined by external structures and circumstances.

The idea that people's sense of reality is dynamic and open to interpretation might seem like an abstract concept. However, this perspective has concrete and important implications for biodiversity conservation efforts because effective conservation is often a negotiation or collaboration among varying stakeholders, who in turn attach very different meanings to habitats, species, natural resources, landscapes, social issues, and decisions. For example, in a conservation project in North Sikkim, India, the state recognized particular areas as sacred groves, which was symbolically important for the Lepcha indigenous group. Like many groups throughout the world, these forests are intimately tied to their identities and ethnicity, and are sites where they assert their indigenous status. The forests hold deep meaning for them, and beyond that, the designation of the land as sacred groves reinforces their legitimacy as indigenous people separate from the larger Nepali majority (Arora 2006).

Since the practice of conservation intersects so directly with people's ideas about land, work, nature, development, health, family, heritage, history, and other deeply held concepts, there are countless other examples of how symbolic meaning operates in relation to conservation objectives throughout the world. The key for conservation planning is to understand and acknowledge the varying symbolic values that differing stakeholders attach to land, waterways, resources, species, or zoning decisions. Many participatory approaches have been developed to identify these issues, but established quantitative or qualitative methods in sociology also can be effectively employed to identify key themes during conservation planning (Newing 2011).

In a practical sense, even language is a salient type of symbol, since words stand in for objects, events, and situations and often communicate embedded or condensed meanings. Language is a powerful tool used to conceptualize and categorize the world. For example, IUCN designates a set of internationally recognized categories for classifying parks. Each label acts as a shorthand symbol that identifies the sets of activities that are allowed and prohibited in particular areas. The adoption of these international categories is an example of the dramatic way that naming and language can produce changes in both understanding and behavior and how symbolic meaning can be transferred across cultures (see Table 8.2 for the list of IUCN categories and the established meanings for each).

Similarly, names are used on a local level to designate what types of use or management are allowed in a given marine or terrestrial area. In this sense, naming is a significant act of power that legitimizes some meanings over others, a point that is critical for conservation practitioners to acknowledge and address. Rules and categories can validate and even legalize the values held by those invested in conservation, however these may not perfectly align with meanings held by other stakeholders. In western Madagascar, for example, researchers interested in the motivations for extreme logging tried to understand the meaning of this act in relation to local concepts of the landscape and viewpoints about culture and economics (Sandy 2006). They found that the local people used entirely different categories for designating types of land and had fundamentally different ideas about ownership and land management regimes than the external groups that were making the official rules about land-management practices.

In another example, conservation practitioners in Guyana were aware that indigenous communities in the Kanuku Mountains used and named land in culturally specific ways, so they engaged in a participatory community resource evaluation with 18 local communities when launching a larger conservation project. The exercise not only helped conservation practitioners understand local terms for landscape features in the area; more importantly, it illuminated how local areas and resources were relevant to the communities in practical, cultural, and symbolic ways. This acknowledgment, in turn, promoted a more meaningful and equitable dialogue between various stakeholders (Stone 2002).

**Table 8.2** IUCN Protected area categorization.

| Category | | Description |
| --- | --- | --- |
| Ia | Strict nature reserve | Protected areas that are strictly set aside to protect biodiversity and also possibly geological/geomorphological features, where human visitation, use and impacts are strictly controlled and limited to ensure protection of the conservation values. Such protected areas can serve as indispensable reference areas for scientific research and monitoring. |
| Ib | Wilderness area | Protected areas that are usually largely unmodified or slightly modified, retaining their natural character and influence, without permanent or significant human habitation, which are protected and managed so as to preserve their natural condition. |
| II | National park | Large natural or near natural areas set aside to protect large-scale ecological processes, along with the complement of species and ecosystems characteristic of the area, which also provide a foundation for environmentally and culturally compatible spiritual, scientific, educational, recreational and visitor opportunities. |
| III | Natural monument or feature | Protected areas set aside to protect a specific natural monument, which can be a landform, sea mount, submarine cavern, geological feature such as a cave or even a living feature such as an ancient grove. They are generally quite small protected areas and often have high visitor value. |
| IV | Habitat/species management area | Protected areas aiming to protect particular species or habitats and management reflects this priority. Many category IV protected areas will need regular, active interventions to address the requirements of particular species or to maintain habitats, but this is not a requirement of the category. |
| V | Protected landscape/seascape | A protected area where the interaction of people and nature over time has produced an area of distinct character with significant ecological, biological, cultural, and scenic value, and where safeguarding the integrity of this interaction is vital to protecting and sustaining the area and its associated nature conservation and other values. |
| VI | Protected area with sustainable use of natural resources | Protected areas that conserve ecosystems and habitats, together with associated cultural values and traditional natural resource management systems. They are generally large, with most of the area in a natural condition, where a proportion is under sustainable natural resource management and where low-level non-industrial use of natural resources compatible with nature conservation is seen as one of the main aims of the area. |

*Source:* IUCN 2021 / International Union for Conservation of Nature.

In practice, the existence of multiple perspectives about the meaning of land, waterways, resources, and species can create significant conflict about the appropriate use or management of a given area. These contested meanings can pose big challenges when designing conservation strategies. Effective conservation requires an understanding of the belief systems of all stakeholder groups and the resulting values they attach to particular places, behaviors, and traditions. Gaining this understanding is a useful way to identify common ground so conservation strategies are relevant and legitimate, since shared meanings can evolve and stakeholders can work to reconcile differing belief systems during a process of interaction. Conversely, neglecting these issues can impair conservation success in both the planning and implementation stages.

### 8.3.2.2  Self and Identity

A key tenet of the SI perspective centers around the fact that interactions and symbolic meanings influence a person's sense of identity and self. Classic thinkers from the SI tradition studied how people adopt differing roles throughout their lifetimes, fashioning themselves within changing social experiences and relationships (Goffman 1956; Mead 1967). People attach significant meanings to particular ideas about themselves and the ways they live and operate in the world. Contemporary SI scholars continue to emphasize that self-image is constructed through an ongoing process of dialogue with ourselves and others, noting that people's sense of self can evolve as they enter new situations or contexts. In addition, people hold multiple ideas about themselves at any given time, so the notion of a "sense of self" actually encapsulates a multitude of beliefs and values (Howard 2000; Wetherell 2009).

People's sense of identity can be a critical factor when designing effective conservation strategies. Those who identify as members of an indigenous group, for example, may view conservation objectives through a lens influenced by this identity, and conservation goals may or may not align with other political rights and agendas that might be important to them. That said, it is important for conservationists not to oversimplify identity and make erroneous assumptions about indigenous beliefs or their connections with the broader world. Instead, it is critical to understand the varied ways indigenous identities may operate across settings (Li 2000; Sawyer & Gomez 2012).

Similarly, conservationists should also try to understand how identity matters to the wide variety of stakeholders who influence conservation outcomes. This is essential to ensure all relevant groups are recognized and have a seat at the table to work out plans and differences. In a proposed project in the Philippines, for example, conservation practitioners designing a large-scale ecotourism venture assumed that creating new opportunities for employment would curtail dynamite and cyanide fishing, a main driver of reef destruction in the Palawan region. The proposed project centered on providing local fishers with an alternative income stream to offset their need to fish using destructive techniques. Interviews with fishers showed that villagers were indeed very interested in working in the emerging ecotourism sector and were enthusiastic about the potential new jobs and sources of income. However, when asked if having the new jobs would stop them from engaging in harmful fishing practices, they said no. Further investigation revealed that fishing was not simply a way to make money but was rooted in a deep sense of self and identity, something they would pursue regardless of other jobs or income streams available.

Designing effective conservation strategies in this context not only required an understanding of the symbolic value of fishing for these groups, but a clear strategy for addressing these motivations in project design. Sociological methods can help elucidate symbolic meaning and people's sense of identity, through surveys that are analyzed quantitatively and through ethnographic approaches such as interviews, oral histories, or observation (Newing 2011; see also Chapter 3).

### 8.3.3 Social Connections

Social networks and **social capital** are closely related concepts built on the basic acknowledgment that humans are social beings. People are members of families, neighborhoods, and communities. They are employed in organizations with colleagues, attend houses of worship, join groups that interest them, and participate in their communities. People create social worlds through continual engagement with those around them. On some level, individuals are defined by who they know and with whom they spend their time. And, as humans engage with each other, patterns emerge and resources of many kinds are exchanged. Social scientists, including sociologists, are interested in the types of connections humans have with others and the consequences these connections have on their lives.

#### 8.3.3.1 Social Networks

Networks exist in both human and natural systems and play an important role in many aspects of life (Castells 2004). Social networks are typically defined as sets of actors who are tied to one another through socially meaningful relationships. The actors can be individuals, groups, or organizations, and the networks can exist at various scales—micro, meso, and macro (Wasserman & Faust 1994; Borgatti et al. 2009). Rather than focusing exclusively on the attributes of any one particular actor, theories of social networks try to explain the connections and dynamics that exist among actors. Some key issues for scholars include how actors are positioned in networks in relation to one another, the strength or weakness of the ties between them, and the patterns of exchanges that emerge from their interactions.

Research has shown, for example, that different types of ties offer different kinds of benefits to people. Having strong ties to other individuals within a tight network can be beneficial in many ways, such as close friends taking care of your children, or community members banding together with law enforcement to watch for poachers at local nature preserves. But, having weak connections to individuals in groups outside of a primary network can also be helpful, for example, if these informal relationships yield useful information or provide access to different sets of contacts than might be available within close, personal connections (Granovetter 1973; J. Scott 2000).

The study of social networks has revealed how authority and responsibility can be both concentrated and dispersed within a network and the ways in which interests among individual actors can converge or conflict. Also important is the variety of characteristics that distinguish one network from another, such as how centralized a network is or the number of connections that members have with other members (see Figure 8.4). Finally, scholars note that networks can be both local and global, operating at a variety of levels and through many different channels (Castells 2004).

Social networks comprise a significant field of study in sociology, offering insights potentially relevant to the practice of biodiversity conservation. Connections among individuals, groups, or organizations can be crucial to conservation efforts, whether they promote the

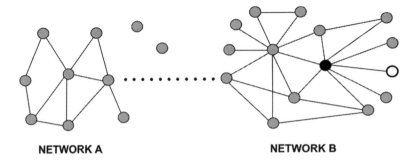

NETWORK A                    NETWORK B

In the diagram above, dots represent actors and lines represent connections between actors. Social networks are often described in terms of three characteristics:

**DENSITY** is a measure of degree of "closeness" of actors within a network compared to other networks. In the example above, Network B is more dense than Network A, because there are more connections among members.

**CENTRALITY** measures the degree to which any particular actor is connected to other actors in their network. An actor with greater centrality has a greater chance of gaining resources and advantages. The actor in Network B represented in BLACK has more connections, and therefore more centrality, than the actor represented in WHITE.

A **STRUCTURAL HOLE** is a gap between two separate networks. In the diagram, there are no actors connecting Network A to Network B. Actors that could connect the two networks (indicated by the dotted line) could be strategically important, given their ability to transmit or control information or actions between the two networks.

**Figure 8.4**   Network characteristics.

exchange of important resources or facilitate collaboration between a variety of stakeholders in a given site. Conservation leaders who are not connected to larger networks may be unaware of important information, including new programmatic trends, new funding opportunities, or potential partners for collaboration. Conversely, organizations or leaders who occupy key positions within important networks may be well positioned to achieve their goals.

Increasingly, conservation efforts have focused on the critical importance of building networks, and a wide variety of interorganizational networks have emerged in recent decades to tackle the social, economic, and political complexities of conservation challenges. For example, in midwestern USA, Chicago Wilderness is a multijurisdictional, public–private partnership of over 250 organizations that coordinate on a range of conservation efforts in four US states. Efforts there have not been free of controversy but demonstrate both the opportunities and challenges of organizational collaboration and communication (McCance 2011).

Similarly, in France, scholars investigated the role that networks played in implementing a European Union-wide conservation program known as Natura 2000. There, efforts to prohibit specific activities threatened traditional resource uses, including common-lands animal grazing by local stakeholder groups. In response, opposing networks of individuals and groups promoted and resisted efforts to harmonize local and national conservation policies and activities, leading the network to reconsider its approach to consultation and negotiation with stakeholders (Alphandéry & Fortier 2002; Osterman 2008; Blicharska et al. 2016).

Although networks are highly relevant to the study of biodiversity conservation, they remain relatively understudied. However, scholars have addressed how conservation governance networks can include multiple actors, ranging from the local level to the international sphere and how organizational networks affect the success of conservation efforts (Brechin et al. 2003; Geisler 2003; Alexander et al. 2016). Sociologists have also developed advanced methods for analyzing social networks that promise new ways to understand how stakeholders and processes can influence the goals, direction, and efficacy of conservation actions (De Nooy 2013; Farrell 2020).

### 8.3.3.2 Social Capital

Social capital can best be described as an intangible asset or resource that individuals and groups accumulate by virtue of their relations with one another. These resources are not physical or material in nature, but instead are formed through social connections, such as the ability to garner beneficial information, or the strategic benefits that might accrue from providing favors to or receiving favors from other actors. The amount of social capital that a person or group holds is dependent on the degree to which social relations or networks can be mobilized. Social capital is therefore quite different from other types of assets, like private property, because it depends on the reciprocal nature of individual and group interactions (Bourdieu 1986). In practical terms, social capital requires a few key elements, including reciprocal transactions, obligations and expectations, enforceable trust, solidarity, and moral obligations (Portes 2000; Putnam 2001). Factors that can increase or decrease the potential for social capital include the degree to which a group or network is open or closed and its level of stability (Coleman 1988).

Seeing social capital in action can be challenging given its intangible nature, but this does not mean it is not a powerful social phenomenon. In places where management of a resource is beyond the control of an individual or even a particular community, such as many forests and fisheries, social capital can be a critical determinant of collective management. In these situations, effectively connecting and working with others allows individuals and groups to collectively manage and protect natural resources (Ostrom 1990; Pretty & Smith 2004). This often requires the trust, reciprocity, and solidarity gained through social capital. Conversely though, social capital can be deployed in ways that work against biodiversity conservation. For example, in Quintana Roo, a town in southeastern Mexico, conservation efforts have been thwarted by gangs, whose established relationships generated social capital that has helped them effectively work together to conduct illegal logging operations (Wilshusen 2009).

Sometimes conservation practitioners intentionally try to mobilize and deploy social capital as part of an overall conservation strategy. For example, social capital is at the heart of many integrated conservation and development projects (ICDPs). ICDPs have gone in and out of favor since emerging in the late 1980s, but through the years they have been employed by many international conservation groups attempting to work on a local level. The stated goals of ICDP strategies vary tremendously, but by definition, all are designed to address both conservation and development goals. Sometimes the express intent of doing the development work is generating good will in a community, which in turn helps conservation organizations attain legitimacy and build working relationships with key stakeholder groups (Margolius et al. 2001). The formal and informal relationships built between conservation organizations and key stakeholders are an example of social capital in action.

In terms of effectiveness, ICDPs have opened the door for many collaborations with communities in areas important for conservation. They also have been the subject of great critique and debate as a conservation strategy. One critique of ICDPs is that they dilute conservation goals in favor of development objectives. Critics contend that the multiple objectives embedded in ICDPs make design, implementation, and monitoring more complicated, arguing that these projects fail to develop clear metrics or approaches for enforcing conservation (Sanderson & Redford 2003). Another critique is that conservation organizations are ill equipped to design or implement development interventions, although many have overcome this barrier by engaging in partnerships with relevant development organizations (Wells & McShane 2004; Gavin et al. 2018).

In response to these concerns, some organizations have moved away from ICDPs in favor of other tools, such as conservation incentive agreements or direct payments for **ecosystem services** (PES). However, the record of ICDPs demonstrates they can be effective in advancing conservation goals when well designed, but lessons learned from them over time underscore the need for explicit attention to the processes of building trust, legitimacy, and reciprocal relations between conservation practitioners, local communities, and governments (Pretty & Smith 2004). Additionally, while some perceive PES to be more clear-cut, this approach often still requires social capital to organize and implement. Given the importance of social capital in a variety of conservation practices, sociologists' approaches to measuring it can be helpful in the design and monitoring of conservation programs (Narayan & Cassidy 2001).

### 8.3.4   Collective Social Action

Collective action in society happens every day in many forms, when people work together to achieve common objectives that are out of the reach of an individual working alone. Collective action can be formal or informal, and can be time limited or can happen over a long period of time. Effective collective action is critical for biodiversity conservation, since no one individual could achieve conservation on his or her own. Groups come together in many ways to advance conservation objectives, including the work of organizations and social movements.

#### 8.3.4.1   Organizations

Organizations are prominent and ubiquitous actors in society, influencing many aspects of our lives. People interact with organizations when they shop, bank, work, and worship. People are born, educated, and placed to rest by organizations. People also turn to organizations to accomplish a variety of other things, including conserving nature and protecting natural resources. When operating effectively, organizations are powerful social mechanisms that can accomplish objectives beyond the reach of individuals alone and, ideally, maximize efficiency.

Organizations can be formal or informal. Formal organizations are consciously created by individuals to accomplish a stated objective. They are typically structured into a set of positions with specific responsibilities, which are in turn linked to stated goals. Positions are then filled with personnel who possess the skills necessary to accomplish key tasks. Formal organizations usually have rules and regulations, or systems governing group members' behavior, and often are recognized by the state. Informal organizations may look similar to formal organizations, but often operate without formal recognition and do not have established systems of coordination or governance.

Organizations can be public or private. Public organizations are extensions of the state or government, operating in the public sector. They are created as a means of achieving collective social goals sanctioned by the government, often to advance public good, and are typically supported by taxes. Private organizations tend to be created to achieve more targeted or narrow goals. Some exist within the market sector and aim to generate profit, engaging in competition with other for-profit organizations. Other private organizations are nonprofits, sometimes called nongovernmental organizations, or NGOs, which exist to promote a specific, documented mission and operate within **civil society**.

Sociologists analyze organizations in terms of their structures and functions, their internal dynamics and practices, how they function on the social landscape in concert with other social groups and actors, and their strategies for engaging the public. Critical issues include whether they are viewed as legitimate, efficient, transparent, durable, or effective (W. Scott & Davis 2007).

Protecting species, habitat, and related ecological services is a difficult task, made even more challenging since **ecosystems** defy political jurisdictions and borders. Given this complexity, organizations play a critical role in advancing conservation goals. Local, national, and international organizations all play a role in conservation, and are sometimes linked through networks (Box 8.3.).

---

**Box 8.3  Applications: A complex web of organizations influencing biodiversity conservation**

In the USA, a multitude of government agencies protect biodiversity, including the Park Service, Forest Service, and Fish and Wildlife Service. Similar agencies exist at the state and local levels. Many NGOs working on conservation issues worldwide have become household names to the public through their fundraising efforts, including the World Wildlife Fund, The Nature Conservancy, Sierra Club, and Conservation International. Other important international organizations include the United Nations (UN) Education and Scientific and Cultural Organization (UNESCO) and the International Union for the Conservation of Nature (IUCN).

International development agencies are also part of the diverse and complex web of organizations involved in or influencing conservation efforts. The UN Development Programme (UNDP), Food and Agricultural Organization (FAO), and Global Environmental Facility (GEF), all members of the UN system, are key actors globally. Bilateral development organizations like USAID, the Canadian International Development Agency (CIDA), Germany's Geselleschaft fur Technische Zusammenarbeit (GTZ), and the Inter-American Foundation (IAF) are among many agencies that offer programming and financial support in many developing countries for conservation-related efforts.

Multilateral funding agencies such as the World Bank and the International Monetary Fund (IMF) as well as regional development banks like Inter-American Development Bank (IDB) also play an important role in structuring conservation initiatives between governments and conservation NGOs. These international organizations have national, regional, and local partners and funding recipients including NGOs, state agencies, rural producer cooperatives, private associations, village councils, and myriad other organizations that can be involved in conservation efforts. Cooperation among these actors working at various levels and scales becomes essential if conservation is to be successfully pursued.

Sociologists studying conservation practices have focused their attention on organizations in two key ways. First, sociologists analyze how organizations operate on the ground and the role they play in conservation efforts. In practice, biodiversity conservation efforts are undertaken by organizations of all kinds: at local, national, and international levels they range from small, grassroots, local NGOs and community-based organizations, to informal conservation clubs and alliances, to state-sponsored task forces, private entities, and government bureaucracies, to large, multinational NGOs sometimes supported by wealthy elite donors, to name a few.

In a perfect world, this cacophony of organizations would work collaboratively to achieve ambitious goals, and in practice, some groups have explicitly joined together to achieve conservation. For example, in Belize, the Association of Protected Area Management Organizations works with smaller, community-based conservation organizations to provide technical support and coordination (see Box 8.4). Partnerships between local, state, and federal agencies as well as nonprofit organizations also have become increasingly common (Brechin et al. 2011; Gavin et al. 2018). For example, in the Democratic Republic of Congo, a partnership between the Congolese Ministry of Forestry Economy, the Wildlife Conservation Society, and Congolaise Industrielle des Bois was created to help protect and effectively manage the Nouabalé Ndoki National Park. The project underscores the value of creating a platform where relevant organizations can come together with other stakeholders to negotiate a conservation strategy that accounts for multiple perspectives, although it has since been criticized by some for exacting too heavy a toll on local communities, including negatively impacting their livelihoods, putting pressure on their resources and ignoring their legal rights (Ayari & Counsell 2017).

---

**Box 8.4   Crossing boundaries: Collaborative governance for nature protection networks in Belize**

In Belize, government officials and leaders of Belizean conservation NGOs developed an elaborate system of collaborative governance of their national parks (Figure 8.5). In 1981, the year Belize gained its independence from the United Kingdom, the Belize Audubon Society (BAS) successfully lobbied the new government to create a protected area system, leading to legislation that created the country's first national parks. The government, however, did not have the financial and human resources to manage the new system so it turned the management over to BAS, which first ran it informally and then engaged in formally contracted, co-management arrangements. Over the last 30 years, the number of national parks has increased in Belize, as has the number of conservation NGOs involved in co-management with the government.

The government of Belize is now the hub of a network of domestic conservation NGOs running the day-to-day operations in 22 of the country's 31 national parks, impacting 75% of the 677,150 acres of the total land base under protection (Brechin & Salas 2011). These arrangements demonstrate the importance of organizations working together in public–private partnerships and how these relationships can shift over time, creating collaborative networks to accomplish goals that the state could not achieve alone. It also underscores the importance of the state in natural resource governance issues (Brechin & Salas 2018).

**Figure 8.5** Left: Steering Committee and representatives of NGO members of the Belize Network of NGOs after a general meeting (2018). Right: Civil society representatives along with the Mayor of Orange Walk Town and the Deputy Chief Environmental Officer receiving a briefing before heading out to monitor pollution in the New River, northern Belize (2019). *Source:* Osmany Salas.

In other scenarios, organizations can end up competing for resources, stakeholder support, and even in their goals and objectives. Furthermore, the goals of conservation organizations can compete with other powerful actors in society when conservation objectives interfere with other social priorities or if the organizations are competing for the same resources but for different purposes. In these scenarios, conservation organizations must navigate the social terrain to maintain their legitimacy and efficacy.

The second way sociologists have studied conservation organizations is by focusing on the practices and global influence of big international conservation NGOs in the last few decades. Critics worry that powerful, well-resourced organizations like Conservation International, World Wildlife Fund, and The Nature Conservancy are able to define conservation agendas without sufficient consultation with important stakeholder groups, such as indigenous peoples or communities (Chapin 2004; Larsen 2016). Many international conservation groups have revisited their approaches for engaging local people in response to these critiques, changing how they operate, although sometimes still come up short in the eyes of critics.

Some scholars worry that conservation agendas may be overly influenced by corporations who donate to and partner with these large, international NGOs, especially corporations with poor environmental practices. Critics of these partnerships argue that they don't change corporate practices, but instead are simply "greenwashing" campaigns that help corporations improve their social and environmental images through association with conservation and environmental nonprofit groups (C. MacDonald 2008; Brechin 2018). Conservation organizations counter that working with corporations is an important way to influence corporate practices, and that this strategy is critical for conservation to succeed because the engagement with corporations encourages more environmentally sustainable practices (O'Gorman 2020). The divergent narratives about the role of large international conservation NGOs underscores the influence and power of these organizations in society. Whether seen as "good, bad, or ugly," they are a force to be reckoned with in societies across the globe (Larsen 2016).

### 8.3.4.2 Social Movements

Social movements are organized, collective action aimed at achieving a goal or creating change in society. They can be motivated by many goals, from reform, to revolution, to resistance. For example, they can aim to address social problems or grievances, give a political voice to those who are disenfranchised, advance a particular political agenda, reform the social order, or conversely, to fight against change and maintain the status quo. One way to think of social movements is to view them as campaigns, performances, or public displays in which ordinary people make claims aimed at others, whether the state, businesses, particular groups in society, or society at large (Tilly 2004).

Scholars note that social movements emerge when the desire for change among individuals is complemented with the concentration of sufficient resources to support a movement, making it viable. Resources in this sense can be viewed in broad terms, including money, leadership, legitimacy, knowledge, media interest, solidarity, and the commitment of participants. Movements can arise from emergent, grassroots efforts or can be organized and supported by formal organizations, sometimes called social movement organizations. For this reason, the study of social movements often overlaps with but also goes beyond traditional organizational theory because of the focus on power and politics.

When thinking about why social movements arise and what makes them successful, it is important to consider the institutional actors involved as well as their logic of operations and governance structures (J. Scott 2000). Similarly, it is important to consider how social movements frame issues as well as their degree of legitimacy, accountability, reliability, and efficiency (Oliver & Marwell 1992). In the last 20 years, attention to the role of agency in social movements has increased, including its role in the mobilization process, how movements are framed culturally, and how a movement's leadership operates (Morris 2000). In addition, the role of transformative events is a key factor in the viability of a given social movement. In this sense, the study of social movements has evolved from looking at static issues like structure, to more dynamic issues, including the roles that creativity, culture, meaning, and morality play in their success (Jasper 2010). Finally, it is important to try to understand social movements at a variety of levels, since they can happen in local, regional, national, and international settings (Keck & Sikkink 1998).

In the USA, social movements in the early twentieth century tended to focus on labor issues, addressing material inequities in society, working conditions, wages, and the political rights of laborers. In the 1960s and 1970s, an era of widespread social unrest led to the emergence of new types of social movements, including those promoting civil rights, women's rights, lesbian, gay, bisexual, and transgender rights, antiwar and peace-building efforts, and environmental protection. These social movements focused less on class divisions and more on the issues of discrimination, identity, quality of life, and self-expression. Aspects of modern environmental movements within the country arose in this context. Globally, social movements are just as diverse and influential.

When thinking about social movements in relation to conservation efforts, two issues are particularly relevant. First, conservation itself can be seen as a social movement. Second, conservation can also be considered as a social process that happens in the context of other social movements, which can be either in sync with conservation goals or at odds with them. Some examples of relevant social movements include activism about land rights, indigenous rights, and women's rights, and movements focused on both environmental justice and climate change.

One of the most famous examples of a land-rights movement that dovetailed with conservation efforts centers around the work of Francisco Alves Mendes Filho, better known as Chico Mendes, who was a rubber tapper, or *seringueiro*, in Brazil. Mendes and his fellow tappers harvested latex sustainably from rubber trees and set up a workers' union to help protect the forests, and their economic way of life, from expanding cattle ranches, with exposure and support from major environmental groups. Mendes's assassination by a cattle rancher in 1988 sparked global outrage. His efforts, however, led to a new category of protected area focused on extractive reserves, which allow for sustainable harvesting of non-timber forest products. This new conservation model provided protection opportunities for traditional and indigenous people and their environments throughout the world.

In other locations, conservation efforts and land-rights movements have clashed. In 1997, local residents revolted against a conservation project inside the Laguna del Tigre National Park in the Maya Biosphere Reserve, taking staff hostage and burning down a seven-building field station because they saw it as competing with their efforts to claim land (Ybarra 2018). Both sides framed the issue as a land grab by the other, underscoring how discursive frames shape public perception of social movements.

Likewise, some women's groups have focused on natural resources management issues because of their interest in providing basic needs for families, such as water, fuelwood, and food. One of the most famous social movements of this sort was the Chipko Movement, which occurred in the mountain forests of India. Using nonviolent methods, rural women literally clasped their hands around trees to stop them from being commercially logged. These grassroots, local actions were aimed at protecting community livelihoods from outside industrial development, and eventually led to policy changes in India that allowed for increased local control over village forests. These historic struggles bolstered environmental movements throughout the world (Guha 2000).

More recently, climate change, which has significant potential impacts on conservation outcomes, has commanded attention on the world stage, and rising and warming seas and changes in temperature and rainfall patterns will have enormous impacts on global conservation efforts. Ecosystems and species already under tremendous stress from global modernization will face additional threats, and protection efforts will need to adjust accordingly (Rands et al. 2010; IPBES 2019). Thus, synergies between social movements focused on climate change and conservation are emerging, aimed at motivating the required economic, social, and political changes necessary to reduce dependence on fossil fuel and to decelerate climate change. Extinction Rebellion, for example, is a newly formed international activist network, sponsoring nonviolent direct action and civil disobedience campaigns focused on climate and ecological issues. In response to movements related to climate change, efforts to protect the use of fossil fuels have also arisen, given that the fossil fuel industry has vested interests in the substantial profits in this sector (Oreskes & Conway 2010; Dunlap & Brulle 2015; Brulle 2020, 2021). Whether in sync with conservation or at odds with it, understanding the goals and intentions of social movements in places where conservation efforts are being planned or implemented is critical.

### 8.3.5  Population

In 1798, British economist Thomas Malthus expressed concerns that exponential human population growth would create a host of problems for societies, including potential resource scarcity, famine, and poverty. At the time that he wrote, the world's population was just under 1 billion people. By the year 2000, the global population had grown to more than 6 billion people. By 2025, world population is expected to increase to over 8 billion people, and current predictions suggest that world population could reach 10 billion by the year 2050 (UN Population Division 2015). Although technology has forestalled Malthus' dire predictions, demographic change remains a central issue in many conservation efforts, given that human populations are often in fierce competition over terrestrial and marine resources and can put direct and indirect pressures on natural systems. In sociology, the subfield of demography offers tools and analytical frameworks to assess the multidimensional concept of population (Hunter 2000).

#### 8.3.5.1  Demography

Demography, which originated from the fields of math and statistics, can be defined simply as the statistical study of human populations. The practice of formal demography uses information about birth rates, death rates, and migration to model overall demographic trends, since together these indicators determine the current and projected size and composition of any particular population. Demographers use these data to generate other indices as well, such as fertility rates, mortality rates, and life expectancies. Measures such as population size, density, distribution, and composition are also important and useful indicators when modeling and predicting demographic changes over time (see Figure 8.6).

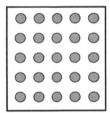 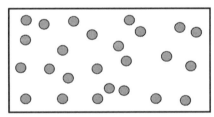 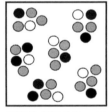

**POPULATION A** *(n=25)*          **POPULATION B** *(n=25)*          **POPULATION C** *(n=30)*

In the diagram above, dots represent individuals in a given population. The bounding boxes represent the size of the land area they inhabit and the different shades of the dots indicate varying characteristics, such as the age or sex of members.

Populations are often described in terms of four characteristics:

**POPULATION SIZE** is the number of individuals in a given population. In the diagram above, Population A and Population B are the same size, with 25 members each. Population C is bigger than both Populations A and B.

**POPULATION DENSITY** is the number of people in a given area. In the example above, Population C is more dense than Population A because it has more people in an equivalent area. Population B is less dense than Population A, because it has the same number of members but they are spread out over a larger area.

**POPULATION DISTRIBUTION** describes the specific placement or distribution of individuals on the landscape. Population A is uniformly distributed, Population B is randomly distributed and Population C is clumped.

**POPULATION COMPOSITION** provides a more in-depth breakdown of the characteristics of a given population, such as their distribution by age or sex. Population C has the greatest diversity of members.

**Figure 8.6**  Population terms.

The practice of social demography builds from the foundation of formal demography but extends the scope of analysis, emphasizing the relationships between demographic phenomena and social, economic, or cultural factors. Thus, social demographers might gather and analyze data about the size and composition of a population, for example, but may then try to link the data to other information such as variations in living arrangements, family structures, levels of education, income, occupation, as well as other attributes including class, race, gender, ethnicity, or religious affiliation (Hunter 2000). Connecting demographic data with information about how a given population uses resources or exerts pressure on the natural environment can be instructive in the practice of conservation. Forecasting future demographic change is also critical to understanding the pressures that a particular habitat, species, or resource might face as populations change.

Understanding the distribution and variation of people on a landscape, as well as how they might be changing, is extremely useful when planning conservation efforts. In 2000, researchers assessed a variety of demographic variables in 25 high-biodiversity areas under significant threat (Cincotta et al. 2000). Using formal demographic analysis, they and others found that a significant portion of the world's population lived within these biodiversity hotspots (Figure 8.7), and that population growth rates and density were higher than world averages in nearly all of these areas, as well as in large, globally significant wilderness areas (Cincotta et al. 2000; Williams 2013). Given these findings, researchers

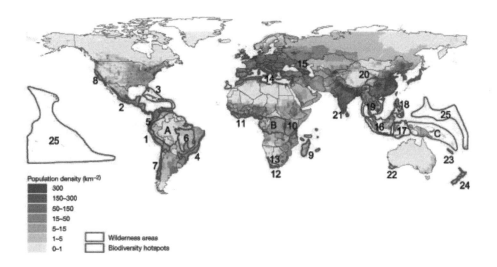

Hotspots: (1) Tropical Andes; (2) Mesoamerica; (3) Caribbean; (4) Atlantic Forest Region; (5) Chocó-Darién-Western Ecuador; (6) Brazilian Cerrado; (7) Central Chile; (8) California Floristic Province; (9) Madagascar; (10) Eastern Arc Mountains and Coastal Forests of Tanzania and Kenya; (11) West African Forests; (12) Cape Floristic Region; (13) Succulent Karoo; (14) Mediterranean Basin; (15) Caucuses; (16) Sundaland; (17) Wallacea; (18) Philippines; (19) Indo-Burma; (20) Mountains of South-Central China; (21) Western Ghats and Sri Lanka; (22) Southwest Australia; (23) New Caledonia; (24) New Zealand; and (25) Polynesia and Micronesia. Major tropical wilderness areas: (A) Upper Amazonia and Guyana Shield; (B) Congo River Basin; and (C) New Guinea and Melanesian Islands.

**Figure 8.7** Population growth in biodiversity hotspots. Figure from Cincotta et al. (2000), with permission.

have been urging conservation practitioners to address population growth in conservation projects, calling for programs that address the high birth rates as well as migration.

There have been calls to address population as a driver of environmental degradation and biodiversity loss since the 1970s, with analyses showing both a general relationship and specific correlations between human population density and species richness (Brashares et al. 2002; Brewer et al. 2013). In response, many conservation organizations and funding agencies, including USAID, created programs focused on family planning over the last couple decades, incorporating fertility issues and integrated health projects into existing conservation programs, or partnering with other groups who focus on population (D'Agnes & Margolius 2007; Oglethorpe et al. 2008).

While many believe these programs have been effective, the approach has been controversial. Critics question whether these programs are a practical or efficient use of resources, whether it is appropriate or even ethical for international conservation NGOs to engage in this type of work, and underscore the challenges of monitoring success (Margolius et al. 2001). Research has also shown that the relationship between population growth and resource use is far from straightforward and that it is simplistic to assume that more people will inevitably put more pressure on ecosystems (see Chapter 5).

The case of HIV/AIDS in sub-Saharan Africa illustrates the diverse ways that demographic changes can affect conservation efforts. HIV/AIDS has decreased the effectiveness of organizations working on wildlife conservation issues, through absenteeism, employee illness, death rates, and increased disease-related financial expenditures (Oglethorpe et al. 2013). In addition, research demonstrated that the mere perception of HIV-related illness can break down channels of communication between an organization and a community, impacting an organization's ability to meaningfully interact with those adjacent to protected areas (Cash 2007). This example underscores how demographic and related social changes, including the social impacts caused by AIDS and other global health issues, remain important to conservation planning.

### 8.3.5.2 Migration

Migration is an umbrella term used to describe the movement of people in or out of an area. It can happen as a result of factors pushing people out of a given location, or can be motivated by factors pulling people into one. According to the UNHCR, the terms *migrant* and *economic migrant* are typically used when people move in order to improve their future prospects (UNHCR 2021). The term *refugee*, however, is typically used to refer to people who move in order to preserve their lives or their freedom, whether in response to persecution, natural disasters, or war. If individuals move for these reasons within their home countries, they are referred to as internally displaced people.

Migration remains a critical issue in conservation, particularly in the context of protected areas. Since protected areas are often the last remaining areas of rich natural resources, they can become magnets for people (Brandon et al. 1998; Oglethorpe et al. 2007; Scholte & De Groot 2010). Migration into the coffee-growing mountains of Sumatra, Indonesia, offers an interesting example. In the Lahat District of Sumatra, researchers worked to understand the factors that had led to an influx of small-scale coffee farmers and the subsequent illegal deforestation of nearly 20% of the forests there (Brechin et al. 1994). They found the deforestation was caused by a chain of events that had started when coffee supplies in Brazil decreased due to frost, causing a sharp rise in both the demand and the price of coffee on the world market. The shift in the market value of coffee, coupled with the introduction of a new fast-growing species of coffee plant that required sun instead of shade, led to high levels of internal migration into an ineffectively administered protected area, as farmers attempted to capitalize on the global trends. Scenarios like this are increasingly common throughout the world as people compete more intensely for resources and both terrestrial and marine protected areas become increasingly attractive as access points for food, land, and raw materials, whether for sustenance or for sale as commodities. Migration remains a challenging force to combat as a source of environmental pressure, because it can be motivated by so many different social forces.

Conversely, it is important to note that people are often displaced by the establishment of protected areas, causing them to leave traditionally inhabited or accessed places. This is not an isolated phenomenon; scholars have documented the significant impact and human cost of displacement as a result of conservation efforts, including impairing people's ability to obtain important resources, diminishing or destroying their livelihoods, estranging them from cultural networks, and inflicting significant emotional and financial distress (Brockington & Igoe 2006; Agrawal & Redford 2009). If enacted forcibly, the practice can also raise basic human rights concerns (see Box 8.5 for a related discussion of fortress conservation).

---

**Box 8.5 Debates: Fortress conservation**

In order to prevent human encroachment into critical areas, some scholars and practitioners advocate "fortress conservation"—promoting strict boundaries between people and nature by whatever means necessary, whether guards, guns, fences, or all three. Some advocates of fortress conservation even support the removal of existing communities from key protected areas, turning the practice of conservation into a driver of migration. The physical displacement and relocation of communities in the name of creating parks or nature preserves, whether voluntary or forced, generates great debate. Scholars refer to the affected groups as "conservation refugees" to convey the gravity of their plight after being disenfranchised from an area (Turnbull 1972; Geisler 2003; Dowie 2009).

Whether they are indigenous groups or simply poor rural families, the social and economic consequences of relocation can be quite extreme, raising questions and concerns about social justice as well as the long-term impacts of these practices. Displacement can result in ongoing conflict between communities and park or reserve personnel. Research shows that many families and communities relocate just outside the boundaries of the established protected area and some attempt to return to the land they have left, potentially threatening the long-term viability of the conservation efforts that the displacement was designed to facilitate in the first place (West & Brechin 1991; Brockington 2002; Wilshusen et al. 2002; Brechin et al. 2003; West et al. 2006).

Alternatives to these approaches include community-based conservation and resource management. Rather than conceiving of people as threats to biodiversity, these approaches attempt to enlist local communities as potential allies in conservation efforts, working to directly engage local communities in conservation activities (Heinen 1996; Agrawal & Gibson 1999; Brechin et al. 2002; Berkes 2004). While often effective, these strategies have challenges as well. Thus, disagreements about whether to pursue fortress-style or community-based strategies generate some of the sharpest and enduring tensions within the world of conservation.

## 8.3.6 Global Social Forces

Social forces are processes and mechanisms that shape individual and societal-level choices, behaviors, or ways of thinking. From an individual perspective, social forces can feel remote or impersonal, because individuals are subject to them even if they did not have a hand in creating them. In this sense, they can be viewed as underlying mechanisms that are a step removed from individual lives even though people are subject to the effects of them. Social forces can be obvious, such as major technological shifts like the rise of the computer, cell phone, and social media, which have changed the way people communicate. Other times, social forces can operate subtly and feel hidden or invisible, such as the way beliefs about science might influence whether people adopt more sustainable practices. Social forces can be structural arrangements or can take other forms, such as events. They can emerge gradually or suddenly, and can operate at many levels from individual to community to global. Whether economic, technological, legal, or cultural, social forces are powerful and salient influences on society that are sometimes adopted and embraced, but also can be rejected and challenged.

### 8.3.6.1 Industrial Capitalism

Industrial capitalism is a major social force in the modern world. Capitalism is an economic system that uses wage labor in production processes and relies on markets to sell and exchange goods, services, and commodities to generate profit. **Industrialization**, on the other hand, is a shift toward mechanization and automation with increasingly rationalized organization and reliance on external energy sources instead of human labor power. Industrialization and mechanization are typically undertaken to accelerate rates of production and/or to reduce labor costs, also with the underlying goal of increasing profit. Capitalism and industrialization are highly interconnected forces that have radically altered the exchange of material between social and ecological systems worldwide (Foster et al. 2010).

Early sociologists analyzed the way industrial capitalism and mass production methods changed society by promoting rapid migration into cities, increasing class stratification, and spawning new forms of bureaucratic organization (Durkheim [1933]1997; Weber [1956] 1978; Marx & Engels 1978). Contemporary sociologists continue to analyze the myriad social effects of industrial capitalism, and in recent decades, some have also turned their attention to its ecological impacts. Environmental sociologists observe that industrial capital increasingly burdens natural systems, as raw materials are extracted and consumed, energy is expended, and ecological systems are taxed by waste and pollution from the by-products of production and the inevitable disposal of postconsumer materials (O'Connor 1997). Concerns mount over the fact that industrial capitalism, in its current form, has the potential to cause huge disruptions to the planet's natural processes (Schnaiberg & Gould 2000).

Within environmental sociology, scholars debate whether industrial capitalism can be made compatible with natural systems. Some sociologists argue that modern industrial capitalism is like a "treadmill of production" (TOP), with a continual drive to accelerate production and consumption in order to generate more profit (Schnaiberg 1980; Gould et al. 2008). From this perspective, firms overlook environmental and social impacts in an effort to maximize revenues for their shareholders. In addition, the state, organized labor, and those with private capital also accept the acceleration of the treadmill because they are heavily invested in economic growth, albeit for different reasons. Thus, those with the power to reform capitalism sometimes have no incentive to do so, although the state does sometimes support conservation policies, especially when countermovements apply pressure (Brechin & Fenner 2017).

The TOP framework raises fundamental questions about whether the constant expansion of the economy is sustainable and proposes that capitalism in its current form will need to be radically reformed to prevent ecological destruction. The framework also highlights the fact that the winners in industrial capitalism are often financial elites, who tend to be personally buffered from the environmental impacts associated with extraction and production processes, since these often take place in locations far from where they live and work. However, people who are negatively affected by the environmental hazards often accept this system, because they believe jobs and livelihoods require an ever-expanding economy through increased production (Gould et al. 2008). Thus, the burdens and costs of industrial capitalism are unevenly divided among groups in society, a perspective shared by those who focus on environmental justice (Taylor 2000; Schlosberg 2007; Martin 2017).

A second approach focuses on what scholars call "ecological modernization" (EM). Those advocating for this perspective view the system of capitalism as both reflexive and responsive. EM scholars argue that market forces will inherently respond to changes in the supply and demand of natural resources, leading to the eventual greening of capitalism as ecological limits are reached and environmental problems grow (Hajer 1995; Mol 2003). Though there are important variations, EM scholars argue that the environmental problems associated with global capitalism are rooted solely in its industrial character, not in its reliance on markets. As a result, they contend that both scientific and technological innovation and adaptation can play a beneficial role in revamping industrial practices, minimizing the disjuncture between industrial capitalism and the natural world (Sonnenfeld & Mol 2000). EM scholars believe that states and civil society can solve environmental problems using key institutions in society, such as markets and the scientific community, leading to a gradual shift in capitalism instead of the radical changes advocated by the TOP theorists (Mol et al. 2009). Since it leaves room for economic growth within the parameters of the planet's ecological limits, the EM approach can be viewed as being in sync with the notion of sustainable development.

In practice, many conservation strategies are designed to counter the TOP, whether by buffering critical areas from resource extraction or mitigating the potentially negative impacts that production processes have on natural systems. Creating parks and reserves, for example, is a strategy used to ensure that particular species or habitats located in key ecological zones are off limits to capitalism's reach. Other conservation strategies attempt to redirect industrial capitalism to particular geographic areas to prevent its impacts in other more critical areas, such as the practice of brokering carbon offsets to mitigate $CO_2$ emissions or safeguarding critical wetlands while allowing or even promoting development elsewhere. Legislation and regulations also prevent the spread of industrial capitalism into critical ecosystems, including appeals to block the construction of large dams, mines, or highways in fragile areas, or campaigns to limit the spread of industrial agriculture.

**Figure 8.8** Left: Coffee farmers raking beans on a coffee plantation in Chiapas, Mexico, in 1999, were early participants in a sustainable coffee initiative launched by Conservation International and Starbucks coffee. Right: Conservation International staff and Starbucks employees visit coffee-growing farms in Sumatra, Indonesia, in 2012. *Sources:* Sterling Zumbrunn/Conservation International (left); Joanne Sonenshine/Conservation International (right).

Reducing or redirecting development may be an obvious goal for protecting particular areas, but in practice, efforts to do this require careful planning. There can be downsides to reducing economic growth in a given area: local people lose development options, companies lose profits, and/or governments can lose state revenue from taxes. Measures to contain or redirect capital may require addressing these losses.

Like EM theorists, however, some conservationists believe that industrial capitalism can be made more sustainable, and work to reform industry standards through the clever use of markets and attempts to mitigate the negative impacts of businesses. One such approach focuses on the certification of products, giving them a "green" or "sustainable" label, to reward business practices that protect biodiversity. For example, Conservation International has worked with corporations to initiate green business practices and guide corporate investment in conservation efforts. The organization partnered with Starbucks Coffee to support coffee farmers who produce more biodiversity-friendly coffee in species-rich tropical forests of the world (Figure 8.8). Similarly, the Forestry Stewardship Council certifies timber products so consumers know that products are harvested in sustainable ways. These efforts work within the overall paradigm of industrial capitalism, attempting to transform its most egregious processes rather than trying to suppress or contain it. Box 8.6 describes how conservation organizations have attempted to use market forces and capitalism to promote conservation, along with some resulting debates.

---

**Box 8.6   Debates: Neoliberal conservation strategies**

Classic liberal economic theories since the eighteenth century have celebrated the ways that market economies give individuals the ability to barter and trade to meet their needs (Smith [1776] 2009). One of the key changes in the global **political economy** since the 1980s, however, has been the promotion of neoliberal economic theory, or **neoliberalism**. Neoliberal perspectives operate under the assumption that free markets and free trade can serve as effective arbiters of social problems and often promote the extension of market logic to realms that are not inherently economic,

**Box 8.6  (Continued)**

such as advocating for the commodification and privatization of social and environmental life (Harvey 2005; Brockington et al. 2008).

Conservation organizations have increasingly adopted neoliberal conservation strategies, moving toward approaches that monetize conservation objectives so they can be bought, sold, or traded using market logic (Büscher et al. 2014). Conservation incentive agreements, for example, are contractual agreements in which organizations pay local communities or organizations cash in exchange for specific conservation-related actions. Similarly, conservation organizations have set up programs that translate the value of natural resources, such as clean air or clean water, into a commodity or product called an "ecosystem service," which can then be bought and sold. Ecosystem services are the beneficial functions that ecosystems provide to people. Payments for ecosystem services (PES) are mechanisms that attempt to measure and convert these services into monetary values. Advocates of these neoliberal strategies argue that cash payments are an efficient and fair way to broker conservation outcomes. They view these strategies as improvements over other conservation approaches, such as environmental education campaigns and ICDPs, because they see them as clear, direct, and enforceable.

However, critics of neoliberal conservation strategies have raised numerous concerns (Apostolopoulou et al. 2021). First, they worry that monetizing conservation objectives subjects the goals of conservation to the logic of profitability, leaving conservation efforts vulnerable to the whims of the market. They argue that conservation efforts can easily be outbid by market players with interests other than the preservation of critical habitats, species, or ecosystems. Second, the neoliberal emphasis on private property and ownership has also generated critiques, since land tenure is often unclear and private property rights do not always exist in many important conservation settings. Third, critics worry that market approaches may unfairly privilege the interests of those with financial capital, including international elites, silencing groups that may be economically or politically marginalized but might still have a legitimate stake in zoning, management plans of particular species, and land and water rights (Igoe & Brockington 2007). Fourth, there is a worry that the emphasis on monetary value can overshadow discussions of the noneconomic value of a resource, suppressing the importance of other legitimate social and cultural logics for conservation (Harmon & Putney 2003). Finally, there is a concern that neoliberal approaches sometimes undermine support for and the capacity of regulatory agencies of local and national governments.

Finally, it is important to note that these agreements do not absolve conservation practitioners from understanding the complex social dynamics in a given region. Regardless of the mechanism used to create incentives for conservation, effective implementation will require that practitioners exercise due diligence to understand relevant social structures, institutions, and power dynamics, as well as how the intervention will be understood by the community or population. The long-term success of these strategies remains to be seen, so conservationists and sociologists alike are watching carefully to determine what works, what doesn't, and why.

### 8.3.6.2 Economic Globalization and World-Systems Theory

Global trade flows and networks have existed for centuries, but recent changes in communication and transportation technologies have quite literally changed the way people and places are connected. Just a few decades have brought new ways to transfer information, changes in financial systems, and the rise of increasingly complex commodity chains that require global networks to operate (Paulson et al. 2003; Grewal 2005). Yet, finding a single definition for **globalization** is challenging, since the term encapsulates a variety of uneven and interconnected sets of processes operating simultaneously. Contemporary definitions tend to emphasize the fact that globalization involves both an intensification in social relations and a shift in spatial relations, connecting local places to extra-local forces in new ways, through new transnational organizations and changing flows of information, capital, commodities, and practices that stretch across the world (Burawoy et al. 2000; Sassen 2007). There are many theories about globalization, some with particular relevance to biodiversity conservation.

World-systems theory (WST) is an economic perspective on globalization that views the modern world economy as a single interconnected system, operating with an unequal exchange of capital, raw materials, and labor between wealthy and poor nations, as well as key classes within societies (Wallerstein 1979; Shannon 1996). In this framework, some nations are classified as core nations, meaning they are able to secure continued streams of profit from global production processes. Other countries are considered periphery nations or semi-periphery nations that serve as sources of cheap raw materials, agricultural commodities, and inexpensive labor for global production processes, leading to their relative disadvantage within the world system. In a nutshell, nation-states are interconnected, but each plays a different role in the global system and gains differently from it. World-systems theorists argue that wealthier, more developed nations have already garnered a disproportionate share of benefits from the world economy and will continue to have disproportionate influence on the global political economic system. This unevenness has implications for natural systems as well.

For decades, world-systems theorists paid little attention to environmental issues or conservation. This perspective has, however, been used to understand how efforts to conserve biodiversity are affected by the overall unevenness of capitalist development (Roberts & Grimes 2002). Given their relative affluence, core nations can pursue higher living standards and maintain natural ecosystems within their own borders by exporting the dirtiest parts of production chains to developing countries. Poorer peripheral countries, on the other hand, have less capital and often lack the technology to pursue clean, energy-efficient development, so they accept polluting industries despite potentially negative environmental or health consequences. These countries also end up heavily exploiting their natural resources, despite gaining few benefits in return, due to the incentives to pollute and over-extract within the global system. Thus, benefits are distributed to certain countries at the risk and expense of others (Wallerstein 1979; Bunker 2005).

In addition to driving ecological destruction through market transactions, the world system also affects conservation efforts through international lending agencies. Some periphery nations, for example, take large loans offered by international lending agencies such as the World Bank or the International Monetary Fund to promote economic development. While these loans are intended to provide an influx of capital to help grow economies and alleviate poverty, they often have the perverse effect of driving natural resource depletion, if governments sell these resources to ensure a steady stream of income to avoid defaulting on the debt. Some international conservation organizations have employed strategies called debt-for-nature swaps to curb this driver of resource destruction, providing debt relief to insolvent nations in exchange for the creation of protected areas or environmental policies. In these projects, large international conservation organizations often act as the intermediary for international lending agencies, governments, and donors enlisted to help pay down the debt, brokering the deal or providing technical support in the form of conservation planning or policy development.

Looking at the world as an interconnected system of nations has led some to claim that the system is designed around principles of "ecological imperialism," where wealthier nations are able to exploit ecosystems in other nations for their own benefit without fair compensation (Bunker 1985; Muradian & Martinez-Alier 2001). Work conducted in Ecuador in conjunction with the Ecuadorian NGO Acción Ecológica, for example, has documented this process of unequal exchange and termed it "ecological debt," arguing that members of wealthy core nations potentially owe billions to poorer nations as reparations for the uneven exchanges with those in the periphery (Martinez-Alier 2002). Interestingly, some attempts by conservation organizations to preserve critical areas of the Global South have also been called ecological imperialism, since periphery nations bear the burden of setting aside land and resources in the name of global conservation at the expense of their own national development. Since the poorer, periphery countries house the highest shares of biodiversity, wilderness areas, and critical habitats, finding ways to strike an equitable balance between global biodiversity conservation efforts and national development is critical.

### 8.3.6.3 Cultural Globalization and World Society Theory

Cultural globalization happens as people are increasingly exposed to other societies as a result of migration, travel, the rise of global media and other communication technologies, new forms of international and transnational organizations, as well as the effects of global capitalism. These diverse forces and processes lead to increased flows of ideas, customs, images, music, symbols, products, advertising, and even people across new geographic scales (Holton 2005; Hopper 2007). Thus, while scholars agree that most cultures have never been entirely isolated, there is now an increasing interpenetration between global and local cultures, due to the new flows circulating more broadly across the globe (Tomlinson 2012).

Although the most visible forms of cultural globalization might be found in the music, food, entertainment, and fashion industries, the concept also encompasses the global movement of ideas and discourses. For example, the term *biodiversity* did not even exist a few decades ago, but is now a concept that is not only used regularly within conservation circles, but is also widely recognized as a concept globally (Machlis 1992; Hannigan 1995).

World society theory has been used to help explain the rise of conservation and environmental policies around the world. World society theory scholars assert that a global perspective on environmental issues has emerged over the last century, and that nation-states around the world enact environmental legislation as they become more plugged into these global debates (Frank et al. 2000). International organizations often play a role in diffusing norms and ideas about environmental protection throughout the world, through their role in promoting parks and protected areas, encouraging state membership in intergovernmental environmental organizations, and promoting environmental impact assessment laws.

However, others point out that the world society theory framework focuses so heavily on the role of international organizations that it fails to acknowledge the role that local and national actors and processes play in promoting environmental change. Domestic social movements, for example, often have pivotal influence on the policies of nation-states motivating local and national policies independent from international actors (Buttel 2000). Furthermore, many emphasize that local people are not merely passive recipients of information from international organizations but respond to and resist or adopt new ideas and influences in the context of their existing social contexts (Escobar 1998).

## 8.4 Future Directions

Biodiversity conservation, from global to local levels, is not an end state or condition but rather a continuous form of social engagement that involves people, including communities, governments, organizations, and other stakeholders of many kinds. The natural sciences provide critical knowledge necessary for protecting ecological systems for the benefit of human and non-human species. But successful conservation also requires a full appreciation of the social, political, cultural, and economic contexts in which conservation takes place. Biodiversity conservation is fundamentally a social and political process. With that in mind, this chapter highlights the importance of understanding broader global contexts, the **positionality** of different stakeholders, and various conduits for collective action.

Looking forward, these three focal areas will continue to be relevant for biodiversity conservation. Increasing human population, continued industrial development linked to globalization, and a changing climate loom as potential social forces that will influence the fate of biodiversity. The conservation community will need to understand these major macro-level social forces in order to adequately plan for the impacts that will follow. Important differences between stakeholder groups also will remain critical factors in conservation success. Deciding the fate of valuable land and natural resources requires full participation by many competing groups and organized interests, each with differing power, resources, and orientations toward biodiversity and conservation. Experience shows that successful conservation efforts cannot rest on the backs of the disadvantaged, and conservation efforts must effectively engage local people and communities in nature protection efforts. The need for clear and transparent collaboration will continue to grow as the scale of conservation projects expands, crossing social and political boundaries. Finally, the future of biodiversity conservation will rely on renewed collective effort, whether this means the emergence of new social movements and better organizational responses to address the needs of our threatened planet, or finding new ways for the conservation community to engage with social movements and issues that may be different from but related to conservation efforts, such as women's rights, land tenure, and antipoverty campaigns.

Learning and self-reflection on conservation projects sometimes focuses on particular case studies, with less investment in analysis that looks systematically across projects, sites, or approaches to elucidate more generally what has worked, what has not, and why. To address the many challenges ahead for conservation, practitioners can adopt social science methods that will help develop what sociologists call "middle-range theories," or theories that are general enough to showcase key issues and themes across projects, but also specific enough to be useful in particular cases. Finally, conservation practitioners would be wise to consider how social justice concerns are increasingly part of the discourse that shapes whether or not conservation efforts are adopted or resisted. Acknowledging and addressing social justice concerns up front in project design and implementation are critical priorities for both ethical and practical reasons. Fundamentally, since people actively shape and create the world we live in, saving biodiversity will require strategies that are sustainable not only for species and habitats but humans as well.

## For Further Reading

1 *Contested Nature: Promoting International Biodiversity Conservation with Social Justice in the Twenty-first Century* (Brechin, Wilshusen, Fortwangler & West, eds. 2003, State University of New York Press, Albany, NY).
2 *Nature Unbound: Conservation, Capitalism and the Future of Protected Areas* (Brockington, Duffy & Igoe 2008, Earthscan Publications Limited, London).
3 Whose knowledge, whose nature? Biodiversity, conservation, and the political ecology of social movements (Escobar 1998, *Journal of Political Ecology* 5: 53–82).
4 *Billionaire Wilderness: The Ultra-Wealthy and the Remaking of the American West* (Farrell 2020, Princeton University Press, Princeton, NJ).
5 Transnational conservation movement organizations: shaping the protected area systems of less developed countries (Lewis 2000, *Mobilization: An International Quarterly* 5 (1): 103–121).
6 Social capital in biodiversity conservation and management (Pretty & Smith 2004, *Conservation Biology* 18 (3): 631–638).
7 *Saving Forests, Protecting People? Environmental Conservation in Central America* (Schelhas & Pfeffer 2009, AltaMira Press, Lanham, MD).
8 *Defining Environmental Justice: Theories, Movements, and Nature* (Schlosberg 2007, Oxford University Press, Oxford, UK).
9 *The Rise of the American Conservation Movement: Power, Privilege, and Environmental Protection* (Taylor 2016, Duke University Press, Durham, NC).
10 Parks and people: the social impacts of protected areas (West, Igoe & Brockington 2006, *Annual Review of Anthropology* 35: 251–277).

## Acknowledgments

The authors thank John Schelhas, Chantal Elkin, the Monks Community Forest, Joanne Sonenshine, Sterling Zumbrunn, and Conservation International for photographs used in this chapter. Many thanks also to several anonymous reviewers for helpful feedback on earlier drafts and to the editors for their assistance and persistence in bringing this volume to press.

# References

Adelman, R.M. (2007) Racial residential segregation in urban America. *Sociology Compass* 1 (1): 404.

Agrawal, A. & Gibson, C.C. (1999) Enchantment and disenchantment: the role of community in natural resource conservation. *World Development* 27 (4): 629–649.

Agrawal, A. & Redford, K. (2009) Conservation and displacement: an overview. *Conservation and Society* 7 (1): 1–10.

Akella, A.S. & Cannon, J.B. (2013) Strengthening the weakest links: strategies for improving the enforcement of environmental laws globally. In: *Transnational Environmental Crime* (ed. R. White), 528–567. New York: Routledge.

Alexander, S.M., Andrachuk, M. & Armitage, D. (2016) Navigating governance networks for community-based conservation. *Frontiers in Ecology and the Environment, Special Issue: Network Governance and Large Landscape Conservation* 14 (3): 155–164.

Alphandéry, P. & Fortier, A. (2002) Can a territorial policy be based on science alone? The system for creating the Natura 2000 network in France. *Sociologia Ruralis* 41 (3): 311–328.

American Sociological Association (2019) Current sections of the American Sociological Association. https://www.asanet.org/communities-sections/sections/current-sections.

Apostolopoulou, E., Chatzimentor, A., Maestre-Andrés, S. et al. (2021) Reviewing 15 years of research on neoliberal conservation: towards a decolonial, interdisciplinary, intersectional and community-engaged research agenda. *Geoforum* 124: 236–256.

Arora, V. (2006) The forest of symbols embodied in the Tholung Sacred Landscape of North Sikkim, India. *Conservation & Society* 4 (1): 55–83.

Ayari, I. & Counsell, S. (2017) *The Human Cost of Conservation in the Republic of Congo*. London: Rainforest Foundation UK.

Bandiaky, S. (2008) Gender inequality in Malidino Biodiversity Community-based Reserve, Senegal: political parties and the "village approach". *Conservation and Society* 6 (1): 62–73.

Bandiaky-Badji, S. (2011) Gender equity in Senegal's forest governance history: why policy and representation matter. *International Forestry Review* 13 (2): 177–194.

Berkes, F. (2004) Rethinking community-based conservation. *Conservation Biology* 18 (3): 621–630.

Besek, J. & McGee, J.A. (2014) Introducing the ecological explosion. *International Journal of Sociology* 44 (1): 75–93.

Besek, J. & York, R. (2018) Toward a sociology of biodiversity loss. *Social Current* 6 (3): 239–254.

Blicharska, M., Orlikowska, E.H., Roberge, J.M. et al. (2016) Contributions of social science to large scale biodiversity conservation: a review of research about the Natura 2000 network. *Biological Conservation* 199: 110–122.

Blumer, H. (1986) *Symbolic Interactionism: Perspective and Method*. Oakland, CA: University of California Press.

Borgatti, S.P., Mehra, A., Brass, D.J. et al. (2009) Network analysis in the social sciences. *Science* 323 (5916): 892–895.

Bourdieu, P. (1984) *Distinction: A Social Critique of the Judgment of Taste*. Cambridge, MA: Harvard University Press.

Bourdieu, P. (1986) The forms of capital. In: *Handbook of Theory and Research for the Sociology of Education* (ed. J.G. Richardson), 241–258. New York: Greenwood.

Brandon, K., Redford, K. & Sanderson, S. eds. (1998) *Parks in Peril: People, Politics, and Protected Areas*. Washington, DC: Island Press.

Brashares, J.S., Arcese, P. & Sam, M.K. (2002) Human demography and reserve size predict wildlife extinction in West Africa. *Proceedings of the Royal Society B: Biological Sciences* 268 (1484): 2473–2478.

Brechin, S.R. (2018) Returning greater integrity to the conservation mission in a post-neoliberal world. In: *The Anthropology of Conservation NGOs: Rethinking the Boundaries* (ed. P. Bille & D. Brockington), 245–250. Gewerbestrasse, Switzerland: Palgrave Macmillan.

Brechin, S.R., Benjamin, C. & Thoms, A. eds. (2011) Networking nature: network forms of organization in environmental governance. *Journal of Natural Resources Policy Research* 3 (3), 211-340.

Brechin, S.R. & Fenner, W.H., IV (2017) Karl Polanyi's environmental sociology: a primer. *Environmental Sociology* 3 (4): 404–413.

Brechin, S.R. & Salas, O. (2011) Government-NGO networks & nature conservation in Belize: examining the theory of the hollow state in a developing country context. *Journal of Natural Resources Policy Research* 3 (3): 263–274.

Brechin, S.R. & Salas, O. (2018) Civil society–government collaborations in Belize, Central America. In: *Conflict and Collaboration: For Better or Worse* (ed. C. Gerard & L. Kreisberg), 120–133. London: Taylor & Francis.

Brechin, S.R., Surapaty, S.C., Heydir, L. et al. (1994) Protected area deforestation in south Sumatra, Indonesia. *The George Wright Forum* 11 (3): 59–78.

Brechin, S.R., Wilshusen, P., Fortwangler, C. et al. (2002) Beyond the square wheel: toward a more comprehensive understanding of biodiversity conservation as a social and political process. *Society of Natural Resources* 15 (1): 41–64.

Brechin, S.R., Wilshusen, P., Fortwangler, C. et al. eds. (2003) *Contested Nature: Promoting International Biodiversity Conservation with Social Justice in the Twenty-first Century*. Albany, NY: State University of New York Press.

Brewer, T.D., Cinner, J.E., Green, A. et al. (2013) Effects of human population density and proximity to markets on coral reef fishes vulnerable to extinction by fishing. *Conservation Biology* 27 (3): 443–452.

Brockington, D. (2002) *Fortress Conservation: The Preservation of the Mkomazi Game Reserve*. Bloomington, IN: Indiana University Press.

Brockington, D., Duffy, R. & Igoe, J. (2008) *Nature Unbound: Conservation, Capitalism and the Future of Protected Areas*. London: Earthscan Publications Limited.

Brockington, D. & Igoe, J. (2006) Eviction for conservation: a global overview. *Conservation & Society* 4 (3): 424–470.

Brulle, R.J. (2020) Denialism: organized opposition to climate change action in the United States. In: *Handbook of US Environmental Policy* (ed. D.M. Konisky), 328–341. Cheltenham, UK: Edward Elgar Publishing.

Brulle, R.J. (2021) Networks of opposition: a structural analysis of US climate change countermovement coalitions 1989–2015. *Sociological Inquiry* 91 (3): 603–624.

Bunker, S.G. (1985) *Underdeveloping the Amazon: Extraction, Unequal Exchange, and the Failure of the Modern State*. Champaign-Urbana: University of Illinois Press.

Bunker, S.G. (2005) How ecologically uneven developments put the spin on the treadmill of production. *Organization & Environment* 18 (1): 38–54.

Burawoy, M. (2005) For public sociology: 2004 presidential address. *American Sociological Review* 70: 4–28.

Burawoy, M., Blum, J.A., George, S. et al. (2000) *Global Ethnographies: Forces, Connections and Imaginaries in a Postmodern World*. Berkeley, CA: University of California Press.

Büscher, B., Dressler, W. & Fletcher, R. (2014) *Nature™ Inc. Environmental Conservation in the Neoliberal Age*. Phoenix, AZ: The University of Arizona Press.

Buttel, F.H. (2000) World society, the nation-state, and environmental protection: comment on Frank, Hironaka, and Schofer. *American Sociological Review* 65 (1): 117–121.

Cash, J. (2007) HIV/AIDS and conservation agency capacity in southern Africa: perceptions of critical impacts, barriers, and intervention strategies. MS Thesis, Department of Society and Conservation, University of Montana.

Castells, M. (2004) *The Information Age: Economy, Society and Culture, Volume II: The Power of Identity*. Oxford, UK: Blackwell.

Chapin, M. (2004) A challenge to conservationists. *World Watch* 17 (6): 17.

Cincotta, R.P., Wisnewski, J. & Engelman, R. (2000) Human population in the biodiversity hotspots. *Nature* 404: 990–992.

Clausen, R. & York, R. (2008) Global biodiversity decline of marine and freshwater fish: a cross-national analysis of economic, demographic and ecological influences. *Social Science Research* 37 (4): 1310–1320.

Coleman, J.S. (1988) Social capital in the creation of human capital. *American Journal of Sociology* 94 (suppl.): S95–S120.

D'Agnes, L. & Margolius, C. (2007) *Integrating Population, Health and Environment (PHE) Projects: A Programming Manual*. Washington, DC: United States Agency for International Development.

De Nooy, W. (2013) Communication in natural resource management: agreement between and disagreement within stakeholder groups. *Ecology and Society* 18 (2): 44.

Dowie, M. (2009) *Conservation Refugees: The Hundred-Year Conflict between Global Conservation and Native Peoples*. Cambridge, MA: MIT Press.

Duncan, O.D. (1959) Human ecology and population studies. In: *The Study of Population* (ed. P.M. Hauser & O.D. Duncan), 678–716. Chicago, IL: University of Chicago Press.

Dunlap, R. (1994) International attitudes towards environment and development. In: *Green Globe Yearbook of International Cooperation on Environment and Development* (ed. H.O. Bergesen & G. Parmann), 115–126. Oxford: Oxford University Press.

Dunlap, R. & Brulle, R. eds. (2015) *Climate Change and Society: Sociological Perspectives*. New York: Oxford University Press.

Dunlap, R., Gallup, G.H., Jr. & Gallup, A.M. (1993) Of global concern: results of the health of the planet survey. *Environment* 35 (9): 7–15.

Durkheim, E. ([1933]1997) *The Division of Labor in Society*. New York: The Free Press.

Elkin, C. (2009) Linking Buddhism and conservation: the case of the Monks Community Forest, Cambodia. Master Thesis, University of London.

Escobar, A. (1998) Whose knowledge, whose nature? Biodiversity, conservation, and the political ecology of social movements. *Journal of Political Ecology* 5: 53–82.

Farrell, J. (2020) *Billionaire Wilderness: The Ultra-Wealthy and the Remaking of the American West*. Princeton, NJ: Princeton University Press.

Field, D.R. & Burch, W., Jr. (1991) *Rural Sociology and the Environment*. Westport, CT: Greenwood Press.

Foster, J.B., Clark, B. & York, R. (2010) *The Ecological Rift: Capitalism's War on the Earth*. New York: Monthly Review Press.

Fowler, F.J. (2013) *Survey Research Methods*. Thousand Oaks, CA: Sage.

Frank, D.J., Hironaka, A. & Schofer, E. (2000) The nation-state and the natural environment over the twentieth century. *American Sociological Review* 65 (1): 96–116.

Gavin, M.C., McCarter, J., Berkes, F. et al. (2018) Effective biodiversity conservation requires dynamic, pluralistic, partnership-based approaches. *Sustainability* 10 (6): 1846.

Geisler, C. (2003) Your park, my poverty: using impact assessment to counter the displacement effects of environmental greenlining. In: *Contested Nature: Promoting International Biodiversity with Social Justice in the Twenty-First Century* (ed. S.R. Brechin, P.R. Wilshusen, C.L. Fortwangler et al.), 217–229. Albany, NY: State University of New York Press.

Goffman, E. (1956) *The Presentation of Self in Everyday Life*. Edinburgh: Doubleday.

Gould, K., Pellow, D.N. & Schnaiburg, A. (2008) *The Treadmill of Production: Injustice and Unsustainability in the Global Economy*. Boulder, CO: Paradigm Publishers.

Granovetter, M.S. (1973) The strength of weak ties. *American Journal of Sociology* 78 (6): 1360–1380.

Grewal, I. (2005) *Transnational America: Feminisms, Diasporas, and Neoliberalisms*. Durham, NC: Duke University Press.

Guha, R. (2000) *The Unquiet Woods: Ecological Change and Peasant Resistance in the Himalaya*. Oakland, CA: University of California Press.

Hajer, M. (1995) *The Politics of Environmental Discourse: Ecological Modernization and the Policy Process*. Oxford: Oxford University Press.

Hannigan, J. (1995) *Environmental Sociology: A Social Constructionist Perspective*. New York: Routledge.

Harding, S.G. (2004) *The Feminist Standpoint Theory Reader: Intellectual and Political Controversies*. New York: Routledge.

Harmon, D. & Putney, A. eds. (2003) *The Full Value of Parks: From Economics to the Intangible*. Lanham, MD: Rowman & Littlefield.

Harvey, D. (2005) *A Brief History of Neoliberalism*. Oxford: Oxford University Press.

Heinen, J.T. (1996) Human behavior, incentives and protected area management. *Conservation Biology* 10 (2): 683–684.

Hoffman, J.P. (2004) Social and environmental influences on endangered species: a cross-national study. *Sociological Perspectives* 47 (1): 79–107.

Holton, R.J. (2005) *Making Globalization*. New York: Palgrave Macmillan.

Hopper, P. (2007) *Understanding Cultural Globalization*. Cambridge, UK: Polity Press.

Howard, J.A. (2000) Social psychology of identities. *Annual Review of Sociology* 26 (1): 367–393.

Humphrey, C.R., Lewis, T.L. & Buttel, F.H. (2002) *Environment, Energy and Society: A New Synthesis*. Belmont, CA: Wadsworth.

Hunter, L. (2000) *The Environmental Implications of Population Dynamics*. Santa Monica, CA: RAND.

Igoe, J. & Brockington, D. (2007) Neoliberal conservation: a brief introduction. *Conservation and Society* 5: 432–449.

IPBES (Intergovernmental Science–Policy Platform on Biodiversity and Ecosystem Services) (2019) Summary for policymakers of the global assessment report on biodiversity and ecosystem services of the Intergovernmental Science–Policy Platform on Biodiversity and Ecosystem Services. Bonn, Germany: IPBES.

IUCN (International Union for Conservation of Nature) (2021) Protected area categories. https://www.iucn.org/theme/protected-areas/about/protected-area-categories (accessed August 1, 2021).

Jasper, J.M. (2010) Social movement theory today: toward a theory of action? *Sociological Compass* 4 (11): 965–976.

Keck, M.E. & Sikkink, K. (1998) *Activists beyond Borders: Advocacy Networks in International Politics*. Ithaca, NY: Cornell University Press.

Larsen, P.B. (2016) The good, the ugly and the Dirty Harry's of conservation: rethinking the anthropology of conservation NGOs. *Conservation & Society* 14 (1): 21–33.

Lewis, T.L. (2016) *Ecuador's Environmental Revolutions: Ecoimperialists, Ecodependents, and Ecoresisters*. Cambridge, MA: MIT Press.

Li, T.M. (2000) Locating indigenous environmental knowledge in Indonesia. In: *Indigenous Environmental Knowledge and Its Transformations: Critical Anthropological Perspectives* (ed. R.F. Ellen, P. Parkes & A. Bicker), 123–146. Amsterdam: Harwood Academic.

MacDonald, C. (2008) *Green, Inc.: An Environmental Insider Reveals How a Good Cause Has Gone Bad*. Guilford, CT: Globe Pequot Press.

MacDonald, K.I. (2010) Business, biodiversity and new 'fields' of conservation: the World Conservation Congress and the renegotiation of organisational order. *Conservation & Society* 8 (4): 256–275.

Machlis, G.E. (1992) The contribution of sociology to biodiversity research and management. *Biological Conservation* 62 (3): 161–170.

Margolius, R., Myers, S., Allen, J. et al. (2001) *An Ounce of Prevention: Making the Link between Health and Conservation*. Washington, DC: Biodiversity Support Program.

Martin, A. (2017) *Just Conservation: Biodiversity, Wellbeing and Sustainability*. London & New York: Routledge.

Martinez-Alier, J. (2002) The ecological debt. *Kurswechsel* 4: 5–16.

Marx, K. (1976) *Capital Volume I: A Critique of Political Economy*. London: Penguin Group.

Marx, K. & Engels, F. (1978) *The Marx-Engels Reader*, 2nd ed. (ed. R.C. Tucker). New York: Norton.

McCance, E. (2011) Networking to conserve biodiversity: the case of Chicago Wilderness. *Journal of Natural Resources Policy Research* 3 (3): 237–250.

Mead, G.H. (1967) *Mind, Self, and Society: From the Standpoint of a Social Behaviorist*. Chicago, IL: University of Chicago Press.

Mills, C.W. (1959) *The Sociological Imagination*. New York: Oxford University Press.

Mol, A.P.J. (2003) *Globalization and Environmental Reform: The Ecological Modernization of the Global Economy*. Cambridge, MA: MIT Press.

Mol, A.P.J., Sonnenfeld, D. & Spaargaren, G. eds. (2009) *The Ecological Modernisation Reader: Environmental Reform in Theory and Practice*. New York: Routledge.

Morris, A. (2000) Reflections on social movement theory: criticisms and proposals. *Contemporary Sociology* 29 (3): 445–454.

Munch, R. (1994a) *Sociological Theory I: From the 1850s to the 1920s*. Belmont, CA: Wadsworth Publishing.

Munch, R. (1994b) *Sociological Theory II: From the 1920s to the 1960s*. Belmont, CA: Wadsworth Publishing.

Muradian, R. & Martinez-Alier, J. (2001) Trade and the environment: from a "southern" perspective. *Ecological Economics* 36 (2): 281–297.

Narayan, D. & Cassidy, M.F. (2001) Dimensional approach to measuring social capital: development and validation of a social capital inventory. *Current Sociology* 49 (2): 59–102.

Newing, H. (2011) *Conducting Research in Conservation: A Social Science Perspective*. New York: Routledge.

O'Connor, J. (1997) *Natural Causes: Essays in Ecological Marxism*. New York: The Guilford Press.

Oglethorpe, J., Ericson, J., Bilsborrow, R.E. et al. (2007) *People on the Move: Reducing the Impacts of Human Migration on Biodiversity*. Washington, DC: World Wildlife Fund & Conservation International Foundation.

Oglethorpe, J., Honzak, C. & Margolius, C. (2008) *Healthy People, Healthy Ecosystems: A Manual for Integrating Health and Family Planning into Conservation Projects*. Washington, DC: World Wildlife Fund.

Oglethorpe, J., Lukas, T., Gelman, N. et al. (2013) *HIV/AIDS and Environment: A Manual for Conservation Organizations on Impacts and Responses*. Washington, DC: World Wildlife Fund–US and Africa Biodiversity Collaborative.

O'Gorman, M. (2020) *Strategic Corporate Conservation Planning: A Guide to Meaningful Engagement*. Washington, DC: Island Press.

Oliver, P.E. & Marwell, G. (1992) Mobilizing technologies for collective action. In: *Frontiers in Social Movement Theory* (ed. A.D. Morris & C.M. Mueller), 251–272. New Haven, CT: Yale University Press.

Omi, M. & Winant, H. (1986) *Racial Formation in the United States: From the 1960s to the 1990s*. New York: Routledge.

Oreskes, N. & Conway, E.M. (2010) Defeating the merchants of doubt. *Nature* 465 (7299): 686–687.

Osterman, O. (2008) The need for management of nature conservation sites designated under Natura 2000. *Journal of Applied Ecology* 35 (6): 968–973.

Ostrom, E. (1990) *Governing the Commons: The Evolution of Institutions for Collective Action*. Cambridge: Cambridge University Press.

Palmer, M. & Finlay, V. (2003) *Faith in Conservation. New Approaches to Religions and the Environment*. Washington, DC: The World Bank.

Paulson, S., Gezon, L.L. & Watts, M. (2003) Locating the political in political ecology: an introduction human organization. *Human Organization* 62 (3): 205–217.

Portes, A. (2000) The two meanings of social capital. *Sociological Forum* 15: 1–12.

Pretty, J. & Smith, D. (2004) Social capital in biodiversity conservation and management. *Conservation Biology* 18 (3): 631–638.

Putnam, R.D. (2001) *Bowling Alone: The Collapse and Revival of American Community*. New York: Simon & Schuster.

Rands, M.R.W., Adams, W.M., Bennum, L. et al. (2010) Biodiversity conservation: challenges beyond 2010. *Science* 329 (5997): 1298–1303.

Roberts, J.T. & Grimes, P. (2002) World-system theory and the environment: toward a new synthesis. In: *Sociological Theory and the Environment* (ed. R.E. Dunlap, F.H. Buttel, P. Dickens et al.), 167–196. Lanham, MD: Littlefield.

Roca, J. (2004) Healthy Communities Initiative Peru Trip Report, Conservation International.

Rudel, T.K. (1998) Is there a forest transition? Deforestation, reforestation and development. *Rural Sociology* 63 (4): 533–552.

Sanderson, S. & Redford, K.H. (2003) Contested relationships between biodiversity conservation and poverty alleviation. *Oryx* 37: 389–390.

Sandstrom, K.L., Lively, K.J., Martin, D.D. et al. (2013) *Symbols, Selves, and Social Reality: A Symbolic Interactionist Approach to Social Psychology and Sociology*. New York: Oxford University Press.

Sandy, C. (2006) Real and imagined landscapes: land use and conservation in the Menabe. *Conservation & Society* 4 (2): 304–324.

Sassen, S. (2007) *A Sociology of Globalization*. New York: Norton.

Sawyer, S. & Gomez, E.T. (2012) *The Politics of Resource Extraction: Indigenous Peoples, Multinational Corporations and the State*. London: Palgrave Macmillan.

Schelhas, J. (2002) Race, ethnicity and natural resources in the United States: a review. *Natural Resources Journal* 42 (4): 723–763.

Schelhas, J. & Pfeffer, M.J. (2009) *Saving Forests, Protecting People? Environmental Conservation in Central America*. Lanham, MD: AltaMira Press.

Schlosberg, D. (2007) *Defining Environmental Justice: Theories, Movements, and Nature*. Oxford: Oxford University Press.

Schnaiberg, A. (1980) *The Environment: From Surplus to Scarcity*. Oxford: Oxford University Press.

Schnaiberg, A. & Gould, K. (2000) *Environment and Society: The Enduring Conflict*. Caldwell, NJ: The Blackburn Press.

Scholte, P. & De Groot, W.T. (2010) From debate to insight: three models of immigration to protected areas. *Conservation Biology* 24 (2): 630–632.

Scott, J.P. (2000) *Social Network Analysis: A Handbook*, 2nd ed. London: SAGE.

Scott, W.R. & Davis, G.F. (2007) *Organizations & Organizing: Rational, Natural and Open Systems*. New York: Routledge.

Shabecoff, P. (2003) *A Fierce Green Fire: The American Environmental Movement*. Washington, DC: Island Press.

Shannon, T. (1996) *An Introduction to the World-System Perspective*. New York: Routledge.

Smith, A. ([1776] 2009) *The Wealth of Nations*. New York: Classic House Books.

Sonnenfeld, D. & Mol, A.P.J. (2000) Ecological modernisation around the world: perspectives and critical debates. *Environmental Politics* 9 (1): 1–14.

Stone, S. (2002) The Kanuku mountains protected area process. Conservation International. https://pdf.usaid.gov/pdf_docs/pnadb870.pdf (accessed May 10, 2022).

Tàbara, J.D. & Pahl-Wostl, C. (2007) Sustainability learning in natural resource use and management. *Ecology and Society* 12 (2): 3.

Taylor, D.E. (2000) The rise of the environmental justice paradigm: injustice framing and the social construction of environmental discourses. *American Behavioral Scientist* 43 (4): 508–580.

Taylor, D.E. (2018) Diversity in Environmental Organizations: Reporting and Transparency Report No. 1. Ann Arbor, MI: University of Michigan Press. https://doi.org/10.13140/RG.2.2.24588.00649.

Tilly, C. (2004) *Social Movements*. Boulder, CO: Paradigm Publishers.

Timko, J. & Satterfield, T. (2008) Criteria and indicators for evaluating socio-cultural and ecological effectiveness in national parks and protected areas. *Natural Areas Journal* 8 (3): 307–319.

Tomlinson, J. (2012) *Cultural Globalization*. Hoboken, NJ: Wiley-Blackwell.

Turnbull, C.M. (1972) *The Mountain People*. New York: Simon and Schuster.

UNHCR (United Nations High Commissioner for Refugees) (2021) 'Refugees' and 'migrants' – frequently asked questions (FAQs). https://www.refworld.org/docid/56e81c0d4.html (accessed January 28, 2021).

UN Population Division (2015) *World Population Prospects: Key Findings & Advance Tables*. New York: United Nations Department of Economic and Social Affairs, Population Division.

Vencatesan, J. (2008) Gender and conservation–some issues. *Current Science* 94 (9): 1120–1122.

Veríssimo, D., Bianchessi, A., Arrivillaga, A. et al. (2018) Does it work for biodiversity? Experiences and challenges in the evaluation of social marketing campaigns. *Social Marketing Quarterly* 24 (1): 18–34.

Wallerstein, I. (1979) *The Capitalist World-Economy*. Cambridge: Cambridge University Press.

Wasserman, S. & Faust, K. (1994) Social network analysis: methods and applications. Structural analysis in the social sciences series, No. 8. Cambridge: Cambridge University Press.

Weber, M. ([1956] 1978) *Economy and Society: An Outline of Interpretive Sociology*. Berkeley, CA: University of California Press.

Weber, M., Mill, C.W. & Gerth, H.H. eds. ([1948] 1998) *From Max Weber: Essays in Sociology*. New York: Routledge.

Wells, M.P. & McShane, T.O. (2004) Integrating protected area management with local needs and aspirations. *Ambio* 33 (8): 513–519.

West, P.C. & Brechin, S.R. eds. (1991) *Resident People and National Parks: Social Dilemmas and Strategies in International Conservation*. Tucson, AZ: University of Arizona Press.

West, P.C., Igoe, J. & Brockington, D. (2006) Parks and people: the social impacts of protected areas. *Annual Review of Anthropology* 35: 251–277.

Wetherell, M. ed. (2009) *Theorizing Identities and Social Action*. London: Palgrave Macmillan.

Williams, J.N. (2013) Humans and biodiversity: population and demographic trends in the hotspots. *Population and Environment* 24 (4): 510–523.

Wilshusen, P.R. (2009) Shades of social capital: elite persistence and the everyday politics of community forestry in southeastern Mexico. *Environment and Planning A: Economy and Space* 4 (2): 389–406.

Wilshusen, P.R., Brechin, S.R., Fortwangler, C.L. et al. (2002) Reinventing a square wheel: critique of a resurgent "protection paradigm" in international biodiversity conservation. *Society and Natural Resources* 15: 17–40.

Ybarra, M. (2018) *Green Wars: Conservation and Decolonization in the Maya Forest*. Oakland, CA: University of California Press.

York, R., Rosa, E.A. & Dietz, T. (2003) Footprints on the earth: the environmental consequences of modernity. *American Sociological Review* 68 (2): 279–300.

# 9

# Conclusion: Toward Better Conversations about Conservation

*Daniel C. Miller, Ivan R. Scales, and Michael B. Mascia*

## 9.1 Introduction

The social sciences comprise a diverse set of concepts, theories, and approaches to understanding what it means to be human and how human societies work. Each of the six classic social science disciplines reviewed in this book—anthropology, economics, human geography, political science, psychology, and sociology—help us understand not only human nature and society, but also human–environment interactions (including the relationships among individuals, groups, and larger societies) with biodiversity and efforts to conserve it. The central contribution of this book is to show how, individually and as a whole, the social sciences are crucial to understanding the causes and consequences of biodiversity loss, as well as possible solutions to this pressing global challenge.

While the social sciences are vital to the science and practice of biodiversity conservation in general, significant variations exist in how they have and might engage with conservation. Diversity within and across the social sciences is a defining feature and is, in our view, a strength. But this diversity can be confusing and even frustrating. We have therefore emphasized consistent organization across the chapters of this book to give the reader a clear sense of the history, major questions, and key conceptual frameworks and analytical lenses for each of the social science disciplines covered, with particular attention to how they relate to conservation. To ensure contemporary relevance, each chapter has concluded by reflecting on future directions for the discipline in conservation research and application. This presentation across the core chapters builds from the review of foundational issues relating to knowledge and methods relevant across all the social sciences presented in Chapter 2.

We hope this approach has helped readers make sense of the diversity in the social sciences and better understand how the social sciences might be relevant to their own interests in conservation. We recognize that these interests are also likely to be manifold. We have written this book with such diversity in mind and expect that readers will include students at different levels, scholars from different disciplinary backgrounds interested in biodiversity conservation, practitioners seeking to build their knowledge of the social dimensions of conservation, and donors and policy makers seeking specific insights from

*Conservation Social Science: Understanding People, Conserving Biodiversity*, First Edition.
Edited by Daniel C. Miller, Ivan R. Scales, and Michael B. Mascia.
© 2023 John Wiley & Sons Ltd. Published 2023 by John Wiley & Sons Ltd.

**Table 9.1**  Principles for enhanced conversations on conservation.

---

1) *Everyone does not need to agree on everything.*
2) *Acknowledge plural and discordant perspectives.*
3) *Seek basic knowledge of different disciplines and ways of knowing.*
4) *Science can inform decision-making, but does not determine it.*
5) *Recognize unequal power relationships and take steps to rectify them.*

---

social science research to inform their decision-making. Different disciplines, conceptual frameworks, and approaches might have resonated to a greater or lesser extent with these different audiences. Accommodating such diverse interests and placing them within larger disciplinary and interdisciplinary contexts has been a major aim of this volume.

In this concluding chapter, we identify key cross-cutting themes as well as areas of disagreement and tension that emerge when examining the chapter contributions in this book as a whole. We also highlight some limitations of the book, including blind spots left by its discipline-by-discipline approach, and focus on a subset of potentially relevant fields of knowledge on the social dimensions of conservation. We then use this synthesis, along with consideration of contemporary global trends, to respond to the broad question: Where to now for conservation and the social sciences? Specifically, we explore questions about the future of conservation social science and its relationship to larger domains of conservation science and sustainability science (Clark & Harley 2020).

In responding to these questions, we sketch out the contours of what might become not only a more fully fledged field of conservation social science but also a more integrated science of conservation that is better able to contribute to the effective, equitable, and enduring conservation of the rich and full variety of life on our shared planet. We do not offer any grand synthesis for any sort of unified field of conservation social science, nor bold prediction of what the future will hold. Instead, we see the social sciences—and conservation more generally—as a forum characterized by creative tensions that should be embraced so as to carefully consider the potential synergies as well as the trade-offs inherent to so much of the conservation enterprise (Hirsch et al. 2011; McShane et al. 2011). At its best, conservation social science can provide an informed, lively forum for conversations about potential synergies among different values, goals, and perspectives, while navigating inevitable trade-offs so as to advance the well-being of humans and the natural world (Table 9.1).

## 9.2  Contributions of This Book

The authors of this volume have carefully distilled information and insights from vast literatures covering the social sciences generally and six classic social sciences specifically. Doing so using a common framework is perhaps the signal contribution of this book. This approach provides a "one-stop shop" for conservation-relevant knowledge and approaches derived from each of the disciplines featured. It also presents a shared vocabulary, including through key terms used across more than one chapter and defined in the glossary, for understanding the different social sciences and their approach or relevance to conservation issues. In these ways, the book also allows for identification of similarities and differences across these disciplines.

A critical contribution of this book is to help move beyond stunted conceptions of the social sciences that are too often found within conservation and the biophysical sciences (Castree et al. 2014). Such conceptions see the social sciences (and the humanities) as an "add on" or, worse, an irritant to the "real" science of conservation. By showcasing their richness and diversity, this volume presents a more complete picture of the social sciences. It has sought to do so in a broadly accessible way so as to help enable more constructive dialogue across the range of social and biophysical sciences in relation to conservation. Only through such dialogue can the social sciences become an equal partner with the natural sciences in conservation scholarship and decision-making. And, we argue, only through such exchange will conservation science, policy, and practice be able to rise to the challenges posed by the complex, rapid, and large-scale social and environmental changes that characterize our contemporary world.

This volume demonstrates the long history of social scientific investigation of human interactions with the environment. It shows how the scientific study of the natural world has inspired development of modern social science and continues to shape its development. In tracing this evolution through to the present, the chapters of this book reaffirm the importance of the social sciences to conservation while also vividly showing that change in perspective and understanding is possible, thereby inviting new ways of thinking about conservation.

There are a number of different ways one might approach the subject of "conservation social science." Our approach has been to start with the classic social science disciplines and present a sweeping overview of their relevance and (potential) contribution to the science and practice of conservation. This approach has its merits and, we believe, fills an important need at this juncture in the development of conservation social science. However, we also acknowledge that there are some limits to such a discipline-by-discipline treatment. Perhaps chief among these is that it underemphasizes the wide range of interdisciplinary fields—conservation marketing, ethnobotany, environmental history, political ecology, science and technology studies, among many others—that are vital to the larger emerging field of conservation social science. Our focus on classical social sciences has also meant we have not given much attention to the environmental humanities, which also incorporate the study of conservation, seeking to "illuminate peoples' complex and divergent understandings of life—human and non-human—on Earth" (Castree et al. 2014, p. 765). Fortunately, other reviews help fill these gaps. Bennett et al. (2017a), for example, provide a useful overview of the broad terrain of conservation social science, including a larger range of disciplines than examined here and exploration of numerous relevant interdisciplinary fields.

### 9.2.1 Major Cross-cutting Themes

**Conservation is an inherently social phenomenon.** This book decisively reaffirms the argument made nearly two decades ago by Mascia and colleagues (2003) and further developed since (e.g. Martin et al. 2016; Bennett et al. 2017a) that biodiversity conservation is inherently social. It is no longer—and indeed never was—sufficient to view biodiversity through the lens of biology and the natural sciences alone. People have been critical in shaping the natural world ever since humans evolved (Scott 2017; Ellis et al. 2021). Humanity's influence is now so extensive that many argue Earth has moved into a new geological epoch where the defining feature is human domination of the planetary system (Folke et al. 2021). Conservation is about protecting biodiversity, but it is people, the values they hold, and the choices they make that will shape what happens to the diversity of life across the globe. And conservation has profound implications for people, who may be affected positively or negatively by various forms of conservation action—and inaction.

**Social science is central to the science and practice of conservation.** This point follows logically from the preceding one. Conservation processes and outcomes are the result of human action or lack thereof. This claim is no longer in doubt (Cowling 2014; Bennett et al. 2017a; Cinner 2018). With their focus on advancing knowledge of individuals, groups, and society, the social sciences are uniquely positioned to examine conservation as a field of study (Balmford et al. 2021). This book shows the many different ways social sciences have and can do so, including from critical and applied perspectives. The in-depth treatment of the relationship of major social science disciplines with conservation provided in these pages complements a number of shorter synthetic perspectives on how to do so that are also worth reading (e.g. Hicks et al. 2016; Bennett et al. 2017b; Reddy et al. 2017; Balmford et al. 2021).

**The social sciences are diverse and unevenly integrated into conservation.** That social sciences exhibit tremendous variety—in their approach to knowledge, the topics they choose to investigate, the assumptions they make, and the methods they use—emerges clearly from the chapters in this volume. We believe it is important to highlight this diversity and to see it as a strength, providing a range of possible ways to advance knowledge related to the science and practice of conservation. With this book, we seek to help the reader make sense of this diversity to allow educated judgment and choice regarding possible disciplines, approaches, and topics of interest for deeper engagement.

Shining a bright light on the different classical social science disciplines reveals how their relationships with conservation have varied. For example, quantitative data and statistical methods have provided a common grammar that has facilitated integration of disciplines like economics in which quantitative methods predominate. The integration of knowledge and approaches from disciplines that emphasize qualitative methods and take a more critical approach, such as social anthropology and many subdisciplines in human geography, has been more challenging. In other cases, like political science, further reasons, including a sense that conservation is not a "proper" topic for the discipline, have inhibited such integration.

How integration of the broad and diverse array of social sciences into conservation might occur—and vice versa—remains an open question, one where tentative answers are being formulated and revised. Here we underscore that each of the social sciences has value for understanding and action in relation to conservation, and that rigorous research can take many forms (e.g. it can be more or less participatory and use qualitative, quantitative, or mixed methods) and still yield important conservation-relevant insights. In so doing, the different social sciences can help understand the various ways conservation is taking place, support conservation in practice, and imagine other ways it could be carried out.

**Scale matters.** The chapters of this book make it clear that understanding conservation requires consideration of multiple spatial and temporal scales. Different social science disciplines emphasize different units of analysis, from psychology's particular interest in the individual to sociology's examination of whole large-scale societies and countries. They also vary in their emphasis on spatial scale, from a concern with households or communities in small geographies (most prominent perhaps in anthropology) through concern with territories within nation-states, as often emphasized in political science, to international, planetary-scale processes, as treated in human geography, economics, and others. The social sciences also consider temporal scale in different ways, from short-term outcomes of market transactions to medium-term results from specific policies, to long-term impacts of slowly evolving social processes. Examining the past—and sometimes the "long" past (Braudel & Wallerstein 2009)—is important across the social sciences. It is important not only to identify potentially generalizable patterns, but, in some cases, to envision or model possible futures (Wyborn et al. 2021).

**Power relations are critical to understanding conservation processes and outcomes.** To understand how conservation processes unfold and with what effect, it is imperative to understand power, including who holds it and how it is exercised. This key social science concept emerges more or less explicitly across the chapters of this volume. Some social scientists emphasize how power is exercised through certain ideas and concepts (e.g. anthropology and human geography) while others focus more on the operation of power through norms and institutions (e.g. economics and political science). Collectively, the chapters draw our attention to the various ways power can be exercised in and through conservation, from rules and regulations shaping how the costs and benefits of conservation action are distributed, to narratives and discourses that shape the very way people think about conservation, which in turn influences how they act.

### 9.2.2   Tensions in Conservation-related Social Science

Looking across the chapters in this volume also reveals a number of important points of tension, both within and across social science disciplines. For example, political science includes the more explicitly normative subfield of political theory as well as subfields that strive for a more objective, "scientific" approach. Similarly, tensions exist relating to the more interpretive and critical emphasis in parts of anthropology and human geography and typically positivist and quantitative approaches in disciplines like economics and psychology. Beyond such general frictions, several other thematic tensions are important to highlight.

#### 9.2.2.1   The Role of Social Science in Relation to Conservation Practice

Tensions exist within and across the social sciences relating to their role in the actual practice of conservation. Roles of scholars (to document, explain, and critique), practitioners (to identify problems and implement solutions), and advocates (to encourage specific goals and actions) underscore the existence of different views of what the ultimate goal of research should be. Many social scientists work *on* conservation rather than *for* conservation (Sandbrook et al. 2013). That is, they study conservation as a social phenomenon with an eye toward understanding the human experience and not (necessarily) to inform or change conservation policy and action. On this view, conservation social science should not just be instrumental but should examine conservation science, policy, and practice to yield more general insights about human behavior and meaning, independent of its utility for conservation (Bennett & Roth 2019; Massarella et al. 2021; Nielsen et al. 2021). Others have highlighted the vital role of social science in informing improved conservation through discussion and debate (e.g. Balmford et al. 2021). Constructive dialogue about the role(s) of social science vis-à-vis conservation will require acknowledging and understanding these different viewpoints.

#### 9.2.2.2 Tolerance for Abstraction and Interest in Generalizability

The social sciences vary greatly in terms of their ambition to draw generalizable conclusions about people and our relationships with one another and with the natural world. At one end of the spectrum are approaches with an explicit focus on hypothesis testing and quantitative methods to produce generalizable explanations. At the other end we find social science disciplines that emphasize the diversity, complexity, and singularity of the human experience and social processes. This range translates into a tension regarding the level of simplification accepted. Different disciplines emphasize different levels of particularity and context and have different interests in generalizing. One can imagine a continuum of approaches ranging from those that embrace and explicitly delve into the complexity of local context to those that seek to draw widely applicable conclusions through abstraction. Choices made on how to approach a given conservation social science issue along this continuum will shape conclusions drawn and assumptions about their applicability and so must be taken seriously. It is our aim to raise this tension so as to enable the reader to be an informed consumer in engaging with social science fields and subfields having different emphases on abstraction and ambitions toward generalizability.

#### 9.2.2.3 Analytical Focus

Do social systems and outcomes result from individuals going about their business and interacting with each other based on their choices, or are larger social structures (e.g. capitalism, nation-states, and cultural norms) the dominant force shaping them? This "structure vs. agency" debate cuts across the social sciences, with certain disciplines and subfields placing a greater emphasis on one side of the argument or another even as theorists in sociology and other disciplines attempt to reconcile these two broad approaches (e.g. Bourdieu 1972; Giddens 1984). This debate pits scholars holding a view of methodological individualism against those emphasizing methodological holism. The former view posits that human actors are the central theoretical and **ontological** feature of social systems, with social structure resulting from the actions and interactions of individuals. This perspective is common in the disciplines of economics and much of political science, for example. Methodological holism, by contrast, holds that human actors are born and socialized into social structures that powerfully shape individual choice and action and that such structures should be the primary analytical focus. Anthropology and human geography are two disciplines where this view is widely held. These varying assumptions about structure vs. agency are important to be aware of and understand because they shape the questions, research designs, methods, and, ultimately, the findings of social scientists studying conservation issues.

#### 9.2.2.4 Conceptual Inconsistencies

To talk productively about any phenomenon first requires naming it. The social sciences study complex, multifaceted, and often hard to pin down phenomena: culture, power, and human well-being, among many. Defining what fits—and what doesn't fit—in the category or process that has been named is the next necessary step. This process of naming and defining is called concept formation. It is essential for social science, allowing us to connect what is happening in the "real world" to the linguistic world in which social science theory and knowledge is developed. But concepts are not static and the same words may have different meanings based on different intellectual traditions across disciplines. For example, some 35 different attributes have been associated with the term "ideology," with different definitions emphasizing some but not others and no single definition covering all of them (Gerring 2001).

Such difference, as highlighted in this book, remains a real tension in the social sciences. It often inhibits dialogue and prevents shared understanding. Synthetic social science research has sought to develop criteria for "goodness in concepts" (Gerring 2001, p. 39) to address this problem. These criteria, such as clarity and coherence, ability to be operationalized in practice through measurement or other indicators, and resonance in ordinary or specialized contexts, can be used to develop and refine concepts across the social sciences so they can form the building blocks for knowledge. But the tension remains, and carefully examining and developing concepts in different disciplinary contexts represents a key challenge for a more robust and meaningful conservation social science.

## 9.3 Enhancing Conversations on Conservation

This book extensively demonstrates that insights derived from social science investigation of conservation as well as social science insights applied to conservation enable a more complete understanding of people and the conservation of biodiversity. But where to now? How might conservation social science further develop as a field of its own and in relation to broader concerns with conservation, the environment, and sustainability? Here we draw from recent literature as well as the findings of this volume to sketch a working response to these vital questions. We do so in four parts. We begin by briefly considering the future of conservation-related inquiry within the social sciences. Second, we explore conservation social science as an interdisciplinary field. Third, we discuss conservation social science in relation to non-social science knowledge, including in the context of a larger conservation science and science of sustainability as well as in relation to indigenous wisdom and other ways of knowing. Finally, we conclude by offering a number of principles that can help facilitate more meaningful and productive conversations across the vast range of interests and perspectives on conservation.

### 9.3.1 Conservation in the Social Sciences

Given the far-reaching importance of biodiversity to humanity and the mounting threats species and ecosystems face across the globe (Steffen et al. 2015; Díaz et al. 2019), there is an urgent need for the social sciences to take conservation more seriously. This is particularly true for social science disciplines that have devoted comparatively little attention to conservation, such as political science, but it holds across the board. We agree with recent calls for dramatically increasing the breadth and depth of social and behavioral science research for the benefit of nature and human well-being (Balmford et al. 2021; Nielsen et al. 2021). Beyond this vital pragmatic need, we argue that knowledge of human behavior, institutions, processes, and outcomes will remain incomplete as long as the natural world and the diversity of life therein is not considered. Conservation itself is also a ubiquitous social phenomenon, affecting the lives of billions of people around the world. For these reasons, examining conservation as an object of study will only help build and strengthen theory and knowledge within the social sciences.

### 9.3.2 Conservation Social Science

The emergence of conservation social science as an interdisciplinary field is a relatively new development. It has been catalyzed by a growing recognition of the importance of the "social dimensions" of biodiversity conservation among conservation professionals and scientists and the need for dialogue and integration across social science disciplines in relation to conservation. The founding and growth of the Social Science Working Group of the Society for Conservation Biology, together with the publication of influential articles by Mascia et al. (2003), Bennett et al. (2017a, 2017b), and others, helped mark out the terrain and give the field direction.

However, while conservation social science is now increasingly recognized as a distinct area of research, its precise contours and boundaries are more difficult to pin down. The field is perhaps best characterized as a noisy, fractious conversation among people with varying **epistemologies**, methodological emphases, and value commitments. As the field continues to grow and evolve, we hope it will embrace this diversity while also seeking opportunities to build more generalizable knowledge through novel integration across the social sciences and the wide range of interdisciplinary fields that have developed to examine human–environment interactions. There is a particular need to develop theories that can be tested empirically using a range of potential methods, quantitative and qualitative, in a variety of different settings. Our hope is that, in doing so, conservation social science will mature into a field that creatively and rigorously illuminates the myriad social dimensions of conservation and human interactions with biodiversity in a rapidly changing world. In this process, it is crucial for next-generation conservation social science to increase its standing and relative power so that it becomes a truly equal partner with the biophysical sciences in the quest to understand—and contribute to addressing—biodiversity conservation challenges, processes, and outcomes. Foundational training in conservation social science, including this book as a key reference, is key to building the knowledge base and epistemic community necessary for the maturation of the field.

### 9.3.3 Integration with Non–Social Science Disciplines and Other Ways of Knowing

The social sciences are not the only avenue to make sense of and understand conservation processes and outcomes. Of course, biology, writ large, is critical to understanding biological diversity, its distribution, and how and why it has changed over time. So, too, are other natural sciences, such as atmospheric science, limnology, and soil science. The discipline of conservation biology emerged to bring a number of different, mostly natural science disciplines together to build knowledge on how to protect and restore the diversity of life on Earth (Soulé 1985). A larger vision of "conservation science" has now developed that encompasses an even broader range of disciplines and places much greater emphasis on the roles of humans and human well-being in diversity conservation (Kareiva & Marvier 2012; Evans 2021).

The arts and humanities, ethics, theology, and other disciplines are also critical to understanding conservation (Castree et al. 2014; Bennett & Roth 2019). So, too, are indigenous and local knowledge (Kimmerer 2013; Hill et al. 2020; Sidik 2022). A more accurate, comprehensive, and dynamic understanding of conservation capable of informing more effective action to maintain and restore biodiversity requires including and valuing these various ways of knowing and forms of knowledge. It also requires seeing conservation in a broader context of other social and environmental issues, not least human development and climate change. In this sense, knowledge on conservation comprises a vital component of broader environmental social sciences (Moran 2010; Castree et al. 2014; Cox 2015), and the burgeoning field of sustainability science (Clark & Harley 2020), among others.

### 9.3.4 Working Principles to Enhance Conservation Dialogue

While indigenous peoples and many religious traditions have long understood and explored humans and nature as inextricably linked, they have been compartmentalized into separate worlds in much of modern science (Kimmerer 2013; Francis 2015; Sidik 2022). Nearly a century ago, the influential ecologist Aldo Leopold worried about this artificial separation, predicting that the fusion of the study of "the human community" with the study of plant and animal communities "will, perhaps, constitute the outstanding advance of the present century" (Leopold 1935 quoted in Meine 2010, pp. 359–360). As this book clearly demonstrates, this fusion is indeed happening. But the vision of a truly integrated science of conservation, where the social and natural sciences are equal partners and draw from other wells of relevant knowledge such as the humanities, law, engineering, and indigenous sources, remains incomplete. To realize this vision—and the broader one of a better world for people and nature that it is meant to support—will require more concerted action, which must start with more constructive conversations. Better conversations are needed among social science disciplines, the social and natural sciences, scientific and other ways of knowing, researchers and practitioners, and various stakeholders. We conclude this book by offering a series of working principles meant to inform these needed conversations.

**Everyone does not need to agree on everything.** Building consensual solutions to conservation challenges is often desirable, but will not always be possible. Synergies among conservation and other societal goals do exist (e.g. Persha et al. 2011), but trade-offs are common and always will be (McShane et al. 2011). Critique and dissent are important as they can help avoid the suppression of marginal voices and reveal new possibilities for addressing interlinked social and environmental problems (Raymond et al. 2022). Social research "for" conservation (often more applied) and "on" conservation (often more critical) are therefore both needed (Sandbrook et al. 2013).

**Acknowledge plural and discordant perspectives.** Perspectives on biodiversity are diverse, within the conservation movement and in society more broadly (Pascual et al. 2021). Enhanced dialogue on the future of conservation, including identifying possible areas of agreement, requires identifying these different views and the interests they represent and really trying to understand them. Doing so will require careful reading of different positions that seek to avoid making unfounded assumptions or creating caricatures of arguments of those holding different views. It will also require distinguishing genuine differences in values versus more superficial differences in interpretation of data.

**Seek basic knowledge of different disciplines and ways of knowing**. Conservation and conservation science are increasingly interdisciplinary (Newing 2010; Kareiva & Marvier 2012; Bennett et al. 2017b). Formal training in more than one discipline or in an interdisciplinary field can be useful but is not necessary to contribute effectively to conservation conversations. However, a basic awareness of other disciplines, methodologies, and worldviews, as provided in this book, is important to help facilitate dialogue. Such awareness can help build respect for the specialized expertise from different disciplines and methods that are necessary to effectively address conservation challenges. Striving for logical clarity and rigor in research, whether using qualitative, quantitative, or mixed methods approaches, can also help.

**Science can inform decision-making, but does not determine it.** Many scientists (natural and social) and the general public often wring their hands because decision-making processes often seem "unscientific" or "political." Yet science (again, both natural and social) does often inform and shape beliefs upon which complex values-based decisions are made even if it is not itself the sole determinant of decision-making outcomes. A more realistic understanding of the role of science in decision-making—as one of many different considerations that a given decision maker might weigh along with, for example, her/his own values, public opinion, interest group pressure, political party affiliation, or a catalyzing event—can lead to more effective conversations and strategies for engaging in decision-making processes. Here, social science has a critical role to play in helping understand these processes and how they might be influenced.

**Recognize unequal power relationships and take steps to rectify them**. Those involved in conservation research, policy, and practice need to imagine others' positions and not be blind to power and the ways in which one's own position in any given circumstance is shaped by the past (Wyborn et al. 2021). There are massive historical and ongoing disparities across multiple aspects of social difference, including gender, race/ethnicity, class, nationality, and disability. Biases in conservation research against women and people of color are increasingly well documented (Miriti et al. 2020; Kothari 2021; Maas et al. 2021) and must be dismantled (Chaudhury & Colla 2021; Rudd et al. 2021). There is an urgent need to diversify conservation conversations so that white, male, and middle-class voices (which include the authors of this chapter) do not predominate as has all-too-often been the case (Bailey et al. 2020; Chaudhury & Colla 2021). Such diversification must include a fundamental shift in power and perspective away from the global north (Kothari 2021) and a clear-eyed analysis of power and injustice that have shaped previous conservation discourse, policy, and action (Kashwan et al. 2021; Raymond et al. 2022).

Embracing these principles and revising them as necessary, we believe, can help foster greater equality, respect, and trust necessary for better conversations among participants in different conservation-related debates. Doing so will often be difficult and will require patience, humility, and imagination. But it is our firm belief that better conversations are necessary for better research, better policies, and better practices capable of transforming currently unsustainable trajectories to a world of mutual flourishing for people and nature.

## References

Bailey, K., Morales, N. & Newberry, M. (2020) Inclusive conservation requires amplifying experiences of diverse scientists. *Nature Ecology & Evolution* 4: 1294–1295. https://doi.org/10.1038/s41559-020-01313-y.

Balmford, A., Bradbury, R.B., Bauer, J.M. et al. (2021) Making more effective use of human behavioural science in conservation interventions. *Biological Conservation* 261: 109256.

Bennett, N.J. & Roth, R. (2019) Realizing the transformative potential of conservation through the social sciences, arts and humanities. *Biological Conservation* 229: A6–A8.

Bennett, N.J., Roth, R., Klain, S.C. et al. (2017a) Conservation social science: understanding and integrating human dimensions to improve conservation. *Biological Conservation* 205: 93–108.

Bennett, N.J., Roth, R., Klain, S.C. et al. (2017b) Mainstreaming the social sciences in conservation. *Conservation Biology* 31: 56–66.

Bourdieu, P. (1972) *Outline of a Theory of Practice.* Cambridge, UK: Cambridge University Press.

Braudel, F. & Wallerstein, I. (2009) History and the social sciences: the longue durée. *Review (Fernand Braudel Center)* 32 (2): 171–203. http://www.jstor.org/stable/40647704.

Castree, N., Adams, W.M., Barry, J. et al. (2014) Changing the intellectual climate. *Nature Climate Change* 4 (9): 763–768.

Chaudhury, A. & Colla, S. (2021) Next steps in dismantling discrimination: lessons from ecology and conservation science. *Conservation Letters* 14 (2): e12774.

Cinner, J. (2018) How behavioral science can help conservation. *Science* 362 (6417): 889–890.

Clark, W.C. & Harley, A.G. (2020) Sustainability science: toward a synthesis. *Annual Review of Environment and Resources* 45: 331–386.

Cowling, R.M. (2014) Let's get serious about human behavior and conservation. *Conservation Letters* 7: 147–148.

Cox, M. (2015) A basic guide for empirical environmental social science. *Ecology and Society* 20 (1): 63. http://doi.org/10.5751/ES-07400-200163.

Díaz, S., Settele, J., Brondízio, E.S. et al. (2019) Pervasive human-driven decline of life on Earth points to the need for transformative change. *Science* 366 (6471): eaax3100.

Ellis, E.C., Gauthier, N., Goldewijk, K.K. et al. (2021) People have shaped most of terrestrial nature for at least 12,000 years. *Proceedings of the National Academy of Sciences of the United States of America* 118 (17): e2023483118.

Evans, M.C. (2021) Re-conceptualizing the role(s) of science in biodiversity conservation. *Environmental Conservation* 48 (3): 151–160.

Folke, C., Polasky, S., Rockström, J. et al. (2021) Our future in the Anthropocene biosphere. *Ambio* 50: 834–869.

Francis. (2015) *Laudato Si [Encyclical Letter on Care for Our Common Home].* Vatican City: The Holy See. Accessed September 2, 2022. https://www.vatican.va/content/francesco/en/encyclicals/documents/papa-francesco_20150524_enciclica-laudato-si.html.

Gerring, J. (2001) *Social Science Methodology: A Criterial Framework*. Cambridge, UK: Cambridge University Press.

Giddens, A. (1984) *The Constitution of Society*. Cambridge, UK: Polity Press.

Hicks, C.C., Levine, A., Agrawal, A. et al. (2016) Engage key social concepts for sustainability. *Science* 352 (6281): 38–40.

Hill, R., Adem, Ç., Alangui, W.V. et al. (2020) Working with indigenous, local and scientific knowledge in assessments of nature and nature's linkages with people. *Current Opinion in Environmental Sustainability* 43: 8–20.

Hirsch, P.D., Adams, W.M., Brosius, J.P., Zia, A., Bariola, N. & Dammert, J.L. (2011) Acknowledging conservation trade-offs and embracing complexity. *Conservation Biology* 25 (2): 259–264.

Kareiva, P. & Marvier, M. (2012) What is conservation science? *BioScience* 62 (11): 962–969.

Kashwan, P., Duffy, R.V., Massé, F., Asiyanbi, A.P. & Marijnen, E. (2021) From racialized neocolonial global conservation to an inclusive and regenerative conservation. *Environment: Science and Policy for Sustainable Development* 63 (4): 4–19.

Kimmerer, R. (2013) *Braiding Sweetgrass: Indigenous Wisdom, Scientific Knowledge and the Teachings of Plants*. Minneapolis, MN: Milkweed Editions.

Kothari, A. (2021) Half-Earth or whole-Earth? Green or transformative recovery? Where are the voices from the Global South? *Oryx* 55 (2): 161–162.

Maas, B., Pakeman, R.J., Godet, L., Smith, L., Devictor, V. & Primack, R. (2021) Women and Global South strikingly underrepresented among top-publishing ecologists. *Conservation Letters* 14: e12797. https://doi.org/10.1111/conl.12797.

Martin, J.L., Maris, V. & Simberloff, D.S. (2016) The need to respect nature and its limits challenges society and conservation science. *Proceedings of the National Academy of Sciences of the United States of America* 113: 6105–6112.

Mascia, M.B., Brosius, J.P., Dobson, T.A. et al. (2003) Conservation and the social sciences. *Conservation Biology* 17: 649–650.

Massarella, K., Nygren, A., Fletcher, R. et al. (2021) Transformation beyond conservation: how critical social science can contribute to a radical new agenda in biodiversity conservation. *Current Opinion in Environmental Sustainability* 49: 79–87.

McShane, T.O., Hirsch, P.D., Trung, T.C. et al. (2011) Hard choices: making trade-offs between biodiversity conservation and human well-being. *Biological Conservation* 144 (3): 966–972.

Meine, C. (2010) *Aldo Leopold: His Life and Work*. Madison, WI: University of Wisconsin Press.

Miriti, M.N., Bailey, K., Halsey, S.J. et al. (2020) Hidden figures in ecology and evolution. *Nature Ecology & Evolution* 4: 1282. https://doi.org/10.1038/s41559-020-1270-y.

Moran, E.F. (2010) *Environmental Social Science: Human–Environment Interactions and Sustainability*. West Sussex, UK: John Wiley & Sons.

Newing, H. (2010) Interdisciplinary training in environmental conservation: definitions, progress and future directions. *Environmental Conservation* 37 (4): 410–418.

Nielsen, K.S., Marteau, T.M., Bauer, J.M. et al. (2021) Biodiversity conservation as a promising frontier for behavioural science. *Nature Human Behaviour* 5: 550–556.

Pascual, U., Adams, W.M., Díaz, S. et al. (2021) Biodiversity and the challenge of pluralism. *Nature Sustainability* 4: 567–572.

Persha, L., Agrawal, A. & Chhatre, A. (2011) Social and ecological synergy: local rulemaking, forest livelihoods, and biodiversity conservation. *Science* 331 (6024): 1606–1608.

Raymond, C.M., Cebrián-Piqueras, M.A., Andersson, E. et al. (2022) Inclusive conservation and the Post-2020 Global Biodiversity Framework: tensions and prospects. *One Earth* 5 (3): 252–264.

Reddy, S.M., Montambault, J., Masuda, Y.J., Keenan, E., Butler, W., Fisher, J.R., Asah, S.T. & Gneezy, A. (2017) Advancing conservation by understanding and influencing human behavior. *Conservation Letters* 10 (2): 248–256.

Rudd, L.F., Allred, S., Bright Ross, J.G. et al. (2021) Overcoming racism in the twin spheres of conservation science and practice. *Proceedings of the Royal Society B* 288 (1962): 20211871.

Sandbrook, C., Adams, W.M., Buscher, B. & Vira, B. (2013) Social research and biodiversity conservation. *Conservation Biology* 27: 1487–1490.

Scott, J.C. (2017) *Against the Grain: A Deep History of the Earliest States*. New Haven, CT: Yale University Press.

Sidik, S.M. (2022) Weaving Indigenous knowledge into the scientific method. *Nature* 601: 285–287.

Soulé, M.E. (1985) What is conservation biology? *BioScience* 35 (11): 727–734.

Steffen, W., Richardson, K., Rockström, J. et al. (2015) Planetary boundaries: guiding human development on a changing planet. *Science* 347 (6223): 1259855.

Wyborn, C., Montana, J., Kalas, N. et al. (2021) An agenda for research and action toward diverse and just futures for life on Earth. *Conservation Biology* 35: 1086–1097.

# Glossary

**Agency**  People's capacity to make decisions and act according to their own choices. When used in relation to non-humans it refers to the capacity to act and affect a process or system.

**Anthropocene**  A proposed geological epoch describing the period in the Earth's history when human activities have significantly altered the planet's biophysical systems. The implication is that human transformation of the planet's ecosystems and climate is now so extensive that it is on a geological scale and will leave various types of evidence in the geological record.

**Anthropogenic**  Caused by, or resulting from, human activity.

**Biodiversity**  The variability among living organisms and the ecological complexes of which they are part, including diversity within species, between species and of ecosystems.

**Biodiversity conservation**  The protection, restoration, and management of **biodiversity.**

**Capitalism**  An economic system characterized by the use of wage labor in production processes and the reliance on markets for the sale and exchange of goods, services, and commodities in order to generate profit.

**Civil society**  A sector of society comprised of formal and informal groups and organizations. These entities are not part of the government (public sector) or markets (private sector) and are instead associated with the people or the citizens of a country or region.

**Common-pool resources**  Resources where one person's use can subtract from the quantity of the resource available to others and where it is often necessary, but difficult, to exclude people from using the resource. These characteristics of "subtractability" and "excludability" mean that common-pool resources are prone to be degraded and are especially challenging to manage and maintain.

**Conservation social science**  The study of the conservation-relevant aspects of human society, including the relationships among humans and between humans and their environment. While all social sciences involve the study of human societies and relationships between individuals and groups within those societies, researchers within the social sciences can differ significantly both in the way they go about studying social processes (see, for example, **qualitative** versus **quantitative** methods) and their research aims. Some conservation social science

*Conservation Social Science: Understanding People, Conserving Biodiversity*, First Edition.
Edited by Daniel C. Miller, Ivan R. Scales, and Michael B. Mascia.
© 2023 John Wiley & Sons Ltd. Published 2023 by John Wiley & Sons Ltd.

primarily seeks to improve biodiversity conservation outcomes. Other researchers are more concerned with revealing and challenging the unequal power relations often involved in biodiversity conservation.

**Culture**  The way of life of a group of people, which can include shared beliefs, values, norms, habits, customs, and practices.

**Decolonization**  The reversal of colonial practices and relationships. Colonialism is the domination of a group by another group. As well as control through military, political, and economic power, colonialism involves the imposition of cultural values and the domination of certain forms of knowledge. When applied to biodiversity conservation, decolonization refers to the process of changing conservation's **discourses** and practices so that they no longer reflect colonial power relations or dominated by western institutions, knowledges, and values.

**Development**  Development is a political term that has a range of meanings depending on the context in which it is used. When used in relation to countries, development refers to Western ideas about the organized attempt to overcome constraints on economic growth through processes of industrialization, urbanization, democracy, and capitalism. More generally, development can be considered as a process of bringing about social change so that people are able to achieve their human potential.

**Discourse/discursive**  The process whereby people form and share knowledge about the world around them (e.g. through speech, texts, imagery, art). Following on from the work of French sociologist Michel Foucault, in many social sciences the term discourse has taken on a more specific meaning and refers to a system of thought and communication that shapes the way individuals and groups perceive and experience the world. Controlling a discourse means being able to shape, influence, and dominate the ways people understand and act in the world. Discourses are thus an important mechanism for exercising power.

**Ecosystem**  A dynamic complex of plant, animal, and micro-organism communities and their non-living environment interacting as a functional unit.

**Ecosystem services**  Ecological processes or functions having monetary or non-monetary value to individuals or society at large. These include supporting services (e.g. productivity or biodiversity maintenance); provisioning services (e.g. food, fiber); regulating services (e.g. climate regulation or carbon sequestration); and cultural services (e.g. tourism or spiritual and aesthetic appreciation).

**Epistemology/epistemological**  The study of the origins, nature, and extent of knowledge. Epistemological questions include: How do we know what we know? What constitutes valid knowledge? What is knowable?

**Ethnography/ethnographic**  The study of a particular human culture or social group through largely **qualitative** methods such as **participant observation** and interviews.

**Globalization**  The expansion and intensification of social relations across space, creating new flows of information, capital, commodities, and practices that stretch across the world.

**Governmentality**  An approach to studying power developed by French sociologist Michel Foucault. Governmentality refers to the process where people's consent to be governed is self-generated, rather than forcibly imposed from the top-down. Individuals internalize certain knowledges and **discourses,** so that their behaviors conform to guidance from powerful institutions and agents. Through a variety of institutions—such as families, schools, and hospitals—individuals learn to regulate their own behavior and that of others.

**Ideology/ideological** A system of beliefs that underpin social and political action. Examples of ideologies include capitalism, communism, and socialism.

**Industrialization/industrialized** A shift toward mechanization and automation, generally accompanied by an increasing reliance on external energy sources rather than human labor power.

**Methodology** An overarching framework for carrying out research that provides the rationale for how to collect, analyze, and interpret data.

**Neoliberalism/neoliberal** A set of economic ideas that promote free markets, individual freedom, and minimal state involvement as the most efficient way to coordinate economic activity and generate economic growth. Neoliberal policies include the privatization of state-run companies and industries, the removal of trade barriers, and the deregulation of financial activities.

**Neo-Malthusian** Ideas about human population growth based on Thomas Malthus' 1798 'An Essay on the Principles of Population' in which he argued that human population growth would inevitably outpace a society's ability to produce enough food and that famine, poverty, and war were thus inevitable. Neo-Malthusians have extended Malthus' ideas to include not only food but natural resources more broadly. Unlike Malthus, some argue for strong birth control measures to restrict population growth in order to avoid poverty and environmental degradation.

**Normative** Based on value judgements and social **norms**. Rather than seeking to describe and explain the world, normative research seeks to change it based on certain beliefs and values (for example ideas of justice and equality).

**Norms** Shared rules and expectations about behavior within a society or group.

**Ontology/ontological** The study of the nature of reality, being, and what exists in the world. Ontological questions include: What is real? What is the real world made of?

**Participant observation** A **qualitative method** whereby researchers immerse themselves in the day-to-day lives and activities of the individuals and groups they are studying, with the aim of developing a deeper understanding of their beliefs, behaviors, perceptions, and practices.

**Political ecology** The study of how political, economic, and social factors shape and are shaped by the environment. Political ecology pays close attention to how politics and power relations lead to the unequal distribution of costs and benefits in natural resource use and environmental change.

**Political economy** The study of the interactions between political and economic processes, including how power relations affect the distribution of wealth between individuals and groups and how wealth affects policy choices.

**Positionality** The social, political, economic, and cultural factors that determine an individual's identity and role in a society. When applied to research it refers to how the researcher's identity and position in society influences, and might bias, research.

**Positivist/positivism** The belief that social processes can and should be studied objectively through sensory experience and direct observation in the same way that the scientific method is used by natural scientists to study the natural world.

**Postmodernism/postmodern** A movement spanning the arts, humanities, and social sciences that emerged as a reaction to the ideas and values of modernism. Modernism is based on a belief in universal values, rationality, progress, and objective truths. It emphasizes humanity's ability to control and improve its condition through innovation, science, and technology. Postmodernism is critical of these ideas and instead emphasizes the subjectivity of human experience and the diversity of values.

**Post-structuralism** An intellectual movement based on, but also critiquing and advancing, structuralism. Structuralism is based on the idea that diverse and seemingly unique social and cultural forms (see **culture**) are part of a coherent system and underpinned by a set of universal structures (e.g. linguistic structures, kinship relations, myths, customs). Both structuralists and poststructuralists pay close attention to the role of language in mediating how humans understand the world. However, while structuralism emphasizes universality and coherence in the structures that underpin societies, post-structuralism stresses the plurality and fluidity of human values, beliefs, and perceptions. Poststructuralists emphasize the importance of **discourse** in shaping the way individuals perceive and act in the world.

**Quantitative research/methods** Research approaches that collect and analyze numerical data and focus on establishing facts about measurable events and processes.

**Qualitative research/methods** Research approaches that focus on the lived experiences and perceptions of individuals and groups. Data are collected through methods such as interviews and **participant observation**.

**Social construction** Something that, rather than being objective and universal, is the product of human interpretation and thus shaped by cultural, geographical, and historical context.

**Social capital** The extent and types of bonds and shared norms among people in a community. Several types have been identified, including bonding social capital, or the emotional connections and supportiveness that develop between similar people; bridging social capital, or the linkages among people and groups that are dissimilar but have been brought together; and linking social capital, or the ability of groups to influence or gain resources from external agencies or authorities.

# Index for *Conservation Social Science*

Note: Page numbers referring to figures are in italic; page numbers referring to tables are in bold; page numbers referring to boxes are not emphasized.

*Conservation Social Science: Understanding People, Conserving Biodiversity*, First Edition.
Edited by Daniel C. Miller, Ivan R. Scales, and Michael B. Mascia.
© 2023 John Wiley & Sons Ltd. Published 2023 by John Wiley & Sons Ltd.

Printed and bound by CPI Group (UK) Ltd, Croydon, CR0 4YY

28/10/2024

14581366-0001